# CLIMATE CHANGE AND THREATS TO HUMAN SECURITY

# CLIMATE CHANGE AND THREATS TO HUMAN SECURITY

*Editor*
**Narottam Gaan**

KALPAZ

*Published by*

Kalpaz Publications
C-30, Satyawati Nagar,
Delhi – 110052
Ph.: 011-47034999, 9811692060
www.kalpazpublications.com
E-mail: kalpaz@hotmail.com

*Distribution Network*
gyanbooks.com
India, USA, Canada, UK, Australia, France



ISBN: 978-93-5722-573-1 (PB)

First Published, 2024

*Printed at:* Gyan Press, Delhi.



**Climate Change and Treats to Human Security**
Editor: Narottam Gaan

## *Dedicated*

*At the Lotus Feet of My Revered Gurudev Paramahamsa Hariharananda with Prostrations and Love*

# Contents


*Preface* 9
*Acknowledgements* 19
*List of Contributors* 21

1. **Introduction: Climate Change and Human Security** **23**
*Narottam Gaan*

2. **Environmental Security and Peace Building: A Plea for a Theoretical Construct** **53**
*Manoj Kumar Mishra*

3. **Warfare: Its Environmental and Human Consequences in the Light of Global Warming** **67**
*Arthur H. Westing*

4. **Climate Change and Threats to Human Security in Bangladesh: An Analysis** **79**
*Md. Rafiqul Islam and Ziaul Haque*

5. **Climate Change, Human Security and Sri Lanka** **115**
*Erandathie Lokupitiya*

6. **Climate Change and Threats to Human Security** **132**
*Dasarathi Bhuyan*

7. **India's Perceptions, Positions and Policies on Climate Change and Human Security** **146**
*Amulya Kumar Tripathy*

8. **India's Role in South-South Cooperation: A Case of Climate Change Negotiations** 164
*Dhanshree Jayaram*

9. **Impact of Climate Changes in Africa : A Review of Human Security and Conflicting Threat** 187
*Saubhagyalaxmi Misra*

10. **Environmental Goods and Services: Challenges and Opportunities for Bangladesh in the Context of Climate Change** 207
*Sumitra Mitra and Saptarshi Basu Roy Choudhury*

11. **Climate Change Debate in Nepal: Policies and Challenges** 224
*Anjan Kumar Sahu*

12. **Climate Change vis-à-vis Human Security in Maldives: Prospects and Perspectives** 240
***Subhashree Patnaik***

13. **Sea Level Rise in Bay of Bengal: Challenges for India** 258
*Ramakrushna Pradhan*

14. **Climate Change Challenges to Human Security in Third World Countries** 280
*Alok Kumar Ray*

15. **Politics of Development in North-East India: Its Impact upon the Bio-diversity, Environment and Human Security** 302
*Sasanka Ghosh*

16. **Reconceptualising Security in the Context of Global Climate Change** 312
*Sonali Mohapatra*

*Index* 327

# Preface

The state centric military based conceptualisation of security is the result of understanding it in terms of walling in what are legitimate occupants within the territorial borders and walling out what are not within the definition of self but can be relegated to the category of 'the other'. Since Westphalia tradition in 1648 'security' till today defined as nothing more than the specialisation by distancing away or walling out the 'other' and boundary making has been reduced to a mere symmetrical labelling in the sense that in a Hobbesian anarchical international system the states are always posited in a situation of being threats to other similar states or being threatened by the latter. Thus perched in an "insecurity of the security" or security dilemma situation like the individualised atomic selfs in a state of nature, the states are always on a military spending spree. Security thus held as synonymous with state, its military institutions, and its associated functions remains impervious to or shows a blithe disregard for the sources of threats not necessarily emanating from the symmetrical states but from asymmetrical not-static sources. Too much securitisation with military forces, techniques and equipments by the state could not enable the latter to give protection to those living within their territorial precincts against those non-state centric threats. Rise of these non-state centric threats to the centre stage of mainstream security analysis sidelining the state centric security constituted the main basic plank of the intellectual fermentation that took place in 80s and 90s in the western world adducing reasons to redefine security. So far underplayed or marginalised under the ontological and epistemological scaffolding of realist understanding of state centric security, the non-state centric security threats included a mélange of sectors like incidence of poverty, economic deprivation, migration, spread of fatal diseases like HIV through pathogens, changing demographic structure, environmental change, climate change, malnutrition, child mortality, and missing women, ethnicity and identity. A paradigm shift took place from state centric security to non-state security threats. The response to the state centric security threats always remains state centric with exercise of military forces both conventional and non-conventional. However, in case of non-state centric security threats the responses may not be necessarily always the traditional state centric approach. But the approach and responses to these non centric non-traditional security threats have the potential to impact international security and national politics as these non-traditional security threats are mostly trans-boundary. So in terms of both source of security threats and responses to these, the paradigm shift deflecting away from state in many cases has become the cornerstone of new security thinking at present.

In the following chapters, the authors have tried to present country wise scenario where climate change has been a threat to human security.

Manoj Mishra in chapter 2 tries to answer the question, which has been of paramount interest among academicians and security analysis experts as to whether the environment should be seen as a security issue. It deals with the various issues ranging from those who believed that the securitisation of the environment is the most important step to secure the survival of humanity, to those who believe that its advocates are simply environmentalists who cynically want to grab the attention of the government. The article further explores into the potential issues, which can be included into the environmental security agenda. It goes on to describe the different schools of thought, which are included in the environmental security approaches. Lastly, it brings the issue of environmental peace building, which is an attempt to broaden the environment and security agenda programme. In this section issues of environmental cooperation are discussed which would prompt stability and peace between states in conflict. In the end, this article makes a sincere attempt to discuss initiatives such as trans-boundary cooperation for environmental conservation (e.g. "*peace parks*"), international river basin management, regional marine agreements and joint environmental monitoring programmes, which could enhance cooperation and ultimately peace between communities and countries.

Arthur H. Westing in chapter 3 dilates upon the impact and devastating consequences of deadly weapons on enmeshed civilians and displaced persons with emphasis on mitigating these disruptions. With continuing advances and modernisation in weaponry, the levels of death, destruction, causalities, and environmental disruption have not been increasing over these years. The person most affected happens to be the people from the Third World countries. The environmental disruptions brought about by wars are more or less unintentional side effects of attack. The damages have been collateral. He points out that the current Afghan war of 2001 provides a good example of the collateral damages done to human beings and environment. He excerpted from history many examples to illustrate how the military forces have targeted the cropland. In the Gulf War of 1991, US forces were environmentally blameless but Iraq caused huge environmental damage by intentional use of liquid oil. Even water infrastructures were targeted. Ghenghis Khan's forces deliberately destroyed the irrigation works of the Tigris River upon which the local agriculture depended. The Second Sino-Japanese War of 1937-1945 provides us with a spectacular example of militarily successful intentional flooding. In order to curtail the Japanese advance, the Chinese in June 1938 dynamited the Huayuankow Dike of the Yellow River near Chengchow. This action resulted in the drowning of several thousand Japanese soldiers and stopped their advance into China along that front. Explosive remnants of war continue to infest battlefields for incredibly long periods. Post war removal of these pernicious remnants is arduous and environmentally disruptive. Another effect more deleterious is the unsettlement of people in Third World countries from war-affected areas to considerably safer places. Their number according to Westing stands at 255 million by 2009. The climate change and its impact in terms of sea level rise will further accentuate the crisis. As conflicts are likely to occur among countries over scarcity of water in trans-boundary rivers, the use of weapons and the post war remnants of explosives may further cause environmental disruptions. If history is a witness, then the terrorists and militants shall not forget to target the river waters and dams and dykes to cause disruption.

Md Rafiqul Islam and Ziaul Haque have explained in chapter 4 how "climate change has emerged as one of the grisly human security challenges of our time with its harrowing effects on a broad range of issues within a paradigm which inseparably links human, environment and socio-economic well-being". They have designed a theoretical build up to show how the traditional concept of security has been inadequate to explain the new kind of threats emerging from environmental disruptions. The security is no longer centered on state and its military institutions but has migrated down to the level of human beings from the state centric notions. In their words climate change will result in extreme weather events, drought, flooding, sea level rise, retreating glaciers, habitat shifts, and the increased spread of life-threatening diseases affecting thousands of millions of people and their communities around the world. Taking the case study of Bangladesh, they have rightly pointed how climate change disrupts the way of life of individuals. The climate change with its devastating consequences have destroyed ecosystem services, diminish environmental resource base – fundamental for human survival – and thus bringing chronic hunger, disease, endemic poverty and loss of livelihoods. Its very geographic location entwined with climate change has increased natural disasters, intensity of flood, inundation of water logging, cyclone and tidal surges during these years with frequency of drought in north-western regions of Bangladesh, river bank erosion, sea level rise and saline water ingress. They have also highlighted the various social effects in terms of increasing poverty, decline in agricultural productivity, chronic hunger and threats to food security and health of the people. The gradual weakening of state capability to meet these challenges has resulted in gradual erosion of social contract in Bangladesh. The migration of people from Bangladesh triggered by environmental disruption and climate change to India as they pointed out has the potential of conflict with the latter.

Erandathie Lokupitiya from University of Colombo has in chapter 5 mentioned how climate change impacts on human security in Sri Lanka have occurred mostly due to temperature and precipitation changes, and sea level rise. The direct impacts of temperature and precipitation changes involve droughts, flood events, landslides, and extreme wind events, etc. Sea level rise has also caused flooding and other impacts in coastal areas. In general, the number of extreme weather events involving temperature and precipitation fluctuations, of which floods and droughts being the largest, has increased over the past four decades. The overall highest frequency, magnitude, and impacts of these were reported during the last decade. Whether such instances are directly related to anthropogenic climate change might be still debatable; however, an overview of the situation with regard to these impacts and their effect on human security is a timely concern.

Agriculture is also affected due to reduced yields and wind-driven soil erosion during droughts in the prominent agricultural areas; soil is dried quickly due to increased evaporation rates with hotter temperatures. Given the low amount of water in the major irrigation tanks, the area under cultivation becomes significantly low during the drought periods, affecting the livelihoods of the producers and food security for consumers of crops, especially rice. In hot temperatures due to global warming, pest and diseases have seriously affected agricultural productivity and fishery industry. Water scarcity has caused drought in many parts of Sri Lanka affecting agricultural productivity as it is mostly depended on agriculture. Sri Lanka has been impacted by climate change, mostly due to extreme weather fluctuations and sea level rise. Climate change has impacted the human security in Sri Lanka through direct and indirect impacts on major economic sectors including agriculture, water, health, and power sector. The country in turn has taken several important policies-regulatory and other measures in addressing the issue. Improper land use and agricultural practices have further aggravated the climate change impacts on the country's major economic sectors.

Dasarathi Bhuyan in chapter 6 has tried to establish and design a model to understand and explain the relationships between hydro-climatic hazards, climate change vulnerability, human security and conflict, on the basis of theoretically informed, comparative empirical research. To map international and national policies for security and adaptation in water resources and hazard management, and develop a policy model for security against hydro-climatic hazards ("hydro-security") in the Indian region has been focused by him. He has dwelt upon vividly how issues of climate change with its disastrous consequences become ultimately the main concerns of human security. He has pointed out how climate change would exacerbate environmental scarcity of resources causing forced distress migration not only within a state but also from state to state. The various implications of climate change would finally cause a threat to international peace and security. In conclusion, he has tried to address the crisis by stressing upon the imperatives for reorientation of science and technology that will replace the all consuming and aggrandising propensities found growing wantonly in all modern human beings. He has searched for a middle path as shown by Buddha and the ancient wisdom of India that will bring an end to the crisis of climate change.

In this chapter 7 Amulya Tripathy has tried to present India's perceptions on the problem of climate change and sustainable development; the kind of negotiating positions that follow from these perceptions; the policies India has undertaken so far and finally, India's possibilities for action that can help contain the threat of climate change. Indians should be concerned about climate change since this phenomenon might have substantial adverse impacts on them. Not all possible consequences of climate change are yet fully understood, but the three main 'categories' of impacts are those on agriculture, sea level rise leading to submergence of coastal areas, as well as increased frequency of extreme events. Each of these poses serious threats to India. However, these are long-term issues. Apart from energy conservation and efficiency improvements, the need to find, develop and exploit non-conventional energy sources, many of them clean and renewables, has long been recognised by the Government of India. In the context of the current debate about climate change, it is necessary to show that far from being inactive, the developing countries, especially India, are taking considerable actions in terms of policies, programmes and projects. Technology transfer can speed up the modernisation process and additional funds can accelerate Government initiatives in energy conservation. However, policies for poverty alleviation must take priority. It is shown that savings in GHG emissions by the poor should not be expected at the expense of development.

Dhanshree Jayaram in chapter 8 has stated that the human-induced process of environmental change that was set off way back during the era of civilisation-building took a perilous turn in the twentieth century due to the industrialisation drive of the Western countries; it is now gradually worsening as the international community awaits a consensual deal. Climate change is one such threat and among all environmental threats, this is one phenomenon, which has gathered maximum amount of attention. One could say that it is not the definition of climate change that needs to be looked into; it is the definition of 'response to climate change' that needs to be transformed so that the 'measures' bear fruits. Geopolitically, climate change is an issue that encompasses all facets of international politics that steer a nation's march towards achieving national comprehensive power. Ironically, the international negotiations themselves have become synonymous with 'conflict' rather than 'cooperation.' The first step towards resolving the conflicts would be re-standardisation of international relations that would create political willingness based on 'complex interdependence.' South-South cooperation could be one of the pillars of such an order as attempts to bridge the differences between the South and the North have failed repeatedly.

In the chapter 9, Saubhagyalaxmi Misra has dealt with the devastating impacts of climate change in African countries. Premised on Homer Dixon's theoretical model she has defined at the outset what the traditional security is and subsequently tried to redefine it in view of the drastic changes in security scenario. The environmental degradation and climate change due to anthropogenic reasons have come as the greatest threat to human beings in recent years. These new changes have challenged the traditional concept of security centered on state. She has shown how the environmental degradation and climate change due to continued adherence to development as based on fossil fuels have impacted water, the natural resources, agriculture, food, health and living condition of the people. The various social effects of the climate change and environmental degradation will be as she depicted are to be seriously felt in terms of decline in agricultural productivity, increasing poverty, health hazards, migration of vulnerable people and growing incapability of states to meet these climate change induced challenges. These will facilitate the conditions for civic violent conflict among states and even within a state challenging the legitimacy of the ruling class. She has illustrated this very glaringly in case of African countries.

Sumitra Mitra and Saptarshi Basu Roy Choudhury in this chapter 10 have discussed that in the context of trade liberalisation in Environmental Goods and Services (EGSs) the developed-developing countries debate persists. The list of EGSs mainly consists of technologically advanced sophisticated products and exports from developed countries.

At present most of the developing countries are net importers of EGSs. Bangladesh being a developing country is no exception. However, economists argue that developing countries will have dynamic benefits from the trade liberalisation of EGSs in the coming years. Studying the opportunities and challenges for Bangladesh, we found that being a labour abundant country it has a potential gain of production shift of manufactured goods from developed countries because labour intensive technologies are environment-friendly. This explanation holds for labour intensive environmental services also. Particular to jute and jute products, Bangladesh has a huge opportunity because jute is one of the Environmentally Preferable Products (EPPs) as defined in WTO negotiations and Bangladesh is the largest producer and exporter of jute in the world. However, there are several shortcomings as well for this country in achieving these potential gains. The smoother access of technology does not always guarantee its absorption. Given Bangladesh's high poverty and poor literacy levels, the technological absorption is a big cause to worry about. Chances are high that the jute export from the country might get refused by developed countries for not following all the international trade norms and standards. Bangladesh also struggles to keep pace with the global trade standards due to the use of unskilled labour, child labour and technologies, which are not environment-friendly. The paper hits upon the possibilities that if India and Bangladesh join hands in the production of jute and jute-related goods then these problems can be overcome.

In chapter 11, Anjan Kumar Sahu has shown how climate change debate in Nepal is a new development. The debate has put the climate change issue at the centre around which political, economic and social issues are defined and shaped. Despite the independent identities of politics, economy and society, the debate shows that climate change has overwhelming presence and the perception is that the climate threat begets socio-economic and political instability. However, the recent climate change policies of the Government of Nepal have been more or less human security oriented and internality in its character. The feeling of insecurity from climate change induced external threats has not been a major policy of policy-makers and political leaders. They are unable to make it a political and public issue and unable to foresee the potential conflict over the water resources of the region. Thus, this paper is an attempt to look into both the internal-external dynamics and human-national security aspects of climate change in Nepal.

In Chapter 12 Subhashree Patnaik depicts how Maldives has always become a referring point in climate change debate. Its vulnerability due to climate change has been predicted to be immense that goes to the extent of extinction of the whole nation. Its geography, economy, agriculture, health, life and livelihood are at stake. Today, when climate change has become a central focus of academic as well as policy discussion, Maldives case seems to be the starting point. Maldives case not only impacts Maldives only but also impacts the human security problem of the neighbouring countries like India, Bangladesh, Sri Lanka, Australia and as far as Germany in terms of refugee problems and business modalities. Therefore, it is not that the sinking state seeks help but neighbouring nations of sinking nation seek help. The paper therefore not only focuses on the impacts of climate change on Maldives's human security but critically predicts its impact on the neighbouring countries in particular and the globe at general. In comparison to its emission level, its endeavour towards emission curb, one can very well judge the intensity of concern Maldives has shown for the cause. One can very well suspect Maldives's concern as individualistic triggered by foolish fear and suspicion, but Maldives has ignited the global climate mitigation process more rigorously. Maldives has been trying to convince the global community that unless we act rigorously and rapidly, not only our globe will lose the Maldives very soon but also will witness the many more unpredicted consequences that can challenge human existence on the earth.

In chapter 13, *Ramakrushna* Pradhan has highlighted that one of the most profound effects of climate change has been the sea level rise due to melting of glaciers and increase in thermal temperature of the sea surface since the mid 19th century. Coastal zones with complex geographical entities encompassing diverse human activities and ecological processes in interconnected maritime and terrestrial spaces are among most vulnerable of all environmental concerns significantly threatening these interdependent systems. The rise in sea levels resulting in disastrous consequences like, intensification of tropical cyclones, larger storm surges, increasing sea-surface temperatures and growing acidification of surface water increasingly endanger the food security, water security, health and livelihoods of coastal population. In this context, this paper attempts to study the sea level rise in Bay of Bengal and its multiple challenges for India. Sea level rise is of great concern to India in view of the damages that occur along the east coast of the country from the cyclones formed in the Bay of Bengal, exacerbation of flooding, beach erosion and salt water intrusion to rivers and ground water aquifers endangering the life and habitats of one-quarter of the population of India living along the coasts. Any increase in frequency and severity of these extreme climate events or change in coastlines as projected is likely to have serious effects and can cause greater harm to the populace. Hence, in the mean time the urgent need is to respond to the coming disastrous through various policy processes like, mitigation, adaptation and protection.

In chapter 14, Alok Kumar Ray has proved that the Third World countries are more prone to the effects of climate change than the other is. Those countries comprise people of diverse cultures, religions, languages and races. Population growth, unstable governments, lesser economic development, poverty, crime, violence and unsustainable utilisation of natural resources are some of their major threats which make them more vulnerable towards climate change and its severe repercussions. In those countries of South Asia, Sub-Saharan Africa and Latin America, people's livelihood options are in direct contrast with environment. Scarcity of resources, its uneven availability among the populace, competitive populist policies of governments to remain in power without looking to the needs of the future generations have already made this region across continents prone to newer and lesser-known threats for their survival from climate change and global warming. Most of the people in this region earn their livelihood through agriculture, which will be severely affected in the coming days due to the negative impact of global warming and climate change. Civil strifes and violence, underground nefarious activities, absence of minimum standard of living condition are serious concern in those areas. Western model of fossil fuel technology based developmental pattern has ruined the socio-economic and cultural pattern of people there. However, the most serious concern is related to industrial development and the corresponding ill effects, which paved the way for changes in the environment. Human ingenuity is capable to sort out the problem to some extent. However, the need is to rely on local models of developments and think for the model of sustainable development. Security is no more related to military aggression. Rather it has become individualistic so to say challenging the authority of western model of nation states.

In Chapter 15 Sasanka Ghosh has outlined the significance of geopolitics and enriched biodiversity of North East Indian region in the process of defining development. Without taking into account this factor the dominant concept of development based on industrialisation, construction of big dams, and projects at the expense of the natural environment on which the local inhabitants depend for their livelihood, which has been superimposed on this region has been the root causes of destruction of the environment and the growing discontentment among the people against the state, thus challenging the process of integration of this region in the mainstream of national politics. The climate change as he has stated will further accentuate the crisis challenging the established notion of security, development and fulfilment of basic human needs.

In Chapter 16 Sonali Mohapatra has described analytically how climate change can usher in a reconceptualisation of traditional concept of security built around state and its military institutions. Her focus is on how climate change challenges the established frame of mind about state centric security.

**Narottam Gaan**

# Acknowledgements

Ideas remain abstruse unless translated into practice. Participation in national seminar sponsored by UGC as keynote speaker on climate change propelled me to think of conglomerating articles from various persons into a book on this, which ultimately finds its reification. This long intellectual stride commenced from the day I conceived of inking a book on this taking not only my own ideas but also ideas and facts from various conferences and seminars abroad and in India I attended, imbibing new thinking, approaches and views from scholars I met, and from books I rummaged through, getting articles, and papers published in various journals of international repute, reached its logical conclusion when all these variant strands of ideas seeped through my in-depth analytical acumen and rigour of scientific scrutiny are blended nicely into a synthesised and pithy tapestry. If the collections from others are likened to bricks, mine is the architectural design, adding blood and flesh to the structure and finally shaping the ultimate attic in the form of this book. But for the help and cooperation from others, it could not have been possible to bring out this book in the present form. My unbounded thanks and gratitude flow from the core of my heart to a litany of scholars, intellectuals and critics who have indirectly or directly helped me in this monumental task. From the fathomless depth of my heart, I always convey my sense of gratification to a host of erudite scholars for the materials, articles and books they have provided me without exhibiting even a shred of evidence of their being hesitant.

I am extremely grateful to my mother and late father and my wife Reena for their inspiration and encouragement and forbearance with my indulgence in this work often neglecting the household chores. My love and blessings are always with my two beautiful flowers Tattwamasi and Yogeswar Krishna who have always spurred me inspiringly to the ethos of trailblazing and emerging shining, and withstood difficulties in the preparation of the book.

From the inner cavern of my heart secrets my reverence and love to Swami Samarpanananda Giri and Swami Sudhananda Giri, two great divine and realised souls for their constant blessings and inspiration. Behind all these and the sole doer, but for whom it is not possible to write such a book are my revered Guruji Paramahamsa Prajnanananda and my ever-worshipped Gurudev Paramahamsa Hariharananda. I deem it a God given privilege to sing in paeans of their Beatific Bliss and Glory for their invisible divine hand for laying in bricks after bricks to design this intellectual attic, which I consecrate at their Lotus Feet.

**Narottam Gaan**

# List of Contributors

**Dr. Narottam Gaan, Ph.D, D.Lit,** Professor and Head, P.G Department. of Political Science, Utkal University, Bhubaneswar.

**Dr. Manoj Kumar Mishra**, Asst. Professor, Malaviya Center for Peace Research, Faculty of Social Science, Banaras Hindu University Banaras.

**Dr. Arthur H Westing,** The author is a forest ecologist (Columbia, A.B., 1950; Yale, M.F., 1954; Yale, Ph.D., 1959), former artillery officer in the US Marine Corps, Research Forester with the US Forest Service, Professor of Ecology at various colleges and universities, Senior Researcher at the Stockholm and Oslo International Peace Research Institutes (SIPRI and PRIO), Lecturer at the European Peace University, consultant to the *UN Environment Programme*, *UN Institute for Disarmament Research*, UNESCO, World Bank, *International Committee of the Red Cross*, *International Union for Conservation of Nature* (IUCN), Republic of Korea, *American Association for the Advancement of Science* (AAAS), and other intergovernmental, non-governmental, and national agencies. He is a member of the *World Commission on Protected Areas* and its Transboundary Specialist Group, of the *International Crane Foundation* and its Korean DMZ Task Force, and of the *DMZ Forum* and its Advisory Board.

***Address:*** Westing Association in Environment, Security, and Education, 134 Fred Houghton Rd, Putney, VT 05346, USA. westing@sover.net.

**Md. Rafiqul Islam,** Asst. Professor, Department of Peace and Conflict Studies, University of Dhaka, Bangladesh.

**Dr. Ziaul Haq,** Asst. Professor, Department of Peace and Conflict Studies University of Dhaka, Bangladesh.

**Erandathie Lokupitiya,** Senior Lecturer, Department of Zoology, University of Colombo, Sri Lanka.

**Dr. Dasarathi Bhuyan,** Lecturer in Political Science, Berhampur University Bhanja Vihar, Berhampur, Odisha.

**Dr. Amulya Kumar Tripathy,** Department of Political Science, DAV Autonomous College, Titilagarh, Odisha.

**Dhanasree Jayaram,** A student of M.S. Geopolitics and International Relations at the Department of Geopolitics and International Relations, Manipal University. She is also a non-resident fellow at the Centre for Air Power Studies, New Delhi. She has been working in the areas of environmental security, geopolitics of climate change and environmental change, geostrategic implications of environmental change, environment and international law and treaties, climate change policy, energy and environment and sustainable development.

**Dr. Saubhagyalaxmi Misra,** Asst. Prof, AMA International University, Bahrain.

**Sumitra Mitra and Saptarshi Basu Roy Choudhury,** Ph.D Scholars at Centre for International Trade and Development, School of International Studies, Jawaharlal Nehru University, New Delhi 110067.

**Anjan Kumar Sahu,** Assistant Professor, Department of Public Policy, Law and Governance, Central University of Rajasthan, Bandersindri, N.H. 8, Tehsil-Kishangarh, Dist-Ajmer, Rajasthan-305801.

**Subhashree Patnaik,** Ph.D Scholar, Department of Political Science, Utkal University, Vanivihar, Bhubaneswar.

**Ramakrushna Pradhan,** Assistant Professor, The Department of Political Science, Ram Krishna Mahavidyalaya, Tripura University, Tripura. He is also a Ph.D research scholar in The School of International Studies, Jawaharlal Nehru University, New Delhi.

**Sri Alok Kumar Ray,** Senior Lecturer-in-Political Science, VHJ Mahavidyalaya, Nikirai, Kendrapara, Odisha.

**Sasanka Ghosh**, Assistant Professor in the Dept. of Political Science, Ram Krishna Mahavidyalaya, Tripura University, Tripura.

**Sonali Mohapatra,** Ph.D Research Scholar, Department of Political Science, Utkal University Vanivihar, Bhubaneswar.

# 1

# Introduction: Climate Change and Human Security

*Narottam Gaan*

The increasing disenchantment was with the prevailing notions of development and security and utter inadequacy of the state apparatus to address the challenges from not only military sector but from economic, cultural, environmental and societal sector in the 1960s, 1970s and 1980s.The growing discrepancy between security of state and security of the individuals created an intellectual upheaval to redefine security and trace back the origin of human security. The concern for individual well being and safety occupied a central position in the multinational World Order Models Project to envisage and construct a new just world order (Mendowitz, 1975).

Among the most prominent protagonists of the idea of human security was in the 1970, the Club Rome group, which produced a series of volumes on the world problematique which were premised on the idea that there is "a complex of problems troubling men of all nations: poverty, degradation of the environment; loss of faith in institutions; uncontrolled urban spread; insecurity of employment; alienation of youth; rejection of traditional values; and inflation and other monetary and economic disruption" (Meadows et. al., 1972 : 10). The report noted that:

> Every person in the world faces a series of pressures and problems that require his attention and action. These problems affect him at many different levels. He may spend much of his time trying to find tomorrow's food.... He may be concerned about personal power or the power of the nation in which he lives. He may worry about a world order...or a war next week with a rival clan in his neighbourhood (*ibid*: 17—18).

Accelerating industrialisation as an ideal type of development, burgeoning population growth, widespread malnutrition, depletion of non-renewable resources and degradation of renewable resources did not come under the rubric of traditional state centric security and need to be understood in their disastrous impact on individuals. This implies that there are alternative ways of redefining development and security. As the report envisions, 'a state of global equilibrium could be designed so that the basic material needs of each person...are satisfied, and each person has an equal opportunity to realise his individual human potential' (*ibid*: 21).

In the 1980s, two other independent commissions contributed to the changing thinking on development and security. The first was the Independent Commission on International Development Issues chaired by Willy Brandt, which in 1980 issued the North South report. In his introduction Brandt wrote: 'Our report is based on what appears to be the simplest common interest that mankind wants to survive and one might even add has the moral obligation to survive. This not only raises traditional questions of peace and war, but also how to overcome world hunger, mass misery and alarming disparities between the living conditions of rich and poor (The Independent Commission on International Developmental Issues, North-South: A Programme for Survival, 1980: 13).

The second commission of 1980s, the Independent Commission on Disarmament and Security Issues authored the famous Common Security report, which also turned attention to alternative ways of thinking about peace and security, while it focused on military issues and the nuances of national security, it acknowledged that in the Third World security was also threatened by 'poverty and deprivation, by economic inequality'. The report also noted 'Common security requires that people live in dignity and peace, that they have enough to eat and are able to find work and live in a world without poverty and destitution (The Independent Commission on Disarmament and Security Issues, 1982: xv and 172).

The end of cold war witnessed a spurt of intellectual activity demonstrating the glare failure of neo realism and realism to account for these upheavals and urging upon the imperatives for redefining security. In 1991 the Stockholm Initiative on Global Security and Governance stressing upon 'Common Responsibility in the 1990s' alluded to "challenges to security other than political rivalry and armaments" and to a wider concept of security, which deals also with threats that stem from failures in development, environmental degradation, excessive population growth and movement, and lack of progress towards democracy' (The Stockholm Initiative on Global Security and Governance, 1991: 17-18). Four years later the Commission on Global Governance's report, our Global neighbourhood, was resonant with the sentiments of the

Stockholm initiatives: 'The concept of global security must be broadened from the traditional focus on the security of states to include the security of people and the security of the planet' (The Commission on Global Governance, 1995: 338). These reports' stress on paramount importance of individual security and welfare of the people over and above the state centric security added new impetuses to the intellectual efforts in the 1990s to articulate explicitly human security with some rigour and candour.

The Pakistani development leader and thinker, Mahbub-ul-Haq, championed the concept of 'human security' as a supplement to his 'human development' concept in the 1993, Human Development Report with considerable impact in 1994, Human Development Report. Central to the formulation of human security is the notion that development thinking and policies must hinge on the welfare of individuals rather than simply the macro economy. He added six types of security to a conventional concern with security from physical violence: income security, food security, health security, environmental security, community/identity security, and security of political freedoms. In the words of Des Gasper, 'Just as human development for him [Haq] was more than a humanising addition of health and education concerns to economic growth, so human security meant securing what is humanly central, not only humanising an existing state security discourse by a concern for the physical security of persons' (Des Gasper, 2005: 223).

The second important intervention on human security was that of Canadian Government and various Canadian academics, which launched a middle powers initiative. To understand the difference between human security and traditional national security conceptions, it is necessary to refer to the Political Scientist Baldwin who has argued that in order to evaluate the debate over the conception of security, it is vital to define the term more closely. This requires at the very least two things: agreement on the root meaning of security, for whom security of which values, how much security, security from what threats and security by what means (Baldwin, 1997: 12-18). Baldwin takes as a starting point—Arnold Wolfer's famous discussion of security as "the absence of threat to acquired values" and modifies this to read a low probability of damage to acquired values' (*ibid* : 13). To understand the UNDP and Canadian Schools' approach to Human Security, to answer the four key questions: Security for whom, security of which values, security from what threats, and security by what means remains important. Haq answers the question of 'security for whom' by holding that human security is not about states and nations, but about individuals and people. Thus, he argues that the world is 'entering a new era of human security'. In which 'the entire concept of security will change dramatically. In this new conception, security will be equated with the security of individuals, not just security of their nations, or put it differently, a security of people, not just security of territory.' He says, 'We need to fashion a new concept of human security that is reflected in the lives of our people not in the weapons of our country.' Whereas the traditional conception of security emphasizes territorial integrity and national independence as the primary values that need to be protected. Human security pertains above all to the safety and well being of 'all the people everywhere in their homes, in their jobs, in their street, in their communities, in their environment.' (Mahbub ul Haq, 1994: 1). Haq listed a myriad of threats to human security out of which drugs, disease, terrorism and poverty are pertinent. Besides, an unequal world order scripted by the Bretton Woods system and the dominant development conception and practices, the yawning gap between the North and the South, and the reliance on arms for security poses the greatest threat to human security.

UNDP's Human Development Report of 1994 called 'Redefining Security: The Human Dimensions' begins with providing an alternative to traditional security and a necessary supplement to human development. It answers the question of 'security for whom' by referring to traditional notion of security, which is premised on protecting the territorial states from external aggression, or protection of national interest in foreign policy and provision of global security from the threat of nuclear catastrophe. What this traditional concept of security ignored was the legitimate concern of ordinary individuals for security in their daily lives. The referent object of security is not the state but the individuals or people. It writes:

> Human security is a child who did not die, a disease that did not spread, a job that that was not cut, and ethnic tension that did not explode in violence, a dissident who was not silenced. Human security is not a concern with weapons—it is concern with human life and dignity. It is concerned with how people live and breathe in a society, how freely they exercise their many choices, how much access they have to market and social opportunity and whether they live in conflict or in peace (UNDPa: 1995: 229).

Along with a sense of choice and surety about the future people should be efficacious and empowered enough to be 'able to take care of themselves: all the people should have the opportunity to meet their most essential needs to earn their own living' (*ibid*: 230). Similarly, the Commission on Human Security conceptualised human security in terms of protecting people's vital freedoms from severe threats, in ways that empower people to fulfill their strengths and aspirations. The focus was on people, considered both as individuals and as communities. To advance the goals of human security, the Commission proposed a framework based upon the protection and empowerment of people. Neither protection nor empowerment can be dealt with in isolation as they are mutually reinforcing. Protection refers to norms, processes, and institutions that are required to shield people from threats. Empowerment allows people to act and develop their capacity (Fouinat, 2004:290-91). The report lists seven components or specific values of human security—economic security, food security, health security, environmental security, personal security, immunity security and political security on the basis of which localised threats can be understood (UNDPa, 1995: 230-234).

The question crops up whether the traditional concept of security founded on state and its military apparatus both nuclear and conventional, military alliance and balance of power could be the right response to the congeries of values to be shielded against and the multifarious threats to humans and living beings to be combated. The missiles cannot plug the ozone hole, nuclear weapons cannot halt the ongoing global warming and weapons of mass destruction cannot check the expansion of desertification. The state equipped with all advanced military weapons and technology cannot remove food insecurity, heath hazards and stall sea level rise and Tsunami like catastrophes. Given the above compendium of threats to living beings and values, a new kind of security agenda is to be designed with new instrumentalities, cooperation between and among a vast range of non-state actors including the state to face the crisis. The place of conventionally entrenched military based force in this new security agenda seems wholly inadequate and irrelevant. The Report advocates the concept of human security with changes in national and international policy focusing on fulfilment of basic human needs, productive and remunerative employment and human rights, preventive diplomacy, and preventive development and the reform of global institutions (*ibid* : 236).

The study on environment and security has evolved over the years: from an early primacy on incorporating environmental and its concomitant upshots into the 'definition of security' to putting a new premium on how environmental degradation can be a cause or magnifier of violent conflict both intrastate and interstate. An emerging trend within this evolution of non-military security threats has been a move toward greater emphasis on the concept of human security. Human security is not in opposition to the earlier trends of redefining security or accounting for the environmental roots of violent conflict. It is an offshoot of these two trends. In a broader sense, human security concerned with security of the people in non-military terms is nothing but an extension of environmental security. The very verbiage used to define the term security in these non-traditional and broader senses is today found not dissimilar to that used to understand and define human security.

Those analysts who have focussed on explicating the environmental causes of violent conflict have also brought the debate closer to the concept of human security [which] offers a third perspective that allows us to move beyond conventional security thinking, appreciates both the local and global dimensions of the many insecurities experienced by real individuals and groups, and identifies useful ways of linking security and development policies (Dabelko et. al., : 48).

While the concept of 'human security' has earlier roots, its recent prominence emanates from the 1994 Human Development Report (UNDP, 1994). Its importance was further advanced by the report of the Commission on Global Governance (CGG, 1995). Both reports tried to shift the direction of the security discussion by focusing on issues of human life and human dignity rather than on weapons and territory.

Lorraine Elliott points out two dimensions of the human security paradigm:

> The first is that the concept of 'human security' provides an antidote to the more conventional focus on states, borders and territorial integrity. The answer to the question 'security for whom' is not the state but the individual and communities, which suggests that even when a state is secure from external threats or internal instabilities, security for its people is not guaranteed. Protecting individuals and communities from the consequences of environmental decline is therefore a security issue. The second dimension is that human insecurity (which includes equity, gender, human rights, and identity concerns) is a central factor in social tensions and political instabilities and conflicts that can...become a feature of state insecurity.... If peoples and communities are insecure economically, socially, politically, environmentally, state security can be fragile or uncertain. Environmental security becomes a distributive equity problem rather than one simply of market failure, externalities or zero sum calculations about access to resources and environmental services (Lorraine Elliott, 2001: 449).

The primacy of state security is strongly embedded in the notion of sovereignty. In its historic meaning, sovereignty implied the security of the sovereign, or 'The Prince'. The emergence of democratic polity and the transfer of power to the 'Citizen' question interestingly the very concept of security rooted in the sovereignty. With sovereignty now ensconcing in the citizenry rather than in the Prince, the notion of security must also be broadened to include the security of the people not the apparatus of the prince. Such a conceptual shift in understanding security does not mince in any way the importance of state security, but it does explicate the need to broaden the term. It is no longer sufficient to hold security of state as synonymous with security of people. Security in conventional parlance understood as synonymous with territorial exclusivity or Prince fails to give security to the individuals and communities. The inadequacy of state security and its brittle impregnability are nowhere felt so grotesquely than in the case of global climate change. Even the most powerful state well equipped with all modern and sophisticated weapons will come to its knees in the face of climate change affecting millions of people in the both developed and developing world. Environmental security and human security meet the same confluence where the need for sustainable development, and new environmental ethics emerging to redefine western pattern of development and its inordinate life style, meet together.

**Climate Change and Human Security**

The question how will changing climate patterns affect both inter-state relations and international and national security in a narrow, geo-strategic sense as well as the well-being and survival of human beings and humankind and thus human, water, health and livelihood security, comes to the fore with the findings of scientists about the imminence and impending danger of climate change looming large on the entire humanity. On 6 April, 2007, the Intergovernmental Panel on Climate Change (IPCC) released the second part of its Fourth Assessment Report demonstrating that the poor of this planet are most likely to suffer the worst effects of climate change. Human security takes on a broader meaning when one considers basic needs for food, water, health—in short, livelihood and a place to live—the issues addressed in the Millennium Development Goals:

> Poor communities can be especially vulnerable, in particular, those concentrated in high-risk areas. They tend to have more limited adaptive capacities, and are more dependent on climate-sensitive resources such as local water and food supplies (IPCC, 2007).

Furthermore, within the larger category of 'the poor' lies the frequently invisible (including within that IPCC summary document) group: women. Worldwide, seventy percent of those living below the poverty line are women (Ulrike Röhr, 2006: 3-7) for whom climate change represents very specific threats to security. When the impacts of climate change are brought home, then women, in their roles as the primary managers of family, food, water and health, must deal very directly with the impacts.

While natural climate variations have existed for millennia, anthropogenic climate change has gradually emerged since the industrial revolution and especially after World War II due to western pattern of development based on the availability of cheap fossil fuels (coal, oil, natural gas) and the dramatic increase in its consumption first primarily in the industrialised countries but now increasingly also in the rapidly growing economies of the BRIC states (Brazil, Russia, India and China), especially of China and India.

**Climate Change, Security and Conflicts**

Ben Wisner and others have established the linkages between climate change and conflict having implications on national and international security (Ben Wisner and others). On 9 January 2004, David King, the UK Government's chief scientific adviser claimed that climate change is a far greater threat to the world than international terrorism. In February 2004, John Reid MP, then British Secretary of State for Defence and now Home Secretary, argued that climate change might spark conflict between nations. He forecast that violence and political conflict would become more likely in the next 20 to 30 years as climate change turned land into desert, melted ice fields and poisoned water supplies. He listed climate change alongside the major threats in future decades, including terrorism, demographic changes, and global energy demand. 'As we look beyond the next decade, we see uncertainty growing; uncertainty about the geopolitical and human consequences of climate change.'... Impacts such as flooding, melting permafrost and desertification could lead to loss of agricultural land, poisoning of water supplies and destruction of economic infrastructure.... More than 300 million people in Africa currently lack access to safe water; climate change will worsen this dire situation (Ben Russell and Nigel Morris, 2006). John Ashton, the UK Foreign Secretary's Special Representative for Climate Change, said at a conference on 'Climate Change: The Global Security Impact', at the Royal United Services Institute on 24 January 2007: "There is every reason to believe that as the 21st century unfolds, the security story will be bound together with climate change (Ben Vogel, 2007). 'Climate change is a security issue because if we don't deal with it, people will die and states will fail,' Ashton concluded. Further, he pointed out that defense and security planners must face a paradox when assessing their responses to the problem. Most security threats in today's world are amenable to some extent to a 'hard power' or conventional reaction, he said, and demand will rise for such responses to climate change-related security problems. 'But there is no hard power solution to climate change—you cannot force your neighbour to change its carbon emissions at the barrel of a gun' (Littlecott, 2007). Sir Crispin Tickell, the former UK permanent representative to the UN, highlighted the environmental factors behind societal collapse. Professor John Mitchell, the chief scientist at the UK Met Office, forecast that the coming decades will see a 30 per cent increase in severe drought. He added that Africa would experience increased desertification, water stress and disease (Tickell, 2003).

Besides the UK, other nations have begun to assess the security implication of climate change. In 2002, the German Ministry for the Environment, Nature Conservation and Nuclear Safety published a commissioned report on climate change and conflicts raised the question whether climate change impacts can increase conflict potentials (Brauch, 2002). In spring 2004 an internal report by Randall and Schwartz for the U.S. Department of Defense on the impact of Abrupt Climate Change on U.S. national security was leaked to the press (*ibid.* 2004).

The British initiative during its Security Council Security Council presidency to put climate change on its agenda for 17 April 2007 has been the most recent attempt to 'securitize' climate change in the context of geo-politics (Press Conference by Security Council President, 4 April 2007).

## Climate Change as a Threat to International, National and Human Security

Climate change poses many new threats and challenges to national security and international stability as well as to human security at other scales. The concept of human security apart from being introduced first by UNDP in 1994 and then developed in a report by the Human Security Commission, co-chaired by Sadako Ogata and Amartya Sen, in its report *Human Security Now* (2003). Its environmental dimension has been addressed by an international team working on Global Environmental Change and Human Security (GECHS) and in several studies by the United Nations University Institute for Environment and Human Security (UNUEHS) (For GECHS, see at http://www.gechs.org/).

In February 1999, during its presidency of the United Nations Security Council, Canada, a founding member of the Human Security Network put human security on the agenda by addressing the impact of armed conflicts on human beings (Dedring, 2007). In March 2005, the UN Secretary General Kofi Annan, in his report *In Larger Freedom (Kofi Annan, 2005)* wrote of human security in a broad sense, the issue was placed on the agenda of the UN General Assembly in the fall on 2005.

Over 17 years, UNDP has incrementally developed a sensitive measure of human development (the Human Development Index – HDI). Earlier studies have shown that the HDI correlates well with measures of disaster risk such as UNDP's disaster risk index (DRI), especially for less developed countries (UNDP, 2004). Preliminary analysis for the year's HDR suggests that climate change poses major obstacles to progress in meeting MDGs and maintaining progress raising the HDI. 'There is a clear and present danger that climate change will roll back human development for a large section of humanity, undermining international cooperation aimed at achieving the Millennium Development Goals (MDGs) in the process (UNDP, 2007).

## Linkages between Climate Change and Security

Some effects of climate change are already evident and will become very clear in the human and climatic short run (2007-20). They will increase and others will manifest themselves in the medium term (2021-50); whilst in the long run (2051-2100), they will all be active and interacting strongly with other major trends: the end of the petroleum economy for many producing and consuming nations, possible financial and economic crisis, a larger population of humans, and a much more urbanised humanity – far in excess of the 50 per cent now living in small to very large cities. All these processes will be accompanied by redistribution of population nationally and internationally (Simon, 2007). Such redistributions typically have significant gender dimensions, for example, extreme event impacts can lead to male out migration in search of work; culminating in an increase in women-headed households – a group often considered particularly vulnerable (Delaney and Shrader, 2000).

The linkages between climate change and security are complex in many ways. To begin with, climate change involves the interactions of many systems such as the atmosphere, hydrosphere, cryosphere, and biosphere that are immensely complex in their own right. Thus, a recurrent theme in IPCC reports is the significance of thresholds and non-linearities. When human systems are added to the mix, complexity escalates (Millennium Ecosystem Assessment (2006). Livelihood security and other aspects of human security interact with geo-strategic (or 'hard') security issues because of the national and regional upheavals that climate stress may put on livelihood systems already vulnerable and incapable of adapting (Wisner et. al., 2004). Worldwide the rural and urban poor are already under stress, and for some groups such as women headed households in Africa, adaptation to climate-induced stress will be very difficult indeed. As Simon points out, climate change has both intermittent but increasingly frequent, extreme impacts (such as, large storms and heat waves) and slow on-set, pervasive, cumulative effects (such as, extinction of life forms and sea level rise) (Simon, 2007). Both kinds of effects may have a role in displacing human populations and disrupting their livelihoods. Some major climate changes may actually occur rapidly.

Some efforts by state actors to mitigate and adapt to climate change may also further stress weak and marginal sections of the population such as indigenous groups and ethnic minorities, increasing discontent and alienation. In particular, large-scale water management and forestry projects in the past have displaced such groups (Art Hansen and Anthony Oliver-Smith, 1982) and without safeguards are likely to do so as states expand mega-projects as part of their national climate adaptation programs. One example is the new dams being built in Guatemala. Such intra-state tension and possible conflict over the distribution of winners and losers in climate adaptation may spill over into regional conflicts, on the basis of recent experience in Darfur. State actor adaptations may also weaken treaties such as regional water basin management and lead eventually to inter-state conflict. For example, Sudan's decision to build a new dam on the Nile could have that result.

One must also consider that the world around us in 2007 has in it a large number of weak and crisis-prone nation states (http://www.crisisstates.com/News/seminars1.htm). It is not likely that such chronic instability will diminish (although details of the pattern may shift geographically) before severe impacts of climate change are felt. Humanitarian intervention in the crises that are likely will become more difficult and run the danger of exacerbating conflict, especially as civilian humanitarian and military relations become more interwoven.

**Climate-induced Conflict Constellations**

WBGU identifies four conflict constellations in which critical developments can be anticipated because of climate change and which may occur with similar characteristics in different regions of the world. 'Conflict constellations' are defined as typical causal linkages at the interface of environment and society, whose dynamic can lead to social destabilisation and, in the end, to violence.

**Climate-induced Degradation of Freshwater Resources and Conflict**

About 1.1 billion people have currently access to unsafe drinking water. The situation could worsen for hundreds of millions of people as climate change alters the variability of precipitation and the quantity of available water. At the same time, demand for water is increasing due to the world's growing population and its mounting aspirations. This dynamic triggers distributional conflicts and poses major challenges to water management systems in the countries concerned. For example, regions, which depend on, melt water from mountain glaciers, which are at risk from climate change, – will require new water management strategies and infrastructures, as well as political efforts to avert national or even trans-boundary conflicts over the distribution of increasingly scarce water resources. However, the countries, which will suffer the greatest water stress, are generally those, which already lack of the political and institutional framework necessary for the adaptation of water and crisis management systems. This could overstretch existing conflict resolution mechanisms, ultimately leading to destabilisation and violence.

There is no dearth of evidence in the past to show as a population growth/demand problem, water scarcity may, indeed, become a source of conflict and instability, but more as a function of supply (Hartmann, 2003). International regimes such as the Nile Treaty are old, inadequate, and fragile. Newer, more forward-looking treaty regimes, such as, the Nile Basin Initiative, are still in their infancy and subject to similar divisive pressures despite the efforts of multilateral agencies like The World Bank. The other tensions and sources of instability discussed above could also place international water management and sharing agreements under pressure. Management of many of the world's 261 internationnal rivers will face severe tests. Furthermore, many coastal freshwater aquifers will suffer salinisation because of sea level rise. In Africa alone:

By 2020, between 75 and 250 million people are projected to be exposed to an increase of water stress due to climate change. If coupled with increased demand, this will adversely affect livelihoods and exacerbate water-related problems (IPCC, 2007).

A team at Keele University and the UK's Centre for Ecology and Hydrology has produced a water poverty index (WPI) Using the WPI they found that already in 2005 a large number of countries were suffering moderate, high, or severe stress. If this is the baseline situation, what will additional stress due to climate change do in many of these places?

**Climate-induced Decline in Food Production and Conflict**

More than 850 million people worldwide are currently undernourished. This situation is likely to worsen in future because of climate change, as food insecurity in the lower latitudes, i.e., in many developing countries, will increase with a temperature rise of just 2°C (relative to the 1990 baseline). With global warming of 2–4°C, a drop in agricultural productivity is anticipated worldwide. This trend will be substantially reinforced by desertification, soil salinisation or water scarcity. In South Asia and North Africa, for example, the areas suitable for agriculture are already largely exploited. This may well trigger regional food crises and further undermine the economic performance of weak and unstable states, thereby encouraging or exacerbating destabilisation, the collapse of social systems, and violent conflicts.

Food and livelihood pressure due to climate change will lead to populist and/or military coups in a number of countries. After the rollout of macro-economic 'structural adjustment' programmes in Africa in the 1980s, one witnessed junior officers in a number of militaries seizing power in the name of workers and peasants who suffered (e.g., in Burkina Faso). This will produce continuing instability in Africa, in particular. Between 1980 and 2001, there were 95 attempted coups in Africa and 33 of them successful. Popular discontent over livelihood security was a contributing cause of many of the conflicts (McGowan, 2003). The same pressures as well as the 'push' provided by conflict will cause considerable population movements and displacement both within countries and internationally (Spanish Environment Ministry, 2006). That, in turn, will increase insecurity – a process that is already occurring. Effects in Africa may include the following:

> Agricultural production, including access to food, in many African countries and regions is projected to be severely compromised by climate variability and change. The area suitable for agriculture, the length of growing seasons and yield potential, particularly along the margins of semi-arid and arid areas, are expected to decrease. This would further adversely affect food security and exacerbate malnutrition in the continent. Local food supplies are projected to be negatively affected by decreasing fisheries resources in large lakes due to rising water temperatures, which may be exacerbated by continued over-fishing. (IPCC, 2007)

Women are responsible for around 70 per cent of household food production in sub-Saharan Africa, often on the basis of informal resource rights. Climate-induced changes in crop and livestock production could threaten those rights, as well as affect the gendered division of labor with negative effects on women's and men's incomes, livelihoods and household security (CIDA).

The climate-food connection will not only affect the poor. In some industrial and industrialising nations, as climate change creates new patterns of food production—new exporting and new importing zones–access to supplies and the energy to import them may become strategic concerns and lead to international conflict.

**Climate-induced Increase in Storm and Flood Disasters and Conflict**

Climate change is likely to result in further sea-level rise and more intensive storms and heavy precipitation. This will greatly increase the risk of natural disasters occurring in many cities and industrial regions in coastal zones. Those risks will be further amplified by deforestation along the upper reaches of rivers, land subsidence in large urban areas and the ever-greater spatial concentration of populations and assets. Storm and flood disasters have already contributed to conflict in the past, especially during phases of domestic political tension, e.g., Central America, India and China. Conflicts are likely to occur more frequently in future, firstly because of regions especially at risk from storm and flood disasters, such as Central America and Southern Africa. Generally they have weak economic and political capacities, making adaptation and crisis management much more difficult. Secondly, frequent storm and flood disasters along the densely populated east coasts of India and China could cause major damage and trigger and/or intensify migration processes that are difficult to control.

**Environmentally-induced Migration and Conflict**

Experience has shown that migration can greatly increase the likelihood of conflict in transit and target regions. It can be assumed that the number of environmental migrants will substantially rise in future due to the impacts of climate change. In developing countries in particular, the increase in drought, soil degradation and growing water scarcity in combination with high population growth, unstable institutions, poverty or a high level of dependency on agriculture means that there is a particularly significant risk of environmental migration occurring and increasing in scale. Most environmental migration is initially likely to occur within national borders. Trans-boundary environmental migration will mainly take the form of south-south migration, but Europe and North America must also expect substantially increased migratory pressure from regions most at risk from climate change. The question as to which states will have to bear the costs of environmentally induced migration in future also contains conflict potential.

**Climate Change Induced Health Hazards and Conflict**

Health impacts of climate change including epidemics and insect outbreaks will have a similar effect, also compounding food and livelihood crises. This is beginning, but it will accelerate in the medium term. As we have seen with SARS and avian influenza, disease does not respect national boundaries in a globalised world. Changing climate may bring many epidemiological surprises as vector habitats change, sometimes quite rapidly.

Projected climate change-related exposures are likely to affect the health status of millions of people, particularly those with low adaptive capacity, as following:

- Increases in malnutrition and consequent disorders, with implications for child growth and development.
- Increased deaths, disease and injury due to heat waves, floods, storms, fires and droughts.
- The increased burden of diarrheal disease; the increased frequency of cardio-respiratory diseases due to higher concentrations of ground level ozone related to climate change and the altered spatial distribution of some infectious disease vectors.

Climate change is expected to have some mixed effects, such as the decrease or increase of the range and transmission potential of malaria in Africa (IPCC, 2007).

**Climate Change Induced Mitigation-adaptation-inequity and Conflict**

Adger et al. raise the issue of 'fairness' in adaptation to climate change (Adger et al., eds., 2006). Mega-projects conceived by nation states as solutions to climate change such as the planting of large scale forestry under the Kyoto Protocol's Clean Development Mechanism (CDM) and the building of more large dams and reservoirs, are beginning to displace large numbers of poor and marginal people, having effects on politics and stability similar to those mentioned above while negatively affecting biodiversity. In the medium term mega-project investments - already a major feature of rapid urban industrialisation in China and India, the world's two most populous countries - will grow rapidly as water shortages intensify. Already 'an estimated 40–80 million people have been forcibly evicted from their lands to make way for dams' (International Rivers Network, *About Rivers and Dams* at http://www.irn.org/basics/ard/).

The social and economic impacts of climate change focused mitigation projects have not yet been fully assessed. Skutsch, for example, highlights the absence of gender impact analysis of CDM projects (Adger et. al. eds, 2006). The 'clean, green' solution of nuclear energy so heavily promoted in the early years of this century and recently adopted as a climate change-fighting strategy by the EU and others has led to a planet-wide race to capture extractive rights and exploit uranium reserves. As seen, the historically high value of uranium has led to violent conflict in the Congo and the renewed use of forced labor. The profound, degenerative human health consequences of working and living in a toxic and radioactive environment are only recently being understood. Epidemics of cancer and other radiogenic health problems now prevalent in areas that hosted the Cold War nuclear machine are certain to expand (Johnston, Barbara Rose, 2007). In addition, a new phase in the growth of nuclear power could exacerbate the problem of 'leakage' of weapons grade material into the hands of non-state actors.

## Climate Change Induced Food–Fuel Conflict and Human Security

Large-scale investments in bio-fuels in the medium and long term as a substitute for green house gas-producing petro-based energy sources may have the perverse effect of taking considerable land out of food production and diverting food grains, thus raising food prices and eroding biodiversity. The maize (corn) price in Mexico has already increased because of demand for grain by US ethanol plants, and protests have erupted (Johnston, Barbara Rose, 2007).

If this occurs, then food and livelihood pressures will increase and the desperation of many rural and urban people will increase national and regional insecurity. Rapid expansion of sugar production in Brazil and Africa palm in Colombia, as energy feed stocks, have been a major source of displacement of small farmers from their lands. In addition, recent research has shown the energy and carbon efficiency of bio fuel production to be variable and often lower than with the burning of fossil fuels (Sauser, 2007).

### Climate Change Induced Disasters Threatening Livelihood–Governance Security

Also in the medium term livelihood security and hence the ability to govern will be undermined by the increasingly frequent occurrence of more and more mega-disasters such as those associated with hurricane Mitch (1998), the Orissa super-cyclone (1999), hurricanes Katrina and Rita, and tropical storm Stan (2005). Even now, the UN reports that over a million people are threatened by erratic weather in five southern African countries (UN, Integrated Regional Information Networks [IRIN]: 5 April 2007). Such catastrophic weather events often produce cascades of secondary physical hazards such as landslides (as Nicaragua during Mitch or in Vargas, Venezuela in 1999) or downstream inundation when dams overflow or dam authorities release large volumes of water to safeguard large dams (as in Mozambique at present and in several recent years). They also trigger changes in social relations, including the exacerbation of unequal gender relations manifested in lack of land and property rights, and rising sexual and gender based violence towards women and girls (Fordham et al., 2006). In urban industrial landscapes such as greater New Orleans, Manila, or Osaka, storms and flooding often bring further complications by damage done to factories, storage facilities, and pipelines. The resulting 'natural-technological' (natech) hazards are very expensive to clean up and can have long-term public health consequences (*ibid.*, 2006). Institutional systems for anticipating or even timely recognition of the 'surprises' that natech hazards may be present in the future and are not yet been developed worldwide:

> Where extreme weather events become more intense and/or more frequent, the economic and social costs of those events will increase, and these increases will be substantial in the areas most directly affected. Climate change impacts spread from directly impacted areas and sectors to other areas and sectors through extensive and complex linkages (IPCC, 2007).

Disasters, development, and conflict have been shown to have complex interactions with one other quite apart from the additional stresses and management challenges likely to accompany increasingly variable weather and more extreme storms (Ben Wisner, 2008).

### Climate Change Induced Sea Level Rise and Displacement Security Threats and Conflicts

Finally, sea level rise in the long term and its collateral impacts on river flow and ocean discharge regimes will cause displacement of many millions of people currently living in coastal areas. For the small island nations of the world, especially the many cultural groups living on coral atolls, entire nations face complete submersion. A recent study calculates that around 634 million people are living less than 10 m above sea level. The study notes that 'of the more than 180 countries with populations in the low-elevation coastal zone, about 70 per cent have urban areas of more than five million people that extend into it.' The authors list of Tokyo, Japan; New York, US; Mumbai, India; Shanghai, China; Jakarta, Indonesia; and Dhaka, Bangladesh (Thomas Wagner, 28 March 2007). One could add other cities as well to the list, for example, Cartagena, Colombia; Lima, Peru; Buenos Aires, Argentina; and Recife and Rio de Janeiro, Brazil. African exposure is high.

Nigerian scholars Ibe and Awosika state: 'This coastal zone consists of four major basins which are bordered on the ocean side by low-lying coastlines which are sandy and muddy in some cases. General beach elevations range from 2-3 m above sea level (Ibe and Awosika, 1991). Most of Africa's major cities are coastal, including Dakar, Senegal; Accra, Ghana; Lagos, Nigeria; Luanda, Angola; Cape Town and Durban, South Africa; Maputo, Mozambique; Dar es Salaam, Tanzania; and Mombasa, Kenya; and Mogadishu, Somalia. The cost of dislocation - which will include salinisation of coastal aquifers and other low-lying fresh water resources — and loss of infrastructure will be difficult for poor countries to bear, especially in Africa (Simon, 2007). The Intergovernmental Panel on Climate Change agrees (IPCC, 1997) predicts:

> Towards the end of the 21st century, projected sea-level rise will affect low-lying coastal areas with large populations. The cost of adaptation could amount to at least 5-10 per cent of GDP. Mangroves and coral reefs are projected to be further degraded, with additional consequences for fisheries and tourism (IPCC, 2007).

Additional negative feedback can be anticipated as degradation of coastal wetland vegetation and coral reefs reduce or remove their protective influence in the face of storms.

**Six Threats to International Stability and Security**

In light of current knowledge about the social impacts of climate change, WBGU identifies the following six key threats to international security and stability, which will arise if climate change mitigation fails (WBGU):

1. **Possible increase in the number of weak and fragile states because of climate change**

Weak and fragile states have inadequate capacities to guarantee the core functions of the state, notably the state's monopoly on the use of force, and therefore already pose a major challenge for the international community. So far, however, the international community has failed to summon the political will or provide the necessary financial resources to support the long-term stabilisation of these countries. Moreover, the impacts of unabated climate change would hit these countries especially hard, further limiting and eventually overstretching their problem-solving capacities. Conflict constellations may also be mutually reinforcing, e.g., if they extend beyond the directly affected region through environmental migration and thus destabilise other neighbouring states. This could ultimately lead to the emergence of 'failing sub-regions' consisting of several simultaneously overstretched states, creating 'black holes' in world politics that are characterised by the collapse of law and public order, i.e., the pillars of security and stability. It is uncertain at present whether, against the backdrop of more intensive climate impacts, the international community would be able to curb this erosion process effectively.

**2. Risks for global economic development**

Climate change will alter the conditions for regional production processes and supply infrastructures. Regional water scarcity will impede the development of irrigated agriculture and other water-intensive sectors. Drought and soil degradation will result in a drop in agricultural yields. More frequent extreme events such as storms and flooding put industrial sites and the transport, supply and production infrastructures in coastal regions at risk, forcing companies to relocate or close production sites. Depending on the type and intensity of the climate impacts, this could have a significant and adverse effect on the global economy. Unabated climate change is likely to result in substantially reduced rates of growth. This will increasingly limit the economic scope, at national and international level, to address the urgent challenges associated with the Millennium Development Goals.

**3. Risks of growing international distributional conflicts between the main drivers of climate change and those most affected**

Climate change is mainly caused by the industrialised and newly industrialising countries. The major differences in the per capita emissions of industrialised and developing/newly industrialising countries are increasingly regarded as an 'equity gap', especially as the rising costs of climate change are mainly being borne by the developing countries. The greater the damage and the burden of adaptation in the South, the more intensive the distributional conflicts between the main drivers of climate change and those most affected will become. The worst affected countries are likely to invoke the 'polluter pays' principle, so international controversy over a global compensation regime for climate change will probably intensify. Beside today's industrialised countries, the major ascendant economies whose emissions are increasing substantially, notably

China but also India and Brazil, for example, will also be called to account by the developing countries in future. A key line of conflict in global politics in the 21st century would therefore divide not only the industrialised and the developing countries, but also the rapidly growing newly industrialising countries and the poorer developing countries. The international community is ill prepared at present for this type of distributional conflict.

**4. The risk to human rights and the industrialised countries' legitimacy as global governance actors**

Unabated climate change could threaten livelihoods, erode human security and thus contribute to the violation of human rights. Against the backdrop of rising temperatures, growing awareness of social climate impacts and inadequate climate change mitigation efforts, the $CO_2$-emitting industrialised countries and, in future, buoyant economies such as China could increasingly be accused of knowingly causing human rights violations, or at least doing so in *de facto* terms. The international human rights discourse in the United Nations is therefore also likely to focus in future on the threat that climate impacts pose to human rights. Unabated climate change could thus plunge the industrialised countries in particular into crises of legitimacy and limit their international scope for action.

**5. Triggering and intensification of migration**

Migration is already a major and largely unresolved international policy challenge. Climate change and its social impacts will affect growing numbers of people, so the number of migration hotspots around the world will increase. The associated conflict potential is considerable, especially as 'environmental migrants' are currently not provided for in international law. Disputes over compensation payments and the financing of systems to manage refugee crises will increase. In line with the 'polluter pays' principle, the industrialised countries will have to face up to their responsibilities. If global temperatures continue to rise unabated, migration could become one of the major fields of conflict in international politics in future.

**6. Overstretching of classic security policy**

The future social impacts of unabated climate change are unlikely to trigger 'classic' inter-state wars; instead, they will probably lead to an increase in destabilisation processes and state failure with diffuse conflict structures and security threats in politically and economically overstretched states and societies. The specific conflict constellations, the failure of disaster management systems after extreme weather events and increasing environmental migration will be almost impossible to manage without support from police and military capacities, and therefore pose a challenge to classic security policy. In this context, a well-functioning cooperation between development and security policy will be crucial, as civilian conflict management and reconstruction assistance are reliant on a minimum level of security. At the same time, the largely unsuccessful operations by highly equipped military contingents, which have aimed to stabilise and bring peace to weak and fragile states since the 1990s show that 'classic' security policy's capacities to act are limited. A climate-induced increase in the number of weak and fragile states or even the destabilisation of entire sub-regions would therefore overstretch conventional security policy.

**Overstretching the Capacities of the Global Governance System**

The greater the scale of climate change, the greater the probability that in the coming decades, climate-induced conflict constellations will impact not only on individual countries or sub-regions but also on the global governance system as a whole. These new global risk potentials can only be countered by policies that aim to manage global change. Every one of the six threats to international stability and security, outlined above, is itself hard to manage. The inter-action between these threats intensifies the challenges for international politics. It is almost inconceivable that in the coming years, a global governance system could emerge with the capacity to respond effectively to the conflict constellations identified by WBGU. Against the backdrop of globalisation, unabated climate change is likely to overstretch the capacities of a still insufficient global governance system.

As the climate-induced security risks of the 21st century have their own specific characteristics, they will be difficult to mitigate through classic military interventions. Instead, an intelligent and well-crafted global governance strategy to mitigate these new security risks would initially consist of an effective climate policy, which would then evolve into a core element of preventive security policy in the coming decades. The more climate change advances, the more important adaptation strategies in the affected countries will become and these must be supported by international development policy. At international level, the focus will be on global diplomacy to contain climate-induced conflicts, as well as on the development of compensation mechanisms for those affected by climate change, global migration policy, and measures to stabilise the world economy. The opportunities to establish well-functioning global governance architecture will narrow as global temperatures rise, revealing a vicious circle: climate change can only be combated effectively through international cooperation, but with advancing climate change, the basis for constructive multilateralism will diminish. Climate change thus poses a challenge to international security, but classic, military-based security policy will be unable to make any major contributions to resolving the impending climate crises (ibid.).

## The Link between Carrying Capacity and Warfare

Today, carrying capacity, which is the ability for the Earth and its natural ecosystems including social, economic, and cultural systems to support the finite number of people on the planet, is being challenged around the world. As predicted, abrupt climate change is likely to stretch carrying capacity well beyond its already precarious limits. As abrupt climate change lowers the world's carrying capacity aggressive wars are likely to be fought over food, water, and energy. Steven LeBlanc, Harvard archaeologist and author of a new book called Carrying Capacity, describes the relationship between carrying capacity and warfare. Drawing on abundant archaeological and ethnological data, he argues that historically humans conducted organised warfare for a variety of reasons, including warfare over resources and environment. Humans fight when they outstrip the carrying capacity of their natural environment (Schwartz and Randall, 2003). Peace occurs when carrying capacity goes up. But such peaceful periods are short-lived because population quickly rises to once again push against carrying capacity, and warfare resumes. Indeed, over the millennia, most societies define themselves according to their ability to conduct war, and warrior culture becomes deeply ingrained. With carrying capacity lowered by abrupt climate change, humanity would revert to its norm of constant battles for diminishing resources, which the battles themselves would further reduce even beyond the climatic effects. Once again, warfare would define human life. As famine, disease, and weather-related disasters strike due to the abrupt climate change, many countries' needs will exceed their carrying capacity. This will create a sense of desperation, which is likely to lead to offensive aggression in order to reclaim balance (*ibid.*).

## Conclusion

Premised on Newtonian and Cartesian epistemology, the western political and economic thinking employed science to the ends of human beings—a commodious and luxurious life. Not only that, it borrowed the insights of science to view nature as the lifeless other to be exploited for the attainment of these ends. This resulted in defining development based on fossil fuel technology and industrialisation, which destroyed the nature with ramifying consequences upon the entire humankind. The result of following the western pattern of development since the beginning of industrialisation has been the accumulation of greenhouse gases in the atmosphere with the devastating consequences as noted above. The more a country is industrialised and considered developed, the more greenhouse gases it has emitted to the atmosphere. On this account, the industrialised North through its centuries of progress on this development trajectory has usurped the resources of the atmosphere at the cost of the developing and underdeveloped countries, and polluted the atmosphere. As the rich industrialised countries have unilaterally and inequitably messed up the atmosphere, they have squarely the historic responsibility of disabusing it of the debris. This is the argument put forth in the climate change debate by developing countries in response to American attempt at bringing the former into the task. Today, the debate remains hanging on a crucial contradiction between the principles of inequity the North have been practicing since industrial revolution as to the uncontested access to the atmosphere denying the same to the South at the cost of nature and future generations, and the preachment of equity principle and meaningful participation to the South on the question of cleansing the atmosphere.

The governments of rich countries have set the wrong targets to tackle climate change using outdated science. A paper published in 2006 by climatologist Malte Meinshausen suggests that if greenhouse gases reach a concentration of 550 parts per million (ppm) carbon dioxide equivalents, there is a 63–99 per cent chance that global warming will exceed two degrees. At 475 parts per million the average likelihood is 64 per cent. Only if concentrations are stabilised at 400 parts per million or below is there a low chance that temperatures will rise by more than two degrees. The IPCC draft report contains similar figures. A concentration of 510 ppm gives a 33 per cent chance of preventing more than two degrees of warming. A concentration of 590 ppm gives a ten per cent chance. The current level of greenhouse gases in the atmosphere is 459 ppm. To give a high chance to humanity of preventing dangerous climate change, what is needed a programme so drastic that greenhouse gases in the atmosphere end up below the current concentrations. However, no government has set itself this task. The EU and Swedish government have established the world's most stringent target, which is 550 ppm. It is of carbon dioxide alone. However, this target gives the human beings a near certainty of an extra 2°C. When other greenhouse gases are included, this translates into 666 ppm, carbon dioxide equivalent. According to last autumn's Stern Report on the economics of climate change, at 650 ppm, there is a 60–95 per cent chance of 3°C of warming.

In his book, Heat and George Monbiot estimated that to avoid two degrees of warming a global emission cut of 60 per cent per capita between now and 2030 is highly required. This translates into an 87 per cent cut in the UK. A recent paper of Climate Change in a journal emphasises the sensitivity of global temperatures to greenhouse gas concentrations remains uncertain. But using the average figure, to obtain a 50 per cent chance of preventing more than 2°C of warming requires a global cut of 80 per cent by 2050. This is a cut in total emissions, not in emissions per head. If the population were to rise from 6 billion to nine billion between now and then, an 87 per cent cut in global emissions per person. If carbon emissions are to be distributed equally, the biggest polluters must make the greater cut: rich nations like US and other western countries. The UK's emissions per capita would need to fall by 91 per cent. But the rich countries appear to quietly have abandoned their aim of preventing dangerous climate change, condemning millions to death. What the IPCC report shows is that the time is to stop treating climate change as an urgent issue. The nations of the world have to start treating it as an international emergency. Since the United States contributes about 25 per cent of the world's $CO_2$ emissions, its own policy could make a large difference.

Global climate change is different from other environmental problems. First, emissions of $CO_2$ and other trace gases are almost irreversible; more precisely, their residence time in the atmosphere is measured in centuries. Most environmental problems are mitigated promptly or unfairly short order when the source is cleaned up, as with water pollution, acid rain or sulphur dioxide emissions. Here, reducing emissions today is very valuable to humanity in the distant future. Second, the scale of the externality is truly global; greenhouse gases travel around the world in a few days. This means that the nation-state and its subsidiaries, the typical loci for internalisation of externalities, are limited in their remedial ability. The poor countries appear to be most vulnerable to the dangerous climate change, mostly a doing of the rich.

Coming to the very root of the crisis, Science provided the ontological foundation to Political Science to look at the nature and the poor who do not have the wherewithal to have a commodious living as 'the other' to be exploited or harnessed for one's end. The entire Earth was looked down upon as a compendium of lifeless raw materials to cater to the materialistic ends of those humans who are propertied or rational in the words of English philosopher Locke, because they have property. To attainment of these goals-wealth production and commodious living by the wealthy few believing in exorbitant consumption and inordinate life style as quintessential of development, the entire Earth has been subjugated and robbed of its bounties at the cost of nature, other humans, non-human living and future generations. That everybody is a self-sufficing and aggrandising part disjointed from the whole has resulted in a mechanical view of the universe. The concept of development based on this paradigm has been the development of one against the other. Woven into this paradigm was the concept of security, which was the security of one or one group, or one state against the other. But today, the devastating consequences of climate change and its pervasiveness impacting the very survival of entire humanity have spurted the traditional security planners into thinking anew what so far they have defined as development and established frame of mind about security in terms of disjointed parts and others as mere objectivities not as inter subjectivities. A profound affirmation of this is found in Eastern insight epitomised in the Indian Upanishads: 'The entire universe is like a family' [*Basudheiba Kutumbakam*]. No man is like an island. Everybody is the inseparable whole. In more mundane terms, as the rich countries and wealthy people have enjoyed, usurped, and destroyed the atmosphere at the cost of the other (humans, nature, living and non-living, and the future generations), they must not only come forward to cut down its consumption, rethink its life style and pattern of development, and follow exemplarily new environmental ethics but also help, support and promote the rest of the world in adapting to the changing planetary system. Martin Parry, a climate scientist with the United Kingdom's Met Office, said destructive changes in temperature, rainfall and agriculture were now forecast to occur several decades earlier than thought. Vulnerable people, such as the old and poor, would be worst affected, and, world leaders had not yet accepted that their countries would have to adapt to the likely consequences. Co-chairman of the IPCC working group, Professor Parry said: 'We are all used to talking about these impacts coming in the lifetimes of our children and grand children. Now we know that it is us.' He added that politicians had wasted a decade by focusing only on ways to cut emissions, and had only recently woken up to the need to adapt. 'Mitigation has got all the attention, but we cannot mitigate out of this problem. We now have a choice between a future with a damaged world or a severely damaged world' (Adam, 2007). How does to meet the challenges of climate change to the very survival of humanity on this earth? The challenges most daunting and requiring urgent attention are related to the very established concept of development, classical notion of security and growth oriented pleasure driven inordinate way of life. Replacement of fossil fuel based technologies seems to be the quintessence to ensure a sustainable world. The Western pattern of development should be redefined urging upon the all other nations to reframe their development on the premises of their own culture and civilisation that would ensure a riddance of western straitjacket. Development is to be understood not as development of one individual, or one nation or a group at the cost of others but as a tapestry in which all living and non-living and the future generations are found to be interwoven. In the similar vein, the conventional notion of security as ever-increasing military buildup by a state against another state needs to be reinvented as the individuals supposed to be secured by the state feel threatened by the apocalyptic climate change. The focal point of security no longer rests with the state but deflects away discursively from it toward the individual human being. The question crops up how the weaponisation forestalls the onset of climate change befalling human beings. It is not to be misconstrued as advocating the demise of state. Rather it adduces for redefining realist security thinking urging upon reshuffling of priorities where individual security remains at the center stage. There is a limit to the growth oriented, pleasure driven and inordinate way of life, which has been the main driving force behind the plundering away of the resources of the earth at the expense of nature and other human and non-human beings. The economy that has been built on this requires to be redefined in order to usher in a sustainable world where basic human needs are to be fulfilled and the difference between needs and wants is to be the guiding principle. A homo-nature balanced development should be premised on a cosmo-centric view of the universe replacing the Newtonian view in which all living and non-living are integrated and integral parts.

## References

A.C. Ibe and L.F. Awosika (1991). "Sea level rise impact on African coastal zones", in S.H. Ominde and C. Juma, eds, *A Change in the Weather: African Perspectives on Climate Change*, 105-12. African Centre for Technology Studies: Nairobi, Kenya [Reprinted with permission by CIESIN at http://www.ciesin.columbia.edu/docs/004-153/004-153.html.

Art Hansen and Anthony Oliver-Smith (1982). Involuntary Migration and Resettlement: The Problems and Responses of Dislocated Peoples Westview Press: Boulder, Co; World Commission on Dams (2000). "People and Large Dams: Social Performance", in Dams and Development at http://www.dams.org//docs/report/wcdch4.pdf. Chris de Wet, compiler (2002). Improving Outcomes in Development-Induced Displacement and Resettlement (DIDR) Projects. Oxford University, Refugee Studies Centre http://www.rsc.ox.ac.uk/PDFs/rrimproving outcomes02.pdf. W. Courtland Robinson (2004). "Minimizing Development-Induced Displacement", Migration *Information Source* http://www.migrationinformation.org/feature/display.cfm?ID=194 .

Ben Russell and Nigel Morris (2006). "Armed forces are put on standby to tackle threat of wars over water", *Independent*, 28 February at http://news.independent.co.uk/environment/article348196.ece.

Ben Vogel (2007). Quoted in "Climate change creates security challenge 'more complex than Cold War'," *Janes.com* at http://www.janes.com/security/international_security/news/misc/janes070130_1_n.shtml.

Ben Wisner (2008). "The Interactions between Conflict and Natural Hazards in an Unstable, Globalizing World: Swords, Plowshares, Earthquakes, Floods, and Storms," in Hans Günter Brauch, et al., eds. *Globalisation and Environmental Challenges: Reconceptualising Security in the 21st Century*, Vol. II, Chapter 15. Springer-Verlag: Berlin (in press); see earlier version in *Journal of Natural Disaster*

*Science* (Kyoto, Japan) 26, 2 (2004): 63-72. at http://www.drs.dpri.kyoto-u.ac.jp/jsnds/download.cgi?jsdn_26_2_3.pdf.

Ben Wisner and others, "Climate Change and Human Security" Simultaneously uploaded to Radix - Radical Interpretations of Disaster at http://www.radixonline.org/cchs.html & http://www.radixonline.org/cchs.doc. Disaster Diplomacy: http://www.disasterdiplomacy.org; Peace Research and European Security Studies (AFES-PRESS): http://www.afes-press.de/html/topical.html.

Ben Wisner, Piers Blaikie, Terry Cannon, and Ian Davis (2004). *At Risk: Natural Hazards, People's Vulnerability and Disaster*. 2nd Edition. On the definition of "vulnerability" at the scale of household livelihoods and its linkage with macro-scale processes. Routledge: London.

Brittany Sauser (2007). "Ethanol Demand Threatens Food Prices", *Technology Review* (13 February) at http://www.technologyreview.com/Energy/18173.

Chris Littlecott (2007). "Climate Change: The Global Security Impact" in a Quatation at February http://www.e3g.org/index.php/programmes/climate-articles/climate- change-the-global-security-impact/.

CIDA (no date).*Gender, Equality and Climate Change.: Why consider gender equality when taking action on climate change?* Canadian International Development Agency at www.acdi-cida.gc.ca/INET/IMAGES.NSF/vLUImages/Climate%20change3/$file/Gender-2.pdf.

Crisis States Programme, Development Studies Institute, London School of Economics: http://www.crisisstates.com/News/seminars1.htm.

David Adam, (2007). "How climate change will affect the world", in a Quatation *The Hindu*, 20 September 2007.

David Baldwin, (1927). The concept of security Review of International studies, Vol-23, 1927, pp. 12-18.

David Simon (2007). "Cities and Global Environmental Change: Exploring the Links", *The Geographical Journal* 173, 1 (March): 75-79 and see chapters 3 and 4 of Nicholas Stern et al. (2007) *Stern Review on the Economics of Climate Change*. London: UK, Department of the Treasury at http://www.hm-treasury.gov.uk/independent_reviews/stern_review_ economics_climate_change/stern_review_ report.cfm.

Delaney and Shrader (2000). Report of the percentage of women-headed households more than doubled in both Honduras and Nicaragua following Hurricane Mitch in 1998, see, Patricia L. Delaney and Elizabeth Shrader "Gender and Post-Disaster Reconstruction: The Case of Hurricane Mitch in Honduras and Nicaragua", LCSPG/LAC Gender Team, The World Bank, Decision Review Draft, Page 24 http://www.gdnonline.org/resources/reviewdraft.doc.

Donella H. Meadows, Dennis L. Meadows, Jorgan Randers and William W.Behruns-III. The limits to Growth (New York; University Books, 1972), P – 10. On the neo-Malthusian roots of "environmental conflict" discourse, see Elizabeth Hartmann (2003) *Strategic Scarcity: The Origins and Impact of Environmental Conflict Ideas*, PhD Thesis, Development Studies, London School of Economics and Betsy Hartmann, "Degradation Narratives: Over-Simplifying the Link Between

Population, Poverty and the Environment," *IHDP Update, Newsletter of the International Human Dimensions Programme on Global Environmental Change*, No. 4, 2002: 6-8 and Elizabeth Hartmann (1998) "Population, environment and security: a new trinity," *Environment and Urbanisation* 10,2: 113-127 — summarised in http://popdev.hampshire.edu/projects/dt/dt27.php.

Francois, Fouinat, (2004). "A comprehensive framework for human security", Conflict, Security and Development 4(3), December: 289-97.

Janos Bogardi and Hans Guenter Brauch (2005). "Global Environmental Change: A Challenge for Human Security - Defining and Conceptualising the Environmental Dimension of Human Security", in Andreas Rechkemmer, ed. *UNEO - Towards an International* Environment, pp. 85-109 Bangkok, October 4-5 2007 at http://www.ids.polsci.chula.ac.th/humansecurity.htm, see also at http://www.gechs.org/.

Geoffrey D. Dabelko, Mark Halle, Steve Lonergan and Richard Matthew, State of the art review on environment, security and development co-operation. Prepared for the Working Party on Development, Co-operation and Environment, OECD Development Assistance Committee, Paris. p. 48.

German Advisory Council on Global Change – http://www.crisisstates.com/News/seminars1.htm. Secretariat Reichpietschufer 60-62, 8. OG D-10785 Berlin.The summary can be downloaded through the Internet from the websitehttp://www.wbgu.de/wbgu_jg2007_kurz_engl.html.

Gasper, Des, (2005). "Securing Humanity: Situating 'Human Security' as Concept and Discourse, in Journal of Human Development 6(2): 221-45.

Hans Günter Brauch (2002). "Climate Change, Environmental Stress and Conflict", in Federal Ministry for the Environment, Nature Conservation and Nuclear Safety, ed. *Climate Change and Conflict*, pp. 9-112. Federal Ministry for the Environment, Nature Conservation and Nuclear Safety: Berlin at http://www.afes-press.de/pdf/Brauch_ClimateChange_BMU.pdf .

Hans Günter Brauch (2004) "Abrupt Climate Change and Conflicts: Security Implications from a European Perspective - Hobbesian vs. Grotian Analyses," Friedrich Ebert Foundation and Carnegie Endowment for International Peace, Washington, DC, 29 March http://www.afes-press.de/pdf/Brauch_Climate Change.pdf On the narrow and broad definitions of 'security', see: Hans Günter Brauch: *Environment and Human Security*," InterSecTions, 2/2005 (Bonn: UNU-EHS); at: http://www.ehs.unu.edu/file.php?id=64; and Hans Günter Brauch: *Threats, Challenges, Vulnerabilites and Risks in Environmental and Human Security. Source 1/2005* (Bonn: UNUEHS) http://www.ehs.unu.edu/file.php?id=63 and Commission on Human Security (2003) Human *Security Now*. Commission on Human Security: New York http://www.humansecurity-chs.org/

International Famine Program (2004) *Ambiguity and Change: Humanitarian NGOs Prepare for the Future*. A report prepared for: World Vision, CARE, Save US, Mercy Corps, Oxfam USA, Oxfam GB and Catholic Relief Services http://famine.tufts.edu/pdf/ambiguity_and_change.pdf .

International Rivers Network, "About Rivers and Dams" http://www.irn.org/basics/ard/

Jeffrey A McNeely (2006) "Bio-Fuels: Green Energy or Grim Reaper?" *The Green Room*, 2 September http://news.bbc.co.uk/2/hi/science/nature/5369284.stm.

Johnston, Barbara Rose (2007) Half-lives and Half-truths: Confronting the Radioactive Legacies of the Cold War. SAR Press: Santa Fe, NM.

Quotations from the IPCC rely on "Working Group II Contribution to the Intergovernmental Panel on Climate Change, Fourth Assessment Report Climate Change 2007: Climate Change Impacts, Adaptation and Vulnerability, Summary for Policymakers, 6 April 2007 http://www.ipcc.ch/SPM6avr07.pdf

Jürgen Dedring (2007) "Human Security and the UN Security Council", In: Hans Günter Brauch, John Grin, Czeslaw Mesjasz, Pal Dunay, Pal Chadha, Navnita Pal, Béchir Chourou, Ursula Oswald Spring, Ursula, P.H. Liotta, Patricia Kameri-Mbote, eds. (2007) *Globalisation and Environmental Challenges: Reconceptualising Security in the 21st Century*. Chapter 48. Hexagon Series on Human and Environmental Security and Peace, Vol. 3. Springer: Berlin (in press).

Kofi Annan (2005) *In Larger Freedom: Towards Development, Security and Human Rights for All*. United Nations: New York http://www.un.org/larger freedom/

Lorraine Elliott, "Regional environmental security: Pursuing a non-traditional approach" In Andrew T.H. Tan and J.D. Kenneth Boutin (Eds), Non-Traditional security issues in Southeast Asia (Singapore: Select Publishing, 2001), p.449.

Mahbub Ul Haq : "New Imperatives Human Security" RGICS paper No – 1 Rajib Gandhi Institute for contemporary studies, Rajib Gandhi Foundation, New Delhi – 1994, p-7.

Maureen Fordham with Madhavi Malalgoda Ariyabandu, Prema Gopalan and Kristina J Peterson (2006*State of the Art in Natech (Natural Hazard Triggering Technological Disasters) in Europe*. DG Joint Research Centre, European Commission and United Nations International Strategy for Disaster Reduction, Ispra, Italy.

Maureen Fordham with Madhavi Malalgoda Ariyabandu, Prema Gopalan and Kristina J Peterson (2006) "Please don't raise gender now – we're in an emergency!" *World Disasters Report 2006*, Chapter 6. International Federation of Red Cross and Red Crescent Societies: Geneva http://www.ifrc.org/publicat/wdr2006/summaries.asp.

Mendowitz, Saul B. ed., 1975. On the Creation of a Just World Order, New York: The Free Press.

Millennium Ecosystem Assessment (2006) *Our Human Planet: Summary for Decision Makers*, p. 6, Box 1.2, "Millennium Ecosystem Assessment Conceptual Framework" http://ma.caudillweb.com//en/Products. Global. Summary.aspx

Neil Adger, Jouni Paavola, Saleemul Huq, M.J. Mace, eds. (2006) *Fairness in Adaptation to Climate Change*. Cambridge, MA: MIT Press; See also: Michael

H. Glantz (1990) "On the Interactions between Climate and Society," *Population and Development Review* 16: 179-200 and Michael H Glantz and Dale Jamieson, co-convenors (2001) "Climate Ethics and Equity," *NCAR Working Group Web Page* (October) http://www.ccb.ucar.edu/ethics/

See: Patrick J. McGowan (2003) "African Military Coups d'Etat, 1956-2001: Frequency, Trends and Distribution..." *Journal of Modern African Studies*, 41, 3: 339–370; at: http://journals.cambridge.org/download.php? file=%2 FMOA%2FMOA41_03%2FS0022278X0300435Xa.pdf &code=aad776a 39cf 997c090 aecae20ceafeb1

Peter Schwartz and Doug Randall, An Abrupt Climate Change Scenario and Its Implications for the United States National Security, October 2003. See: "Press Conference by Security Council President, 4 April 2007" http://www.un.org/News/briefings/docs//2007/070404_Parry.doc.htm

Sir Crispin Tickell (2003) "Risks of conflict: Population and Resource Pressure", In: Hans Günter Brauch, P.H Liotta, Antonio Marquina, Rogers, Paul; Selim, Mohammed El-Sayed, eds. *Security and Environment in the Mediterranean. Conceptualising Security and Environmental Conflict*, pp. 13-18. Hexagon Series on Human and Environmental Security and Peace, vol. Springer: Berlin-Heidelberg.

See: Spanish Environment Ministry (2006) *Conference on Desertification and Migration*. 25-27 October 2005, Almeria, Spain http://www.sidym2006.org/eng/eng_ponencias_conclusiones.asp

The commission on Global Governance, our Global Neighbourhood (New York: Oxford University Press, 1995) p. 338.

The Independent Commission on Disarmament and security Issues; common security (New York: - Simon and Schuster, 1982)

The Independent Commission on International Development Issues, North-South: A Programme for Survival, 1980. Cambridge, MA: The MIT Press.

The Independent Commission on Disarmament and Security Issues, Common Security: A Blueprint for Survival, 1982. New York: Simon and Schuster, 1982, pp. XV and 172 respectively.

The Stockholm Initiative on Global Security and Governance, 1991. "Common Responsibility in the 1990s" paper published by the Prime Minister's office, Government of Sweden, Stockholm, pp.17-18.

Thomas Wagner, (28 March 2007) "Major Cities Warned against Sea Level Rise" *Independent (South Africa) On Line* http://www.iol.co.za/index.php?set_id=1&click_id=31&art_id=nw20070328024424684C606123

See: Ulrike Röhr (2006) "Gender and climate change," *Tiempo* 59 (April): 3-7 http://www.tiempocyberclimate.org/portal/archive/pdf/tiempo59high.pdf .

UNDP (1994) Human Development Report: New Dimensions of Human Security. UNDP: New York.

UNDP (2004) Reducing *Disaster Risk: A Challenge for Development*. Geneva: UNDP, Bureau for Crisis Prevention and Recovery http://www.un.org/special-rep/ohrlls/ldc/Global-Reports/UNDP%20Reducing% 20Disaster% 20Risk.pdf
.

UNDP (2007) *Human Development Reports* http://hdr.undp.org/ ; see also UNDP, UNEP, World Bank, ADB, AfDB, GTZ, DFID, OECD, and EC (2003) ***Poverty and Climate Change: Reducing the Vulnerability of the Poor through Adaptation***. World Bank: Washington http://www.undp.org/climatechange/adap01.htm.

UNDP (2007) "Poverty Eradication, MDGs and Climate Change" http://www.undp.org/climatechange/adap01.htm.

United Nations Office for the Coordination of Humanitarian Affairs - Integrated Regional Information Networks (IRIN) (5 April 2007) "Southern Africa: Extreme weather threatens over a million people" http://www.reliefweb.int/rw/RWB.NSF/db900SID/SHES-6ZYQST?OpenDocument.

# 2

# Environmmental Security and Peace Building: A Plea for a Theoretical Construct

***Manoj Kumar Mishra***

## Introduction

In the post cold war era *'Redefining security'* writes David A. Baldwin has become something of a cottage industry (David A. Baldwin, 1997: 1). More interestingly arguments about whether the environment should be seen as a security issues range from those who believe that the securitisation of the environment is the most important steps to securing the survival of humanity. Those who believe that its advocates are simply environmentalists cynically attempting to grab part of the governmental attention and spending that traditionally attached to security issues. It has been called by Myers as *'ultimate security'* and a *'pollution'* of true security by Deudney (Myers, 1993 and Deudney, 1990).

The discourse lies between these two extreme positions. The environmental issue has been probably received more attention than any of the other domains in the broadened security concept and there is a large and growing literature devoted to the intellectual and policy implication of the concept of environmental security. Approaches ranges from the more specific focus on the linkages between environmental change and violent (deadly) conflict to the possible role of environmental conservation, cooperation and collaboration in peace building.

Environmental security has been defined as concerning 'the maintenance of the local and planetary biosphere as the essential support system on which all other human enterprises' (Buzan, 1991). Define thus; it clearly falls within the scope of security issue, since it encompasses the same dangers that focused so much attention on the issue of nuclear weapons during the cold war. Robert Kaplan arguing that accelerating environmental collapse will trigger increase in crime and ethnic conflict that will plunge the world into conflict-ridden anarchy.

That the environment emerged on the broadening security agenda in the past two decade is hardly surprising. Off late it has been observed that issues such as degradation or desertification threaten many developing countries than they are by the threats of external military forces. Tunisia has reflected this reality in its national security doctrine by defining its struggle against desertification as a key part of its national defense efforts to contain and reverse the advances of the Sahara (Sheehan, 1999). Some are of the view that the advent of environmental security is an example of post materialistic world. Societies can be seen as operating in terms of 'hierarchy of needs', so that once the basic needs are met, there is a movement up the hierarchy, and as this happens, new 'needs' emerge. In this sense, a 'hierarchy of security needs', may be operating, so that developments such as economic development, economic globalisation and the end of cold war have encouraged the development of an environmental security 'need' that was previously seem to be present (Sheehan, 2006: 100).

**Human Versus The Natural Environment: The Never Ending Debate**

If we look from a theoretical perspective, there are two possible kinds of environmental security threats. The first are those that emanate from the natural environment, earthquakes, cyclones etc. The second are those from those are the result of human agencies affecting the natural environment such as green house gas emissions resulting in climate change. There seems to be a consensus that it is the latter category i.e. the proper concern of environmental security. These are threats resulting from the way that the process of civilisation has come to involve a manipulation of nature that has reached self-defeating proportions because of massive population increased and the enormous growth of economic activity in the last century.

According to Barry Buzan, Weaver and De Wilde (Buzan, Waever and De Wilde, 1998). The environmental debate is really preserving existing levels of human civilisation. The real concern is whether the ecosystems needed to preserve and further develop human civilisation are sustainable. So the environmental security is thus not about humanity struggle with nature as such, but a problem of coping with the dynamics of its own post industrial cultures. Environmental security addresses threats to humanity posed by human impact upon the 'natural' environment. It is about the effects of atmospheric change, deforestation, chemical pollution, soil erosion, etc. These are problems primarily induced by human activity.

They may be the result of deliberate acts, or they may be caused by the unintended side effects of the pursuit of other objectives. It is not about dealing with natural disasters such as earthquakes as hurricanes. It is the results, not of humanity's struggle against the unpredictable effects of nature, but of humanity's struggle against itself. Environmental threats as Prins suggests that unlike military threats are 'threats without enemies (Prins, 1993).

To look at environmental security is such a way is that it does not make or require any link with war or violent conflict between human groups, as do the arguments of some other interpretations of environmental security threats. Environmental degradation is seen as a security problem because it affects negatively on the health and well-being of human beings. It represents direct existential threats to human in certain respects and therefore encourages a mindset that sees human is welfare and welfare of natural systems as being crucially linked (Mische, 1989).

**The Environmental Security Agenda**

A number of issues can potentially be included as environmental security concerns. These are some of the categories given below (Sheehan, 2006: 101—102)

(a) Environmental security concerns include disruption of ecosystems, such as climate change, loss of biodiversity, deforestation, desertification, depletion of the ozone layer, and the various frames of pollution. These factors the natural environmental balance if local to global ecologies is being damaged by the activities of the human population.

(b) The second concern is issues related to energy. This is related to the depletion of natural resources, different kinds of pollution involved in the storage and transportation of chemicals, oil and nuclear materials, and the problem of energy scarcities and uneven distribution.

(c) The third category, which can be included under the environmental security agenda, is the effects of human population growth, which include major population movements, excessive consumption and waste, pollution and oversees of limited natural resources.

(d) The issues related to food scarcities and uneven food distribution, including famines and the loss of fertile soils and water resources in many parts of the world.

Various scholars like Simon Dalby and others criticise many of the environmental security issues discussed above. For Dalby it is the pressures of globalisation and the dynamics of post modern capitalism, that are the culprit, rather than population growth *per se*, because of 'the accelerating processes of globalisation that are interconnecting the world's economics and cultures in ways that often operate to undercut traditional economics and challenges the sustainability of agricultural and survival practices' (Simon Dalby, 1998).

Gwyn Prins is of the opinion that environmental security is like 'threats without enemies', which is not the case. The basic logic of environmental security is that, globally, humanity is living beyond the carrying capacity of the planet. 'Carrying capacity' is defined as 'the total patterns of consumption that the Earth's natural systems can support without undergoing degradation' (Buzan, Waever and De Wilde, 1998:81). Carrying capacity is influenced by numbers, technology and lifestyle. The highly developed countries of the 'West', their advanced consumer life styles make greater demands upon the planet resources than the lifestyles of the developing countries. The world's poorer countries are not the most to blame for global environmental problems, in that their lifestyle are far less demanding on global resources, but they are the most vulnerable to its consequences.

This leads to one major implication in teams of identifying the providers of security. In the cold war era discourse, the provider of security was clearly the state, and the instrumental providers were the armed forces. But when damage to the environment *per se* becomes part of the security problematic. It remains partly and crucially a responsibility of the state to seek to maximise security in this area, and yet it also becomes possible for NGO's communities, scientists and even common people to take on an important responsibility for enhancing environmental security for society as a whole.

**Can The Environment be Securitised?**

There have been various efforts to include environmental issues on the security agenda by a number of scholars. Arguing in this context Peter Gleick has linked the 'environment' to a number of other sectors, including the military and societal sectors, where security thinking embraces problems of population growth, transnational pollution, widespread poverty, and inequitable social system (Peter Gleick, 1991).

Those scholars who are against securitisation of the environment are of the view that it is too amorphous a concept to be theorised. Marc Levy argues that 'the environment', 'can refer to anything in which something takes place or which affects what people do; in other worlds, almost anything at all' (Marc Levy, 1995: 37). He therefore argues that the term should be reserved for those issues that involve ecological feedback and equilibrium or are critical to the sustenance of human life (Marc Levy, 1995: 38).

In 1991, Daniel Deudney resented a set of arguments against the securitisation of the environment. Deudney said that environmental degradation and threats to military security were fundamentally different issues, for four main reasons (D. Deudney, 1991).

(1) They are different kinds of threats. Human beings are killed by many kinds of like, accidents, illness, aging, but these are not seen as security threats. If everything that could cause death of human beings were labelled a 'security threat', the team would lose any analytical usefulness it possessed.

(i) The scope, subject matter and source of threats to environmental and military security are different. Environmental threats may not be purely 'national' in character in terms of their causes, their consequences, or their solutions. This is not the case in traditional security threat perception.

(ii) The degrees of intention of threats between environmental and military security differ greatly. Threats of violence are generally international, planned, organised, deliberate acts by human activities. Environmental threats, in contrast, are largely international. They are side effects of other activities.

(iii) Finally, the organisations that provide protection from violence differ greatly from those engaged in environmental protection. To link organised violence and environmental habitability might create a conceptual confusion, leading to a 'militarisation of our thinking about the relationship between humanity and environment' (Graeger, 1996: 111).

Deudney further argues that there is a risk of stirring up unhelpful nationalistic feelings. The national security mindset creates an inappropriate 'us/them' attitude. The kind of zero-sum thinking characteristic of the national security mindset is simply not appropriate for dealing with environmental protection issues. There is no specific 'enemy' as such in the sense of a hostile 'other'. If there is a culprit at all, it is 'us'.

**Environmental Security Approaches: Different School of Thought**

There seems to be a problem in securitising the environment as it remains controversial in many quarters. The traditional military analysts have largely ignored the ideas, as security still remains a conservation realist account. At least three distinct components of the environmental security spectrum can be identified.

(i) The first group is composed of military security specialists who have become aware of environmental issues in the form of resources scarcity problem. This group has simply focused on natural resources as a potential course of future interstate conflict and war. The Toronto School (Homer Dixon, 1994), associated with Thomas Homer-Dixon has sought empirical evidence for this kind of environmental security approach. The Toronto school puts emphasis on the links between renewable resource scarcity and violent conflict. It concentrates upon issues such as scarcities of cropland, forest, fish stocks and water. The concept of environmental scarcity as defined by the Toronto School of thought is of three types:

(a) Supply-induced scarcity is caused by degradation and depletion of an environmental resource for example, the erosion of cropland.

(b) Demand-induced scarcity results from population growth within a region or increased per capita consumption of resources, either of which heightens the demands for a resource.

(c) Structural scarcity arises from an unequal social distribution of a resource that concentrates it in the hands of a relatively few people, while the remaining population suffers from serious short ages (Homer-Dixon and Blitt, 1998).

What is interesting of the Toronto school is how environmental scarcities contribute to violence in developing countries? This Toronto School tries to draw from a wide range of case studies that under certain circumstances, scarcities of renewable resources, such as cropland and water, do produce conflict instability. They further argue that environmental scarcity rarely contributes directly to interstate conflict, though the 1969 war between El Salvador and Honduras is described by Homer-Dixon as a first class example of an ecologically driven conflict (Homer-Dixon, 1991:76).

The Toronto school are of the opinion that how alarmed the international community should be about the linkage between environmental issues and national and international conflict. Environmental issues often play much more than a minimal role in the generation of conflicts, but at the same time, they should not be over-sectionalised or exaggerated in their effect (Homer-Dixon and Blitt, 1998: 1). The lasting contribution of the Toronto school has made an important and extremely valuable contribution in terms of their detailed empirical research into case studies that is characteristic of their approach, for example, the contribution of water shortages to political violence and feelings of social and political alienation in the Gaza Strip (Kelly and Homer-Dixon, 1998).

The second group of thought is the Swiss Peace Foundation's Environment and Conflicts Project (ENCOP). Environmental conflicts are defined in the project as traditional conflicts that are induced or sparked by environmental degradation (Libisezewski S., 1992). Maldevelopment (associated with increases in poverty and inequity, natural resources ever exploitation, environmental degradation) and environmental discrimination (when access to natural capital for certain groups is systematically restricted by the conscious activities of distinct and often powerful actors with control over those environmental resources) were as factors contributing to violent conflicts.

ENCOP identified contexts that were most susceptible to environmentally following: dry lands, mountain areas with low-land verses high-land interactions, transboundary river basins, areas degraded by dams and mines, tropical forests, and sprawling metropolises. ENCOP's research linked these susceptible contexts to the triggering factors of maldevelopment and environmental discrimination to generate seven major types of conflict in which environmental stresses play a role, even if not necessarily the primary or predominate one ethno political conflicts, demographically caused migration conflicts, international water conflicts and global environmental conflicts, such as ozone depletion and global warming (Baechler, 1998).

The third school of thought as expounded by Paul Collier and A. Hoeffler, first at the World Bank and then as leaders of an Oxford University research project, was it is argued that natural resource abundance rather scarcity was critical to the incidence of civil war. They proposed that, holding the level of grievance constant, more rebellions are likely in countries with an abundance of natural resources. These natural resources are often an aim and sometimes the primary goal of emergent movements. The looking of natural resources is hypothesised to be pivoted in financing, for military purchases of arms, labour and food (Paul Collier and A. Hoeffler, 2000).

Several environmental and human rights groups supported this 'honey pot' prospective in their own research and analysis, for example, on cases of conflict over diamonds (Renner M, 2002). Some researchers have found that the implementation of peace arguments is likely to be more difficult in countries abundant in natural resources (Stedman S., 2001). While others have suggested that natural resources dependence poses difficulties in post war peace-building efforts (Doyle M. Sambanis N., 2000).

The conclusion, which emerges from the above debate within different school of thought, is that environmental domain lends itself to a post realist approach to security. Because, as Jessica Matthews noted in 1989, many of the key issues are regional or even global in scope and cannot be neatly confined by sovereign territorial boundaries, so that traditional realist approaches to international relations distort, misunderstand, or ignore those (Matthews, 1989).

In determining whether the environment is truly a security issue, the element of threat is important. To state this in another way as Neville Brown has put it, just as the threat of attack constituted the main reason for employing security strategies in the past, so continuing acts of environmental degradation can be seen to present 'ecological threats to stability and peace' (Brown, 1989). In the worst-case of this kind of thinking, environmental collapse is equated with the effects of a nuclear holocaust, producing a total breakdown of the natural system.

## Environmental Peacebuilding: Broadening the Environment and Security Programme

Since the beginning of this decade and the last, the apocalyptic warnings of water wars and environmental refugees have been slowly giving way to a growing hopes that the environment more specifically, environmental cooperation could prompt stability and peace between states in conflict. According to this theory, initiatives such as transboundary cooperation for environmental conservation (e.g., peace parks), international river basin management, regional marine agreements, and joint environmental monitoring programs could enhance co-operation between communities and countries (Alexander Carius, 2000).

Some scholars have increasingly focused their attention on the potential for using environmental threats as a common aversion to stimulate conflict resolution and collaborative behaviour towards longer-term peace. This is possible by identifying the conditions under which environmental cooperation best facilitates conflict transformation and peace building, and which forms of negotiation or state holder participation have been particularly successful.

## Approches to Environmental Peacebuilding

Most ecological peace initiatives fall into one of three partly overlapping categories (Carius and Dabelko, 2004; Conca, Carius and Dabelko, 2005).

- Initiatives to prevent conflicts that are directly related to the environment,
- Efforts to initiate and sustain a dialogue on transboundary environmental cooperation between parties to a conflict, and
- Initiatives that seek lasting peace by promoting conditions for sustainable development.

## Preventing Conflict Directely Related to the Environment

Environmental cooperation could play a role in preventing violence that erupts due to the uncontrolled exploitation of natural resources, the destruction of ecosystem or the devastation of livelihoods based on natural resources. Most of the research that establishes a link between environment degradation and violent conflict focuses on two solutions—reducing the pressure on resources on which people we economically dependent; and strengthening the institutional capacities to respond to environmental challenges.

Environmental cooperation could also help assuage the anger of victims of environmental injustice, who are often already socially and economically disadvantaged. Latent environmental problems may combine with material insecurity and perceived marginalisation to create an explosive situation, especially where ethnic identity determines access to political and economic opportunities. The most heavily polluted industrial regions in the post-Soviet Baltic States, for instance, have a largely ethnic Russian population, generating a volatile mix of burgeoning ethnic and national identity, mounting social discrimination and environmental mismanagement. Active environmental cooperation could help alleviate an important source of this festering discontent.

**Using Enviromental Cooperation as a Platform for Dialogue**

A second approach to environmental peace building seeks to address conflicts that have no specific environmental cause. These initiatives attempt to create peace by bringing conflicting parties together to reach cooperative solutions to common environmental challenges. Countries have found that environmental issues are one of the few topics on which they can sustain an ongoing dialogue.

One of the most complex, unresolved conflicts in the highly unstable Caucasus region is Armenia and Azerbaijan's dispute over Nagorno-Karabakh. In the fall of 2000, the government of Georgia convinced Armenia and Azerbaijan to set up a trilateral biosphere reserve in the southern Caucasus region. Although Armenia and Azerbaijan are not yet ready to cooperate directly with each other, the agreement calls for them to create national biosphere reserves, which will ultimately be integrated into a single conservation area. A similar attempt has been made in Kashmir, over which India and Pakistan have been fighting since the end of British colonial rule. Some international environmentalists have proposed establishing a peace park in the Karakoram mountains lying between India and Pakistan in the joint management of this unique glacial region in which more soldiers are estimated to have died from the cold and altitude than from enemy fire (Haider, 2005), could help defuse this deadly border conflict.

**Promoting Sustainble Development to Achieve a Durable Peace**

A third approach is based on the premise that long-term and comprehensive sustainability is a prerequisite for a lasting peace. The joint management of shared resources can be not only a way to keep both parties talking, but also the key to negotiating a resolution. For example, even if water scarcity is not the cause of the conflict between Israelis and Palestinians, a solution to their shard water problem is necessary for a peaceful resolution.

**Making Environmental Peacemaking a Reality**

Environmental peacemaking initiative can also reduce tension and violent conflict between countries, which can bring about tangible environmental, economic, and political gains. Environmental peace-making strategy offer the chance to craft a positive, practical policy framework for cooperation that can engage a broad community of stakeholders by combining environment, development and peace related concern.

Government and other actors have not pursued enough peace-oriented cooperative activities on environmental problems. It is only in few occasions they have just begun to share experience and knowledge about environmental peacemaking through peace-and-conflict assessments of environmental projects and programmes. Without such knowledge, the international community may be missing powerful peace-making opportunities in the environmental domain.

The challenge, before the international community, is to gather evidence-however partial and indirect-that more aggressive environmental peacemaking strategy could create opportunities. This might be used to make governments, inter-governmental organisations, social movements, and other actors to be more aggressive about environmental co-operation and peacemaking.

A novel way to build environmental peace building in the 21st century is globalisation. Although its effects are complex and by no means entirely healthy for ecological sustainability, but globalisation has the ability to move political dynamics out of narrow interstate settings and into a broader society-to-society context and is an important and healthy sign. This new social space holds much of the potential for environmental peacemaking. It is well worth finding out whether these changes create opportunities to build peace, lessen environmental insecurity, and break of the zero-sum logic that often plagues international relations (Conca, Carius and Dabelko, 2005: 157).

## References

Archarya, Amitav. 1998. 'Collective Identity and Conflict Management in South-East Asia'. A. Adler and M.Barnett, eds, in *Security Communities*. Cambridge: Cambridge University Press, pp. 198-227.

Baechler, G. 1998. 'Why Environmental Degradation Causes Violence'. *Environ. Changer Secur: Project Rep.*, Swiss Peace Foundation, Zurich, Switz.

Balwin, David A. 1995. 'Security Studies at the End of the Cold War'. *World Politics 48,* pp. 117-41.

———. 1997. 'The Concept of Security'. *Review of International Studies 23,* pp. 5-26.

Booth, Ken (ed.). 1991a. *New Thinking about Strategy and International Security*. London: Harper Collins.

———-. 1995. 'Human Wrongs and International Relations'. *International Affairs 71*. pp. 103-22.

———-. 1997. 'Security and Self: Reflections of a Fallen Realist', in Keith Krause and Michael Williams, eds, *Critical Security Studies.* Minneapolis: University of Minnesota Press, pp. 83-119.

Booth, Ken and Smith, Steve,(eds). 1995. *International Relations Theory Today.* Cambridge: Polity.

Brandt Commission.1985. *Common Crisis: North-South Co-operation for World Recovery.* Cambridge, Mass: MIT Press.

Brown, Neville. 1989. 'Climate, Ecology, and International Security'. *Survival* 31 pp. 519-32.

Bulloch, John, and Darwish, Adel. 1993. *Water Wars: Coming Conflicts in the Middle East.* London: Victor Gollancz.

Buzan, Bary. 1983. *People, States, and Fear.* Brighton: Harvester Wheatsheaf.

———. 1984. 'Economic Structure and International Security: The Limits of the Liberal Case'. *International Organisation* 38, pp. 597-624.

———. 1991(a). *People, States and Fear.* 2nd ed. Boulder: Lynne Rienner.

———. 1991(b). 'Is International Security Possible?', in Ken Booth, (ed.), *New Thinking about Strategy and International Security.* London: Harper Collins, pp. 31-55.

———. 1996. 'International Security and International Society',. in Rick Fawn and Jeremy Larkin, (eds). *International Security after the Cold War: Anarchy and Order Reconsidered.* London: Macmillan, pp. 261-87.

Buzan, Barry, Jones, Charlesm and Little, Richard. 2993. *The Logic of Anarchy: Neorealism to Structural Realism.* New York: Columbia University Press.

Buzan, Barry and Segal, Gerry. 1994. 'Rethinking East Asian Security'. *Survival 36,* pp. 3-21.

Buzan, Barry and Waever, Ole. 1997. 'Slippery, Contradictory? Sociologically Untenable? The Copenhagen School Replies'. *Review of International Studies* 23, pp. 241-50.

Buzan, Barry, Waever, Ole and De Wilde, Jaap. 1998. *Security: A New Framework for Analysis.* Boulder: Lynne Rienner.

Carius, Alexander. 2005. 'Environmental Peacebuilding: Conditions for Success' *Environmental Change and Security Program Report.* 12. pp. 59-79.

Collier P. and Hoeffler A. 2000. 'Greed and Grievance in Civil War'. *World Bank Policy Res. Pap.* 2355, Washington, DC.

Conca, Ken, Akexander Carius and Geoffrey D. Dabelko. 2005. 'Building Peace through Environmental Cooperation', in World Watch Institute (ed.), *State of the World 2005: Redefining Global Security* (pp. 144-59). Washington, DC: World Watch Institute.

Dalby, Simon. 1992. 'Security, Modernity Ecology: The Dilemmas of Post Cold War Security Discourse', *Alternatives* 17, pp. 95-134.

———. 1998. 'Human Security: Environmental Dimensions of a Contested Concept', paper presented at the 1998 Meech Lake Conference on Taking Human Security Seriously.

Deudney, Daniel. 1990. 'The Case against Linking Environmental Degradation and National Security', *Millennium: Journal of International Studies 19,* pp. 461-76.

———. 1991. 'Environment and Security: Muddled Thinking', *Bulletin of the Atomic Scientists* 47, No. 3, pp. 22-28.

Deudney, Daniel and Matthews, Richard (eds). 1999. *Contested Grounds: Security and Conflict in the New Environmental Politics.* Albany: State University of New York Press.

Doyle M., and Sambanis N. 2000. International Peace Building: A Theoretical and Quantitative Analysis. *Am. Polit. Sci. Rev.* 94(4): 779-802.

Galtung, Johan. 1971. 'A Structural Theory of Imperialism', *Journal of Peace Research 8,* pp. 81-118.

Gleditsch, Nils Peter. 1992. 'Democracy and Peace', *Journal of Peace Research 29,* pp. 369-76.

Gleick, Peter H. 1991. 'Environment and Security: The Clear Connections', *Bulletin of the Atomic Scientists 47,* No. 3, pp. 16-22.

———. 1993. 'Water and Conflict: Fresh Water Resources and International Security', *International Security 18,* pp. 79-112.

Graeger, Nina. 1996. 'Environmental Security?' *Journal of Peace Research 33,* pp. 104-16.

Haidar, Suhasini. 13 June 2005. 'Indian PM's Bid for Peace Mountain' CNN. Available online at http://edition.cnn.com/2005/WORLD/asiapcf/06/13/siachen.india/index.html

Held, David and McGrew, Anthony. 2001. 'The Great Globalisation Debate: An Introduction', in David Held and Anthony McGrew, (eds), *The Global Transformations Reader.* Cambridge: Polity Press, pp.1-45.

Homer-Dixon, Thomas. 1991. 'On the Threshold: Environmental Changes and Acute Conflict', *International Security 16,* pp. 76-116.

———. 1994. 'Environmental Scarcities and Violent Conflict: Evidence from Cases', *International Security 19,* pp. 5-40.

Homer-Dixon, Thomas and Blitt, Jessica (eds) 1998. *Ecoviolene: Links Among Environment, Population, and Security.* Oxford: Rowan and Littlefield.

Homer-Dixon, Thomas and Percival, Valerie. 1996] *Environmental Scarcity and Violent Conflict: Briefing Book.* Toronto: University of Toronto Project on Environment, Population, and Security.

Kaplan, Robert D. 1994. 'The Coming Anarchy.', *Atlantic Monthly* 273, No. 2, pp. 44-76.

Kelly, Kimberley and Homer-Dixon, Thomas. 1998. 'The Case of Gaza', in Thomas Homer-Dixon and Jessica Blitt (eds), *Ecoviolence: Links Among Environment, Population, and Security.* Oxford: Rowman and Littlefield, pp. 67-107.

Khagram, Sanjeev and Ali, Saleem. 2006. 'Environment and Security', NO 31: pp. 395-411, The Annual Review of Environment and Resources is online at http://environ.annualreviews.org

Krause, Keith and Williams, Michael. 1996. 'Broadcasting the Agenda of Security Studies: Politics and Methods', *Mershon International Studies Review,* 40, sup. 2, pp. 229-54.

———.1997. 'From Strategy to Security: Foundations of Critical Security Studies', in Keith Krause and Michael Williams (eds), *Critical Security Studies.* Minneapolis: University of Minnesota Press, pp. 33-59.

Leitenen, Kari. 1997. 'The Concept of Environmental Security', unpublished paper. Tampere, Finland.

Levy, Marc A. 1995. 'Is the Environment a National Security Issue?' *International Security* 20, pp. 35-62.

Libiszewzewski, S. 1992. *What is an Environmental Conflict?* ENCOP Occas. Paper: 6. http://www.isn.ethz.ch/pubs/ph/details.cfm?lng=en&id=236

Matthews, Jessica Tuchman. 1989. 'Redefining Security', *Foreign Affairs* 68, pp. 162-77.

Mische, Patrocoa. 1989. 'Ecological Security and the Need to Reconceptualise Sovereignty', Alternatives 14, pp. 389-427.

Myers, Norman.1989. 'Environment and Security', *Foreign Policy* 74, pp. 23-41.

————, 1993. *Ultimate Security.* New York: W.W. Norton.

Pettiford, L. 1996. 'Changing Concepts of Security in the Third World', *Third World Quarterly* 17, pp. 289-306.

Peirages, D. 1997. 'Demographic Change and Ecological Security', *Environmental Change and Human Security Project Report,* No. 3, pp. 36-46.

Porter, Gareth and Brown, Janet W. 1991. *Global Environmental Politics.* Boulder: West View.

Prins, Gwyn (ed.) 1993. *Threats without Enemies: Facing Environmental Insecurity.* London: Earth Scan.

Rener, M. 2002. *The Anatomy of Resource Wars.* World Watch, Pap. 162, Washington, DC.

Rochlin, James F. 1997. *Redefining Mexican 'Security': Society, State, and Region Under NAFTA.* Boulder: Lynne Rienner.

Rogers, K.1997. 'The Intrastate Security Dilemma: Ethnic Conflict as a "Tragedy"', *Journal of Peace Research* 36, pp. 183-202.

Rothschild, Emma. 1995. 'What Is Security?' *Daedalus* 124, pp. 53-98.

Schelling, Thomas C. 1958. 'The Strategy of Conflict', *Journal of Conflict Resolution 2*, pp. 203-64.

————. 2000. 'The Evolution of the Concept of Security since 1945', in Michael Sheehan, (ed.), *National and International Security.* Aldershot, pp. 1-31.

Sheehan, Michael. 2006. *International Security: An Analytical Survey.* Viva Books Pvt. Ltd.: New Delhi.

Sorenson, Theodore. 1990. 'Re-Thinking National Security', *Foreign Affairs* 69, pp. 1-18.

Starr, Joyce R. 1991. 'Water Wars', *Foreign Policy 82*, pp. 17-36.

Stedman, S. 2001. *Implementing Peace Agreements in Civil Wars: Lessons and Recommendations for Policymakers*. Int. Peace Acad. Policy Pap., Ser. Peace Implement., StanfordUniversity at http://www.ipacademy.org/PDF_Reports/Pdf_Report_Implementing.pdf

Ullman, Richard. 1983. 'Redefining Security', *International Security* 8, pp. 129-53.

Westing, Arthur H. (ed.). 1986. *Global Resources and International Conflict: Environmental Factors in Strategic Policy and Action*. Oxford: Oxford University Press.

———, (ed.) 1988. *Cultural Norms, War, and the Environment*. Oxford: Oxford University Press.

———, 1989. 'The Environmental Component of Comprehensive Security', *Bulletin of Peace Proposals* 20, pp. 129-34.

# 3

# Warfare: Its Environmental and Human Consequences in the Light of Global Warming

*Arthur H. Westing*

Here I address the impact of warfare (armed conflict) on the environment, and the effects of such affect the enmeshed civilians, with special emphasis on the generation of displaced persons, and then conclude with some suggested approaches to at least mitigating those disruptions. I set the stage for my exposition with ***five*** rather uncomfortably disquieting facts:

***First***, we must not lose sight of the fact that wars are by their very nature *deadly* and *destructive* — that is to say, they are by specific intent highly detrimental to an opponent's armed forces, people, and land.

***Second***, we must recall that wars have occurred without substantial, if any, interruption throughout the long sweep of human history, thereby establishing them among the characteristic endeavors of our species. Arthur Koestler had it right in suggesting (1978) that, 'The most persistent sound which reverberates through man's history is the beating of war drums.' Indeed, to provide some modern perspective, a few dozen wars have been in progress in the world at any one time during the past half century.

***Third***, we must recognise that despite ever continuing advances in weaponry through the ages, the actual levels of death, destruction, human casualties, and environmental disruption have *not* been increasing over the years, only in the efficiency with which they are being carried out. Rather, those levels have been linked all along to the *objectives*, the *will*, and the *tenacity* (or reluctance to capitulate) of the parties to the conflict.

***Fourth***, we must be aware that only a relatively small and more or less steady fraction of the huge and ever growing numbers of persons being displaced from the rural reaches of the Third World—and, in the process, perhaps half of them fleeing to cities—are doing so in order to escape the dangers of war or persecution. What *is* driving those ever-growing numbers of migrants from their homes is an inability to subsist on their increasingly degraded resource base, and must thus be thought of as *environmental refugees* seeking a better life.

***Fifth***, we must understand that the ***global warming*** currently in progress exacerbates the ever more precarious balances that now exist both within the global biosphere and within the global sociosphere—and, importantly also the balance between those two tightly interlocking spheres. The ensuing civil unrest to be expected is in turn likely to lead to increased incidences of armed conflict both within and between nations.

**Environmental warfare**

***Levels of environmental battlefield disruption:*** It is true that the environmental disruption brought about by wars is in some instances a more or less *unintentional* side effect of attacks being aimed primarily at ostensibly military targets such as enemy combatants, fortifications, naval vessels, transportation systems, factories. And the like or to stretch a bit the notion of what constitutes a military target, also cities. Such actions result in greater or lesser ancillary or *collateral* (although usually quite predictable) environmental disruptions. A good example of such collateral environmental disruption is provided by World War I in 1914-18, during which the farmlands and woodlands of France and Belgium received extraordinary levels of incidental battlefield damage, their assisted recovery occurring just in time to have it happen all over again a generation later, during World War II in 1939-45. And the substantial environmental and social disruptions being brought about by the US rural bombing campaign in the current Afghan War in 2001–provides us with a good up-to-the-minute example of such collateral effects (see below).

The next level of environmental disruption—a level often resorted to—is *intentional* attacks directed against the enemy's countryside—against its farms, plantations, orchards, fields, forests, and water systems—and if sufficiently intense, referred to as a scorched-earth strategy. An *additional* level of intentional environmental attack is sometimes distinguished, one in which a relatively small input of triggering energy is employed for the purpose of causing a disproportionately greater amount of environmental devastation by the release of so-called *dangerous forces* (for example, from attacks on major dams, chemical factories or nuclear plants).

Before continuing I should perhaps make it clear that I am *not* here dealing with the environmental impact of the military sector of society during either peacetime or wartime aside from direct wartime battlefield depredations, dwelling especially on those that have had widespread, long-lasting (long-term), or severe effects. I am also *not* dwelling upon the often-monumental *direct battlefield casualties*, either military or civilian. Although before continuing, several examples of at least such battlefield fatalities would be instructive. Thus, during World War II in 1939-45, there was an estimated 21 million military plus 30 million civilian fatalities, together representing 4 per cent of the global population of the time. The most severely hit countries were Poland, thereby losing 18 per cent of its population; Yugoslavia and the USSR each 11 per cent; Germany 9 per cent; and Austria 5 per cent. Then during the Korean War in 1950-53 there were an estimated 800 thousand military plus 600 thousand civilian fatalities, with North Korea thereby losing 6 per cent of its population, and South Korea 3 per cent. And finally, during the Second Indo-China War (or Viet Nam Conflict) in 1961-75, there were an estimated 1.2 million military plus 700 thousand civilian fatalities, with North Viet Nam losing 3 per cent of its population, and South Vietnam 5 per cent.

***Examples of environmental disruption:*** Some examples from throughout history of *intentional* environmental disruption will clarify the actions to which I have just been alluding. I shall be picking and choosing from the long sweep of military history to help make the point that serious environmental disruption is neither a new phenomenon nor an uncommon one.

***Attacks on the land:*** We learn from the *Old Testament* that Philistine crops (corn, grape, olive) were destroyed by their Israelite opponents as long ago as perhaps the 12th Century BCE. Half a millennium later, Herodotus tells us that when the Scythians retreated before the Persian army of Darius the Great (*ca* 512 BCE, they withdrew, choking up the wells and springs along the way, leaving the whole countryside bare of forage, and destroying all that grew on the ground. In the same era, Thucydides notes that the Spartans, in their ultimately successful efforts to defeat Athens during the Peloponnesian War of 431–404 BCE, made a continuing practice of attacking their enemy just before harvest time in order to destroy the annual grain crop before it could be reaped.

Moving along to the 13th Century CE, Ghenghis Khan, routinely waged *la guerre a l'outrance* (a war of extremes). Thus, in conquering Cathay (now northern China) everything in the open country was annihilated or driven off, crops trampled and burned, and herds appropriated or slaughtered. And several years later, in overrunning what is now Afghanistan, Ghenghis's forces trampled and burned whatever crops they came upon, together with razing the accompanying villages. Here I might jump ahead for a moment to the Afghan War of 1979-1989, in which the Soviet forces employed similarly relentless scorched-earth tactics, including the systematic destruction of crops and killing of livestock.

Dropping back to the Thirty Years' War of 1618—48, in which the Protestant armies of the German states and Sweden waged an ultimately successful war against the Catholic armies of the Holy Roman Empire, the German and Swedish forces time and again systematically decimated the agricultural fields of central Europe. Then I might move on to the Chinese Rebellion of 1850—64 in which the Tai Ping (Great Peace) movement was ultimately unsuccessful in overthrowing the Manchu Dynasty. Among other approaches, the Manchu forces employed brutal large-scale scorched-earth tactics in order to starve the rebellious regions into submission.

As early as 1629, the British Virginia colonists tried to maintain the upper hand with their hostile Native North American neighbors by continual incursions to destroy their crops. And in General George Washington's most carefully planned campaign of the US War of Independence of 1775—83, in 1779 he went a long way towards neutralising opposition from the Iroquois allies of the British via the systematic destruction of their food supplies and crops. Subsequent successful subjugation of native North Americans during the 19th Century consistently involved destruction of the natural resources (livestock, croplands, fruit trees) upon which they depended, being especially successful against the Navahoes in 1864. During the US Civil War of 1861—65 Union forces (in such proudly touted Union campaigns as General William Tecumseh Sherman's March to the Sea in Georgia and General Philip Henry Sheridan's ravaging of the Shenandoah Valley in Virginia) systematically confiscated food stores and draft animals and destroyed farms. In the Philippine War of 1899—1903 the USA meticulously destroyed villages, storehouses, crops, and draft animals throughout entire recalcitrant provinces (to the great distress of Mark Twain, among others).

Moving on to World War II of 1939—45, the USA attacked German grain crops with incendiary attacks. Not long after, the Korean War of 1950—53 saw the USA directing some of its most successful air attacks against North Korea's major irrigation dams for the express purpose of disrupting the enemy's production of its staple food, rice. And in pursuit of the Vietnam Conflict of 1961—75 the USA routinely carried out elaborate policies of systematic forest and crop destruction programs with widespread, long-lasting, and severe effects — the former primarily to deny its Viet Cong enemy cover and concealment, the latter primarily to deny it rice and other field and tree crops, salt, and even medicines. I might add here that a government study found that to deny a Viet Cong soldier of his food required just over 100 similar civilian deprivations. And, as one now knows, the massive use by the USA of Agent Orange as an anti-plant chemical warfare agent has led to originally unforeseen human medical tragedies for so many exposed Vietnamese, Americans, and others.

As to the Gulf War of 1991, in which US forces were for once rather environmentally blameless (although there has been some concern expressed over residual depleted uranium in the battlefields from its armor-piercing munitions). *But* in which the US opponent, Iraq caused substantial amounts of environmental havoc through the intentional release of huge amounts of liquid oil on both the land of Kuwait and into the Persian Gulf, as well as extensive oil smoke into the regional atmosphere.

As a more current example, the diplomatically complex and seemingly intractable *Afghan War of 2001–* is offered here in some detail. This war was launched by the USA with support from the United Kingdom and Australia (although without United Nations Security Council authorisation) presumably prompted by the just prior al-Qaida attacks on the USA in 11 September 2001. Doing so with the stated dual purposes (a) of overthrowing the government of the Islamic Emirate of Afghanistan controlled by the Taliban (an Islamic movement of Afghan [Pashtun] Sunni Muslims established in the early 1990s and known for its strict interpretation of the restrictive Muslim Sharia law; and representing the main rebel force in this war). And (b) of preventing al-Qaida (a militant worldwide movement of Sunni Muslims that grew out of the Soviet-Afghan War of 1979–88, its goal being the establishment of a pan-Islamic caliphate throughout the Muslim world; and here allied with the Taliban rebel forces) from enjoying a haven and a base of operations in that country.

It did not take long to oust the existing government and oversee its replacement by what became the Democratic Islamic Republic of Afghanistan. Then in late 2001, the United Nations Security Council established an International Security Assistance Force (ISAF) to protect Kabul and its new government, with the management of ISAF two years later assumed by the North Atlantic Treaty Organisation (NATO), with 42 participating countries, but overwhelmingly dominated by the USA. Outside greater Kabul rebel forces, composed primarily of the Taliban plus some al-Qaida troops, continued year after year to fight ISAF, controlling much of the countryside and with no clear winner in sight. In 2011, the al-Qaida leader Osama bin Laden was killed in Pakistan by the USA, and finally in 2012 an agreement was reached by ISAF to withdraw most of its troops by mid-2013. Although with the USA then having bilaterally agreed to a 'strategic partnership' with the Afghan government for prevailing over the still to a considerable extent Taliban-controlled country. The Post-conflict and Disaster Management Branch (now named the Disasters and Conflicts Programme) of the United Nations Environment Programme (UNEP) has been working in Afghanistan since 2002, and in 2009 the United Nations Security Council established the United Nations Assistance Mission in Afghanistan to coordinate all non-military United Nations activities in the country.

The Afghan War of 2001–is discussed below under three headings: (a) its environmental impact; (b) its social impact; and (c) its economic impact. (The 2009 UNEP report entitled *UNEP in Afghanistan: Laying the Foundations for Sustainable Development* has provided much of the information being presented below).

*Environmental impact of the Afghan War of 2001* : Of the 33,000 square kilometers of Afghanistan's pre-war woodlands (important for building materials, firewood, wildlife, and water conservation), military actions have destroyed about twenty thousand; and rampant illegal logging is occurring in much of the remaining woodlands. War damage has substantially degraded much of the 300,000 square kilometers of rangelands to unpalatable and eroded scrublands (those debilitated grazing lands now supporting only about half of the 35 million pre-war livestock and previously significant numbers of wildlife). With much of Afghanistan's wild nature having been disrupted and degraded by the war, its natural flora and fauna are all now under serious threat. Experience suggests that assisted agricultural recovery is likely to take between five and ten years, whereas unassisted (natural) environmental recovery must be measured in decades. Finally, it is unfortunate that security considerations have prevented the establishment of the 15 per cent or so of the country as protected nature reserves, a fraction that every country should be morally obligated to maintain in order to contribute to its overall environmental integrity and to minimise further plant and animal extinctions. Indeed, biological considerations suggest that 50 per cent more or less would have to be protected to prevent further losses of biodiversity, whereas several major environmental organizations routinely suggest a politically feasible goal of 10 per cent or 12 per cent. In 2010, the states parties to the 1992 Convention on Biological Diversity [UNTS 30619] agreed that 17 per cent be the goal for the earth's land areas and 10 per cent for its ocean areas.

*Social Impact of the Afghan War of 2001* : Three-quarters or more of the *circa* 30 million Afghani citizens live in the countryside dependent for their livelihoods upon agricultural, pastoral, or other natural resource dependent pursuits, and the lives of perhaps one-third of those rural dwellers have been seriously disrupted by the war. More than 25 per cent of the country's farmlands have been prevented by the war from being used, and a significant fraction of those areas remaining in active cultivation are being used for the illegal production of poppies (see below).

An estimated 13,000 Afghani civilians were killed during the first decade of the war, and many more wounded. As to the military casualties during that time (the summation of those armed forces that are (a) killed in action, (b) wounded in action sufficiently seriously to require hospitalisation, (c) subsequently dying of wounds, and (d) missing in action) the armed forces of the USA suffered about 14,000 killed in action and perhaps twice that number wounded in action. To those numbers must be added the casualties—(a) of the other 41 members of the United Nations International Security Assistance Force (ISAF), together perhaps amounting to one-third as many as those of the USA, (b) of the large contingent of US civilians employed by the USA, and (c) of allied Afghan Government forces (perhaps a very large number, although one that has not been reliably estimated). Similarly, no reliable numbers of casualties seem to be available for the Afghan rebel forces (Taliban plus al-Qaida).

*Economic impact of the Afghan War of 2001 :* About two-thirds of Afghanistan's infrastructure has been destroyed. The partial collapse of national and local institutions, and with it the rule of law, have led to the mismanagement of the country's natural resources (forests, rangelands, soils, waters, wildlife). For the USA alone, the monetary cost of the first decade of the war has been calculated by its Congressional Research Service to come to about US$ 468 billion. For the remaining 41 ISAF countries, as a first approximation, perhaps one-quarter that amount was spent by them. For the opposing rebel forces (Taliban and al-Qaida), the monetary cost during that period is estimated to have come to perhaps US$ 3 billion, sadly enough apparently obtained primarily from the sale of Afghan grown poppies destined for the global opium black market. The extensive wartime destruction of both the Afghan infrastructure and environment has driven the inhabitants to exploit the country's resources unsustainably for their immediate survival, and thus at the expense of long-term economic, cultural, and underlying environmental well-being for the society.

Before leaving attacks on the land, the matter of ***self-inflicted scorched-earth:*** events could be mentioned. Two major examples from World War II of 1939—45 follow. One of them was carried out by the Soviet Union in 1941 to impede the German invasion. And for another, in 1944 when Germany was occupying Norway, and fearful of a major Soviet attack from the Kola Peninsula to the northeast, German forces literally laid waste to a huge fraction of the northern part of the country in order to impede that expected (though never realised) attack, in the process displacing some 50,000 or more of the local mainly rural inhabitants.

***Attacks on the water infrastructure:*** Some *levees, dikes,* and *dams* lend themselves well to attack. For example, in subjugating Mesopotamia (now Iraq) in the 13th Century, Ghenghis Khan's forces deliberately destroyed the irrigation works of the Tigris River upon which the local agriculture depended. Then during the Franco-Dutch War of 1672—78, the Dutch in June of 1672 were for a time successful in preventing the French from overrunning their country by cutting dikes to create what was termed the Holland Water Line. They did this despite vehement objections from the local inhabitants to be displaced thereby. And I might add that the following winter was uncharacteristically cold, permitting the French to advance over the frozen waters.

The Second Sino-Japanese War of 1937—45 provides us with a spectacular example of militarily successful intentional flooding. In order to curtail the Japanese advance, the Chinese in June 1938 dynamited the Huayuankow Dike of the Yellow River near Chengchow. This action resulted in the drowning of several thousand Japanese soldiers and stopped their advance into China along that front. In the process, however, the flood waters also ravaged major portions of Henan, Anhui and Jiangsu Provinces, including over 4,000 cities and villages. Immense areas of cropland were inundated and their topsoil washed away, with several hundred thousand Chinese drowning in the process and several million left homeless. Indeed, this act of environmental warfare appears to have been the single most devastating human act in the full history of our species. The River was not to be re-tamed for a decade, and the three devastated provinces may not yet have been fully rehabilitated. World War II in 1939—45 was also witness to huge areas of intentional rural flooding with devastating consequences, especially by British forces in the Ruhr Valley of Germany and by German forces in the Netherlands.

***Explosive remnants of war:*** A word is in order on the often staggering numbers of *land mines, cluster bomb units* (CBUs), and *other live ordnance* that together continue to infest battlefields for incredibly long periods of time. Those explosive remnants lead, of course, to some abandonment of the affected lands. But it is a sad fact that worldwide they currently result in thousands of casualties per year from past wars, most often to those farmers, herders, and loggers who feel compelled to return to their traditional homes, fields, and forests. I must also note that any post-war removal of those lastingly pernicious remnants is arduous, expensive, environmentally disruptive, and above all dangerous.

Perversely on the *plus* side, when explosive remnants (or other factors) prevent or impede the return to civil uses of former theatres of military operation, this permits floral and faunal recovery of those lands (one important example of such recovery being the Korean Demilitarised Zone [DMZ], which has been off limits for over half a century now).

## Refugees, Environmental and Other

Each year, growing numbers of rural people in the poorer (the Third World) countries of the world leave their homes of long standing in order to migrate to locations of less danger or greater opportunity, as already noted, perhaps half of them to cities. In 2009, about 255 million (or almost 4 per cent of the global population) being in that incredibly unenviable status. Of those 255 million displaced persons, an estimated 27 million

(11 per cent of them) remained within the bounds of their country, a group I refer to as *internal refugees* (but who in diplomatic language are known as 'internally displaced persons'). The remaining 228 million found their way into adjacent countries and beyond, the *cross-border refugees* (or in diplomatic language, comprised primarily of, 'international migrants', with the rest being 'official refugees'). So the question arises of what fraction of those many, many refugees is fleeing from the numerous wars always in progress and what fraction is doing so for other reasons. The answer turns out to be quite straightforward.

In looking back over recent decades, one sees that the frequency of wars, both major and minor, has been fluctuating only mildly, that the number of despotic regimes worldwide has actually been in slow decline, and that political rights and civil liberties have been improving at least somewhat forever greater numbers of people. Furthermore, the frequency of natural disasters (earthquakes, volcanic eruptions, floods, hurricanes, and the like) has at least to date held quite steady—although, of course, I must suggest here that ***global warming*** may well soon lead to increasing numbers of adverse weather effects, for example, with associated increases in civil unrest particularly in the Third World.

By contrast, human numbers, especially in Africa and elsewhere in the Third World, continue their relentless increases, unavoidably in step with environmental strains and resultant ever less adequate means of subsistence for so very many millions of people. In other words, as already suggested earlier, those hapless individuals fleeing from danger (whether that be from war or from persecution) represent a significant although relatively constant number of refugees—roughly 14 million. Whereas those fleeing from a no longer sustainable resource base—and thus properly referred to as *environmental refugees*—currently represent some 240 million persons, a value, moreover, increasing by perhaps 8 million per year—yes, by 8 million per year.

**Some Implications**

Thinking about the implications of what I have been saying—and to look ahead a bit — I should like for us all to consider a number of actions that would go toward mitigating *three* areas of concern, namely, battlefield disruption of the environment, the ever increasing numbers of refugees, and the exacerbations to be expected as ***global warming*** proceeds:

***First***, as to wartime *environmental disruption* and accompanying human consequences, I have tried to demonstrate that, although of serious concern, this has *not* over the centuries increased in magnitude to any significant extent. Nonetheless, the excesses perpetrated during two modern wars have, in fact, awakened the world to the perniciousness of military environmental disruption: namely, the Second Indochina War (or Vietnam

Conflict) of 1961—75 and the Gulf War of 1991. Therefore, important among our ameliorative efforts must be to convince the relatively few recalcitrant nations to adopt the ***1977 Protocol I** on International Armed Conflicts* [UNTS 17512], an epochal treaty that not only establishes restraints on attacking agricultural areas, dams and nuclear electrical generating facilities. Even more importantly, makes illegal those means of warfare that may be expected to cause the natural environment *widespread*, *long-lasting*, and *severe* damage—a novel addition to international humanitarian law that had been instigated by US actions in the Viet Nam Conflict.

Would that the ***1977 Protocol II** on Non-international Armed Conflicts* [UNTS 17513] were a stronger instrument, owing to its specific reference to the now substantially greater numbers of non-international (internal) wars than international wars. But, nonetheless, the latter also deserves every country's support for it does provide for certain restraints on the destruction of agricultural areas and on the release of so-called dangerous forces (such as those that could result from an attack on a dam, chemical factory, or nuclear power plant).

Regarding *explosive remnants of war*, it again becomes a high-priority mission for us to convince the few remaining hold-outs to adopt the *1997 Land Mine Convention* [UNTS 35597] which renders illegal the use of—indeed, the very possession of—anti-personnel land mines. Moreover, the more recently enacted *2008 Cluster Munition Convention* [UNTS 47663] is a similarly laudatory effort at restraint of those pernicious weapons, and thus its widespread adoption is to be supported.

For all four of those multilateral treaties I have singled out it is important to recognise that they are not only *proscriptive* in their outlawing of an action, but as well *normative* in serving to introduce and establish a relevant standard.

***Second***, as to *civil environmental disruption* and accompanying human consequences, I must suggest that as tragic (and, in wishful theory, as avoidable) as both the incidental and especially the intended environmental disruptions from warfare may be, it is the environmental depredations arising from the *civil sector of society* that not only dwarf all those of the military sector, but as well keep growing in magnitude. So, to curb the ever increasing flow of environmental (and other) refugees to the cities of the world, we must recognise *two* fundamental approaches: On the one hand, to curb the *per capita* environmental impact of humankind; and, on the other, to address the overwhelming tragedy of the world's unconscionable but nonetheless ever rising global *over-population*.

***Third***, as to ***global warming*** (and associated ocean acidification), the exacerbations to be expected from that direction are sure to be detrimental to both the military and civil issues I have been describing.

On the one hand, the predicted changes in climate and sea level will make life for many in the rural reaches of the Third World even more precarious and less sustainable than it is at present, or even impossible, making for the generation of even greater than current increases in environmental refugees. And on the other, the resulting civil unrest could lead to an increase in the numbers of internal wars, with all of their attendant adverse social and environmental consequences. Therefore, finally, one additional challenge that we should in all good conscience accept (for the just-mentioned and other poignant reasons) is to work toward stabilising the climate—striving for this through that seemingly elusive goal of a *sustainable energy-consumption* future.

## References

Westing, A.H. 1976. *Ecological Consequences of the Second Indochina War*. Stockholm: Almqvist and Wiksell, 119 p. + 8 plates. In Japanese: Tokyo: Iwanami Shoten Publishers, 220 pp. + 8 plates. 1979.

Westing, A.H. 1980. *Warfare in a fragile world: Military impact on the human environment*. London: Taylor and Francis, 249 pp. Available (15.8 MB) at: http://books.sipri.org/product_info?c_product_id=269

Westing, A.H. (ed.). 1985. *Explosive remnants of war: Mitigating the environmental effects*. London: Taylor and Francis, 141 p.

Westing, A.H. 1989. Herbicides in warfare: The case of Indochina. In: Bourdeau, P. et al. (eds). *Ecotoxicology and climate: With special reference to hot and cold climates*. Chichester, UK: John Wiley, 392 pp. 337-57.

Westing, A.H. 1990. Our place in nature: Reflections on the global carrying-capacity for humans. In: Polunin, N., and Burnett, J.H. (eds). *Maintenance of the biosphere*. Edinburgh: Edinburgh University Press, 228 pp: p. 109.

Westing, A.H. 1992. Environmental refugees: A growing category of displaced persons. *Environmental Conservation* (Cambridge, UK) 19(3): 201-07.

Westing, A.H. 1994. Population, desertification, and migration. *Environmental Conservation* (Cambridge, UK), 21(2): 110-14, 109.

Westing, A.H. 1997. Environmental protection from wartime damage: the role of international law. In: Gleditsch, N.P. et al. (eds). *Conflict and the environment*. Dordrecht: Kluwer Academic Publishers, 598 pp. 535-53.

Westing, A.H. 2001. A Korean DMZ park for peace and nature: Towards a code of conduct. In: Kim, Chae-Han (ed.). *The Korean DMZ: Reverting beyond division*. Seoul: Sowha Publishing Company, 359 pp. 157-91. In Korean: Same citation. 2000.

Westing, A.H. 2001. Peace, culture and ethics: Recent history of conservation values in peace and war. In: Tolba, M.K. (ed.). *Our fragile world: Challenges and opportunities for sustainable development*. Oxford: EOLSS Publishers and Paris: United Nations Educational, Scientific and Cultural Organisation (UNESCO), 2263 pp. 865-72.

Westing, A.H. 2003. Environmental dimension of the Gulf War of 1991. In: Brauch, H.G. et al. (eds). *Security and environment in the Mediterranean: Conceptualising security and environmental conflicts.* Berlin: Springer Verlag, 1134 pp. 523—34 + 1003 —1089 *passim.*

Westing, A.H. 2013. *Arthur H. Westing: Pioneer on the Environmental Impact of War.* Berlin: Springer, 150 p.

## Notes

* This contribution updates and expands an address the author gave at Columbia University on 1 October 2010. The data on numbers of wars are from the *Stockholm International Peace Research Institute* in collaboration with the University of Uppsala *Institute of Peace and Conflict Research.* The data on numbers of displaced persons are from the *International Organisation for Migration* (Geneva), based in part on data from the *United Nations High Commissioner for Refugees* (Geneva). The data on political oppression are from *Freedom House* (Washington). The author thanks Carol E. Westing for relevant insights.

** Upon completing his undergraduate training in botany (Columbia, B.A., 1950), the author was commissioned as an officer in the US Marine Corps, followed by about a year of front-line combat in Korea. His subsequent training was in forestry (Yale, M.F., 1954) and physiological ecology (Yale, Ph.D, 1959). In a two-year research position with the US Forest Service, the author investigated the efficacy of killing unwanted trees with herbicides (subsequently to be named 'Agent Orange'), followed by 25 years in academia (including Purdue University, the University of Massachusetts, Hampshire College, and the European Peace University), professing tree physiology, forest ecology, and environmental conservation.

An abiding concern for the integrity of the biosphere, together with the author's past wartime experiences of drastic environmental disruption, led to on-site investigations, especially on behalf of the *American Association for the Advancement of Science* (AAAS), of the long-lasting environmentally destructive US strategy throughout the Viet Nam Conflict of denying its Viet Cong enemy both cover and sustenance. Although, as it eventually turned out, to little military effect, but with significant societal and diplomatic repercussions. After leaving academia, the author was for eight years Director of the *United Nations Environment Programme* (UNEP) project on 'Peace, Security, and Environment', which was headquartered at the *Stockholm International Peace Research Institute* (SIPRI) and *International Peace Research Institute Oslo* (PRIO), its outcomes including an early environmental assessment of the Gulf War in 1991.

# 4

# Climate Change and Threats to Human Security in Bangladesh: An Analysis

*Md. Rafiqul Islam and Ziaul Haque*

## 1. Introduction

Today, climate change has emerged as one of the grisly human security challenges of our time with its harrowing effects on a broad range of issues within a paradigm, which inseparably links human, environment and socio-economic well-being. Extreme weather events, drought, flooding, sea level rise, retreating glaciers, habitat shifts, and the increased spread of life-threatening diseases, exacerbated by climate change, affect thousands of millions of people and their communities around the world. These conditions disrupt the way of life individual and community feel safe and secure, and prevent them from reaching their full potential and achieving enduring freedom from both fear and want. They destroy ecosystem services, diminish environmental resource base–fundamental for human survival–and thus bring chronic hunger, disease, endemic poverty and loss of livelihoods. Indeed, climate change presents an all-encompassing threat, even though, the social impacts of climate change vary across the regions of the world. Rising global temperatures will jeopardise the bases of many people's livelihoods, especially in the developing regions, increase vulnerability to poverty and social deprivation, and thus put human security at risk. Particularly in weak and fragile states with poorly performing institutions and systems of government, climate change are also likely to overwhelm local capacities to adapt to changing environmental conditions and will thus reinforce the trend towards general instability that already exists in many societies and regions (WBGU, 2008:1).

However, Bangladesh bears the biggest brunt of a crisis not caused by her but mainly by the external sources (Ali, 1999: 114). It is frequently cited as one of the countries hardest hit by climate change due to her geographic location, low elevation, high population density, poor infrastructure, high levels of poverty and critical dependency on natural resources. It remains the most vulnerable country in the world to tropical cyclones and the sixth most vulnerable to floods. More than 68 million people have been directly affected over the last eight years, and millions of lives and livelihoods are threatened by frequent weather-related disasters. With low-lying lands, coastline areas and floodplains occupying 80 per cent of the country, Bangladesh is highly exposed to both disasters and sea level rise. Of its 155 million inhabitants, half live below the poverty line and over a third suffers from malnutrition and hunger (Global Humanitarian Forum- GHF, 2009: 17). It is estimated that climate change could affect more than 70 million people of Bangladesh. It is increasingly recognised, therefore, that the adverse impacts of climate change will severely limit many problem-solving capacities of Bangladesh and thus will put additional stress on overall human development of the country. Against this backdrop, the purpose of the study is to explain the transition from concerns about security to concerns about human security. And from a human face of climate change, it will explore the ways in which climate change poses significant risks to human security in Bangladesh.

## 2. Conceptual Framework

### *2.1. Security a relative concept*

Security is a perpetual concern in human life since it is an inseparable part of his existence and survival. Therefore, security is a relative concept. The quest for security never ends as the 'feeling of being safe' is relative to the threats against which the protection must be ensured. Security as Smith posits it is an 'essentially contested concept' (Tadjbakhsh and Chenoy et.al. 2007: 10). *English Oxford Dictionary* defines security: 'The condition of being protected from and not exposed to danger; safety...Freedom from care, anxiety or apprehension; a feeling of safety or freedom from an absence of danger' (Gary and Murray et. al., 2001: 585–610). But the International Relations (IR) theorists' interpretation was narrowly founded (Buzan, 1991: 14) and understood by a dominant 'classical view'–mainly a state-centric focus, meaning to retain the ability to maintain national and territorial integrity in an anarchic international system. This Realist School summarised the traditional view of security as a derivative of power. Thus, Walter Lippman (1943) in his classical formulation posits that 'a state is secure to the extent to which it is not in danger of having to sacrifice core values if it wishes to avoid war and is able, if challenged, to maintain them by victory in such a war' (cf. Romm,

1993: 5). Security, therefore, is taken to be about the pursuit of freedom from threat and the ability of states and societies to maintain their independent identity and their functional integrity against forces of change, which they see as hostile. According to Arnold Wolfers, 'security, in an objective sense, measures the absence of threats to acquired values, in subjective sense, the absence of fear that such values will be attacked' (cf. Prabhakaran Paleri, 2008:45). This classical view had been on ascendancy during the cold war shaping the entire mode of security edifice. However, with the winding down of cold war and bipolar politics, this traditional modelling of security became outmoded. And with an increasing interplay of many non-traditional aspects within the scope of security perception and the vivid change in the nature of threats the states exposed vulnerable to, a 'paradigm shift' in the security mindset had then been noticed. This necessitated re-thinking of the concept of security and a search for trade-off links with this age-old question and its essential ingredients. Since then traditional notion of security has been broadly debated expanding the concept and including many issues within this process of evolution. The concept of 'human security' has been the latest significant development in its sophistication and present discourse of security studies.

### *2.2. The concept of human security*

The concept of 'Human Security' evolved at a time of great international shift (Shahrbanou Tadjbakhsh, et.al. 2007: 1) when the traditional role of state broadly transformed and traditional notion of threats was contested. Human security concept marked, therefore, an end of the old chapter and beginning of a new one. It was the first development in that direction of 'new manner of thinking' when the world was at a crossroads and a new roadmap was in high demand. The *UNDP Human Development Report* 1994, therefore, first coined the term 'human security' as a way to go from here (UNDP, 1994). Thus, this report broadened the concept of traditional security and defined human security as 'safety from chronic threats such as hunger, disease, and repression as well as protection from sudden and harmful disruptions in the patterns of daily life' (UNCHS, 2003). Marking this change in international system, Kofi Annan in one of his landmark speech notes:

> Geopolitical patterns, economic trends, technological change and other developments are severely straining the system of collective security that has been in place for the past 60 years. Today's threats–familiar ones with added potency, and some entirely new dangers–are borderless, highly connected and capable of crippling, and even destroying societies everywhere (Annan, 2005).

Thus, human security addresses the major changes in international system and the genesis of which was conditioned by new possibilities and new threats. To simply put human security debunks the question of security from its traditional conception of the safety of states from military threats to concentrate on the safety of people and community. Once the referent object of security is changed to individuals, it then proposes to extend the notion of safety to a condition beyond mere existence (survival) to life worth living, hence, well being and dignity of human beings. But, there is no universal definition of human security. Hybrid reports, analyses and critiques prevent reaching a consensus except the agreement about the shift of focus–from a state-centred approach to a human-centred one. *UNDP Human Development Report 1994* notes that human security is *people-centred.* It is concerned with how people live and breathe in a society, how freely they exercise their many choices, how...they live in conflict or in peace (UNDP, 1994: 23). Barry Buzan, a reformer, has echoed this line of reasoning when he looks at security from all angles going from micro to macro, also addressing the social aspects of security and how people or societies construct or "securitize" threats (Marianne Stone, 2009: 2).

The UN Commission on Human Security, co-chaired by Sadako Ogata and Amartya Sen defined human security as concerned with safeguarding and expanding people's vital freedoms. It requires both protecting people from critical and pervasive threats and empowering people to take charge of their own lives (Ogata, et. al. 2003: IV). However, the Commission warns that human security and national security are not mutually exclusive, as Ogata remarked, 'human security does not seek to supplant state security, but rather to complement it' (Ogata, 2003:2). The commission posits that security between states remains a necessary condition for the security of people, but national security is not sufficient to guarantee peoples' security (cf. Sadako Ogata, et.al. 2003: 3). The Ecological Institute in a discussion on human security summarised that 'security is a much wider concept today than it was just some decades ago. Today, the concept is applied not only to new and diffuse military threats such as global terrorism but has been broadened to include additional areas such as food, water, health and other issues, of which many are intimately related to the environment and to its protection (Ecological Institute, 2011). Former UN Secretary-General Kofi Annan points out that human security joins the main agenda items of peace, security and development. According to him:

Human security, in its broadest sense, embraces far more than the absence of violent conflict. It encompasses human rights, good governance, access to education and health care and ensuring that each individual has opportunities and choices to fulfil his or her potential. Every step in this direction is also a step towards reducing poverty, achieving economic growth and preventing conflict. Freedom from want, freedom from fear, and the freedom of future generations to inherit a healthy natural environment — these are the interrelated building blocks of human - and therefore national - security (Annan, 2000).

Human security, therefore, is an integrative concept, unites a number of disparate thoughts and is concerned with reducing and removing the insecurities that plague human lives. So, it has many components but two broad categories could be identified there—'freedom from fear' and 'freedom from want'. As the UNDP reports posit it—'the battle of peace has to be fought on two fronts'. The first is the security front where victory spells freedom from fear. The second is the economic and social front where victory means 'freedom from want' (UNDP, 1994: 24). Within these two broad threads, human security concept increasingly implicates to include economic security, food security, health security, environmental security, personal security, community security and political security. Though, these categories seem to be discreet but they all are fundamental to protect the vital core of all human lives in ways that enhance human freedoms and human fulfilment. Threats to each category are interlinked and interconnected and they are mutually linked in a domino effect.

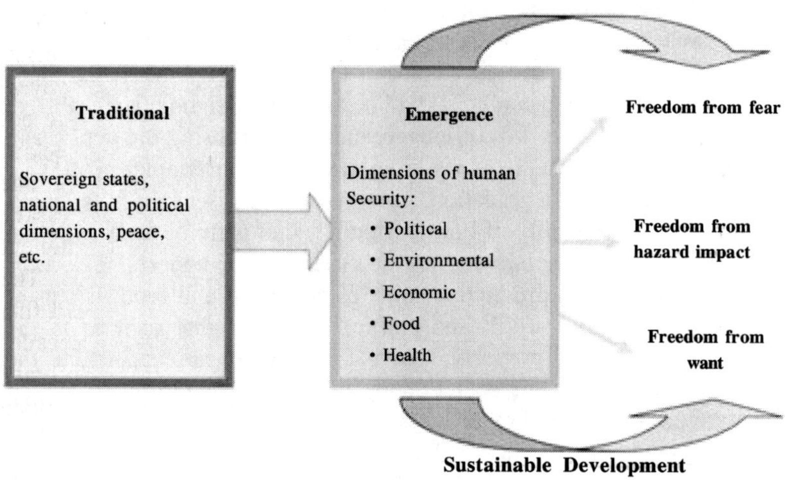


**Figure: Evolution of the Security Paradigm.**

(*Source*: UN University and IEHS, 2008).

This figure shows the nature of shifting of the concept of military based security system to the human centric security system.

## 3. Global Environmental Changes and Human Security

Environmental degradation has become a human security concern because the prospects for human security are deeply affected by local and global processes of environmental change. According to the International Institute for Environment and Development (IIED), the environmental security aspect of human security arises out of our own damage to our local environments and to our planet (Saleemul Huq, et. al.). Indeed, the constraints imposed by local environmental conditions and their natural variability are powerful determinants of the security of individuals and societies. Thus, the Global Environmental Change and Human Security (GECHS) assert that human insecurity is a function of many social processes. It causes some people to be more sensitive and less able to prepare for and respond to sudden and incremental environmental changes (Matthew, et. al., 2010: 1). A study puts it simply—'environmental change reveals the connections—as well as the frictions—between the security of individuals and communities and the security and sustainability of ecosystems and species, including humanity' (Matthew, et. al., 2010: 4). Robert D. Kaplan identifying the environment as part of a terrifying array of problems that will define a new threat to our security in the 21st century commented that:

The political and strategic impact of surging populations, spreading disease, deforestation and soil erosion, water depletion, air pollution, and, possibly, rising sea levels in critical, overcrowded regions like the Nile Delta and Bangladesh—developments that will prompt mass migrations and, in turn, incite group conflicts—will be the core foreign-policy challenge from which most others will ultimately emanate, arousing the public and uniting assorted interests left over from the Cold War (Kaplan, 1994).

Environmental degradation, thus, poses new security challenges that undermine the quality of life, disrupt it social patterns and aggravate the existing socio-economic and political tensions. It undermines access to basic needs such as productive soils, clean water, and food; it puts at risk enshrined human, civil, and political human rights such as to the means of subsistence, property, and nationality. It can undermine the provision of economic and social opportunities required to foster human security; and in these and other ways it can undermine people's ability to pursue the kinds of lives they value. It may also be an indirect factor in the generation of violent conflicts (Matthew, 2010: 21). Thus, global environmental degradation and its sweeping impacts on many fundamental aspects of human life and living have made it a significant part of human security assessment and policy.

## 4. Climate Change and Human Security

The security link of climate change is irrefutable. Once John Ashton, UK Foreign Secretary's Special Representative for Climate Change, said, there is every reason to believe that as the 21st century unfolds, the security story will be bound together by climate change (Ashton, 2007). True,

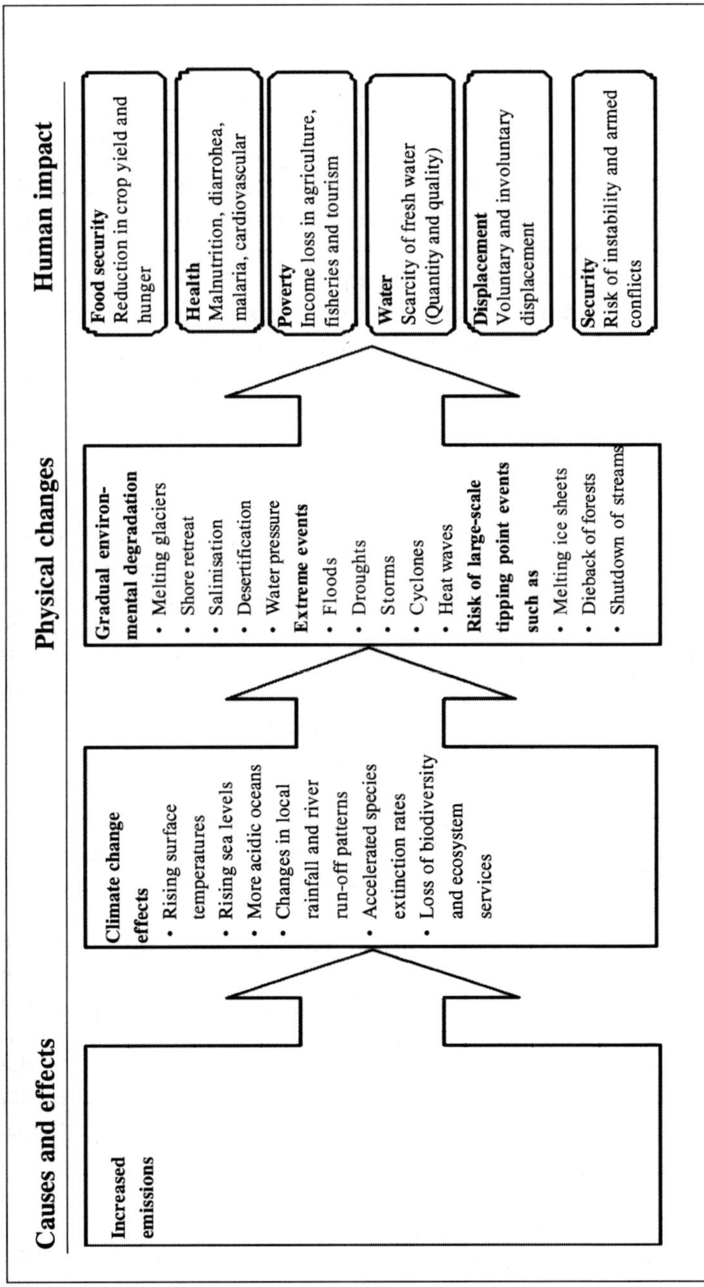


**Figure: The Links from Increased Emissions to Human Impact**

(*Source*: Global Humanitarian Forum, 2009).

climate change is now increasingly recognised as a major human security issue that poses significant global threats. It unfolds an ominous threat to human society making it the greatest emerging humanitarian challenge of our time. Rising sea levels, desertification, changes in local rainfall and river run-off patterns with increased precipitation, salinisation of river deltas, reduce access to fresh and safe drinking water. This in turn affects health negatively and poses real threat to food security in many developing countries. According to one count currently over 2.8 billion people live in areas of the world prone to more than one type of the physical manifestations of climate change—floods, storms, droughts, sea level rise (GHF, 2009: 15). Studies suggest that many of these people are not resilient to those extreme weather patterns and climate variability. They are unable to protect their families, livelihoods and food supply from negative impacts of seasonal rainfall leading to floods or water scarcity during extended droughts. Climate change is multiplying these risks. A publication of the 'Global Humanitarian Forum' reveals that the world experiences over 400 weather-related disasters per year leaving a frightening toll in their wake: almost 90 million people requiring immediate assistance due to personal injury, property loss, exposure to epidemics, disease or shortages of food and fresh water (GHF, 2009:12). Recognising the climate change as a multiplier of human impacts and risks, the study also summarised the following postulation about climate change (GHF, 2009: 3-22):

- Hundreds of thousands of lives are lost every year due to climate change. This will rise to roughly half a million in 20 years. Over nine in ten deaths are related to gradual environmental degradation due to climate change.
- Economic losses due to climate change currently amount to more than one hundred billion US dollars per year, which is more than the individual national GDPs of three quarters of the world's countries.
- Over half a billion people are at extreme risk to the impacts of climate change, and six in ten people are vulnerable to climate change in a physical and socio-economic sense. More than 300,000 die due to climate change every year—roughly equivalent to having an Indian Ocean tsunami annually.

Through a complex set of effects, climate change affects human health, livelihoods, safety, and society. The following figure shows climate change impacts on people:

## 5. Climate Change and Bangladesh

### *5.1. Geography a paradox*

Bangladesh has become one of the 'frontline states' (GHF, 2009: 17) of climate change due to its location in ecologically and geographically sensitive area. As the climate change unfolds her, unique geographic setting has become a 'paradox' making the country worst victim of climatic variability. The country has very low and flat topography except in the northeast and southeast regions. About 10 percent of her land is hardly 1m above sea level rise and one-third under tidal excursions (cf. Matthijs, 2010: 18). It has contrasting weather with the Himalayas and the Khasi-Jaintia Hills to the north and the Bay of Bengal and northern Indian Ocean to the south (Ali, 1999: 110). Both of these varied geographical settings, controls, modify and regulate the weather and climate of the country. The low-lying deltaic plain is mainly fed the mighty rivers namely the Ganges, the Brahmaputra and the Meghna, which together forms the largest river system in the world (Elhance, 1999: 156). These rivers not only traverse landmass across South Asia but support and nourish food banks on which thousands of millions of people rely for their daily subsistence. These rivers have large number of tributaries and distributaries. The country popularly known as a 'land of river' has 230 rivers covering a length of about 24,140 km. One-study counts that these rivers drain about 85 per cent of the total volume of water brought into Bangladesh (Siddiqi, 139). The average annual sediment load carried by the rivers to the Bay of Bengal is around 2 billion tons annually (Viles et. al., 1995: 294).

The volume of water that is carried by this river system from the catchments of the Himalayas and sediment load it discharge to the Bay of Bengal profoundly shape the morphological and hydrological equilibrium and provides lifeblood to the active functioning of the delta. Therefore, any climate variability such as decreased rainfall and drought seriously hampers the habitat it buttresses and the geo-physical and hydrological features it maintains. Given the massive concentration of population on these river basins, any sudden climate-related onset will jeopardize the life and means of support of thousands of millions of people. Climate change is expected to exacerbate many of these existing vulnerabilities with increasing the frequency of floods, storm surges and droughts. The Inter-Governmental Panel on Climate Change (IPCC) in its Fourth Assessment Reports (AR4) has identified the completely Asian mega-delta specially Bangladesh as a key concern. Besides, Bangladesh has a distinct 710 km coastline which includes coastal plain islands, tidal flats, estuaries, neretic and offshore waters. According to Population Census 2001, this coastal area represents an area of 47,211 km$^2$, 32 per cent of the country's geographical area, wherein 35 million people, i.e., 28 per cent of the country's total population live at 6.85 million households (BBS, Population Census 2001). The coastal zone covers 19 districts out of 64 of which 12 districts meet the sea or lower estuary directly. The average population density in the coastal area, according to a study, is currently 890. Study projects that a 1m rise in sea level will inundate about 17.5 per cent of the landmass mostly in the western coast. Moreover, much of the coastline is often devastated by severe tropical cyclones and tidal surges claiming huge loss of lives, property and livelihoods. Available verified data projects that since 2000 the country has experienced more than 70 major disasters. Tropical cyclones, local storms, floods and droughts, have killed 9,000 people and caused damages of more than $5 billion. One-fifth of the country is flooded every year, and in extreme years, two-thirds of the country has been inundated (GHF, 2009: 17).

**Table: Vulnerability of Bangladesh to Climate Change**

**Most vulnerable countries to floods or cyclones (Deaths/100,000 people exposed to floods or cyclones)**

| | *Floods* | | | *Tropical cyclones* | |
|---|---|---|---|---|---|
| 1. | Venezuela | 4.9 | 1. | Bangladesh | 32.1 |
| 2. | Afghanishtan | 4.3 | 2. | India | 20.2 |
| 3. | Pakistan | 2.2 | 3. | Philippines | 8.3 |
| 4. | China | 1.4 | 4. | Honduras | 7.3 |
| 5. | India | 1.2 | 5. | Vietnam | 5.5 |
| 6. | Bangladesh | 1.1 | 6. | China | 2.8 |

* Of major flood affected countries reporting an average of over 200 deaths/year.

*Source* : UNDP (2004), A Global Report : Reducing Disaster Risk : A Challenge for Development http://www.undp.org/bepr

IPCC Fourth Assessment Report (AR4) has observed the following trends and climatic anomalies in Bangladesh (IPCC, 2007):

- In Bangladesh, the average temperature has registered an increasing trend of about 1°C in May and 0.5°C in November during the 14-year period from 1985 to 1998.
- The annual mean rainfall exhibits increasing trends. Serious and recurring floods have taken place in 2002, 2003, 2004. While frequency of cyclones originating from the Bay of Bengal has decreased since 1970, the intensity has increased.
- Salt water from the Bay of Bengal has penetrated 100 km much inland along the tributary channel during the dry season
- The precipitation decline and droughts has resulted in the drying up of wetlands and severe degradation of ecosystem.

Indeed, climate change is stress maker. It unfolds major threats by adding significant stress to Bangladesh's physical and environmental resources, human ability and economic activities. From a human impact lens of climate change, the following section of the study will map out the major threats climate change poses to Bangladesh.

## 5.2. Climate Change and Human Security Issues

### 5.2.1. Threats of increased natural disasters

Bangladesh will remain vulnerable to frequent natural disasters because of her disaster-prone geographical location and climate change is expected to worsen the case. Global warming is projected to increase the frequency and intensity of tropical storms, floods and landslides. In 2007, the number of reported disasters confirmed the global upward trend in natural disaster. Asia is the region hardest hit and most affected by natural disasters. According to a count, 37 per cent of the 2007 year's reported disasters occurred in Asia accounting 90 per cent of all the victims and 46 percent of the economic damage. Moreover, from 1970 to 2005, 171 disaster events have been recorded.

**Top 10 Natural Disasters in Bangladesh for the Period 1900 to 2011 Sorted by Numbers of Killed**

| *Disaster* | *Date* | *No. Killed* |
|---|---|---|
| Drought | 1943 | 1,900,000 |
| Epidemic | 1918 | 393,000 |
| Storm | 12-Nov-1970 | 300,000 |
| Storm | 29-Apr-1991 | 138,866 |
| Storm | Oct-1942 | 61,000 |
| Storm | 11-May-1965 | 36,000 |
| Flood | Jul-1974 | 28,700 |
| Storm | 28-May-1963 | 22,000 |
| Storm | 24-May-1985 | 15,000 |
| Storm | Jun-1965 | 12, 047 |

(*Source*: EM-DAT, the International Disaster Database, Center for Research on the Epidemiology of Disaster CRED, 2011)

### 5.2.2. Intensity of Floods, Prolonged Inundation and Water Logging

Floods remain a fact for many in Bangladesh with around a quarter of the country inundated in a normal year (GoB, 2009: 8). Indeed, floods have become unpredictable in terms of onset and scale. Climate change is expected to increase the risk of more frequent and severe floods through higher river flows resulting from heavier and more erratic rainfall in the Ganges-Brahmaputra-Meghna river system during the monsoon and increased melting of the Himalayan Glaciers (cf. Matthew Walsham,

**Table: Different Types of Flood Occurring in Bangladesh**

| *Type of Flood* | *Causes of occurrence* | *Time/duration* | *Tentative affected area* |
|---|---|---|---|
| Flash Flood | Run-off during exceptionaly heavy rainfall occurring in neighbouring upland areas | Pre monsoon months of April and May | The foot of the Northern and eastern hills of Bangladesh |
| Rainwater flood/ Monsoon Flood | Heavy rainfall occurring over flood plane and terrace areas within Bangladesh. | April-May, June-August | In the south-western part of the country |
| River Flood | Snow melt in high Himalayans, Heavy monsoon rainfalls over the Himalayans, the Assam Hills, the Tripura Hills and the Uppar Brahmaputra and Ganges Flood plains | April-May and June-September | Catchment areas of three major rivers. |
| Coastal Flood | In case of important cyclones the entire coastal belt is flooded. Coastal areas are also subjected to tidal flooding | Tidal flood occurs from June to September | South western coastal areas. |

(*Source*: Planning Commission, GoB and UNDP Bangladesh, 2009: 32).

2010:10). This combined effect of upstream flows, precipitation and terrestrial run-off will result in over flooding, causing water logging and prolong flood almost every year.

The coastal floodplain is only 3-5 meter from the mean sea level. Studies predict that a vast coastal area, approximately 18 per cent of total land, would submerge by 1 metre sea level rise. In the last 25 years, Bangladesh has experienced six severe floods, with the 1988 and 1998 floods alone causing 2000 - 6,500 and 1,100 deaths, respectively and displacing as many as 45 and 30 million people (GoB, 2009:9). A recent severe flood in 2007 alone inundated around 32,000 km$^2$ affecting 16 million people in around 3 million households. Around 45,000 houses were severely damaged with 1.12 million hectares of cropland affected and an estimated loss of BDT 5.8 million in livestock (Disaster Management Bureau, 2007).

**Top 10 Natural Disasters in Bangladesh**
**for the Period 1900 to 2011**
**Sorted by Numbers of Total Affected People**

| *Disaster* | *Date* | *No. Total Affected* |
|---|---|---|
| Flood | Jun-1988 | 45,000,000 |
| Flood | July-1974 | 38,000,000 |
| Flood | 20-Jun-2004 | 36,000,000 |
| Flood | May-1984 | 30,000,000 |
| Flood | 22-July-1987 | 29,700,000 |
| Drought | July-1983 | 20,000,000 |
| Flood | July-1968 | 15,889,616 |
| Storm | 11-May-1965 | 15,600,000 |
| Storm | 29-Apr-1991 | 15,438,849 |
| Flood | 5-July-1998 | 15,000,050 |

(*Source:* EM-DAT, the International Disaster Database, Center for Research on the Epidemiology of Disaster CRED, 2011).

### *5.2.3. Tropical Cyclone and Tidal Surge*

Bangladesh coastal zone is a geographical 'death trap' (Shamsuddoha, et.al. 2007:16) since the coastal line is highly vulnerable to tropical cyclones, storm and tidal surges. Of 80 tropical cyclones (McBride, 1995) formed in the world's waters every year usually about 6.5 per cent does occur in the North Indian Ocean (Neumann, 1993). 'Bangladesh Climate Change Strategy and Action Plan' counts that on average, tropical cyclones hit Bangladesh, every three years. Accompanied by high winds and storm surges of up to seven metres, these cyclones caused extensive damage to houses and loss of lives and livelihoods. Tropical cyclones in 1970 and 1991 killed 500,000 and 140,000 people, respectively (GoB, 2009). Other study found that nearly one million people have been killed in Bangladesh by cyclones since 1820. A UNDP report titled 'Reducing Risk of Natural Disasters—A Development Challenge' reveals a grime face of this reality. The reports project that during 1980 to 2000 cyclone caused death of 250 thousand people worldwide, of which 60 per cent were in Bangladesh (Shamsuddoh et. al., 2007: 16). The country faced 48 major cyclones between 1,584 and 2007. During a storm in May 1991, about 140,000 people died, and damage and displacement caused an estimated loss of US$2.4 billion (WEDO, 2008: 49).

Available data shows that in Bangladesh over 5 million people live in areas highly vulnerable to cyclones and storm surges. According to a count roughly 55 per cent of the coastal population lives within 100 km of the 710 km-long coastal belt of Bangladesh a huge majority of whom are low-income agricultural workers; 70 percent of whom are landless and relatively asset-poor (WEDO, 2008: 49). While Bangladesh has made significant progress in terms of disaster preparedness measures and early warning systems, compared to the intensity of the cyclones, it is far behind. In November, 2007 Bangladesh was stuck with a 4th category cyclone Sidr which affected 30 districts out of 64 impacting the lives and livelihoods of 8.7 million people, and damaging nearly 1.5 million houses and some 4.1 million trees (WEDO, 2008: 47). In 2009, cyclone Aila alone affected 3.9 million people and the loss of about 150,000 livestock, breaching 1,742 kilometres of embankments and displacing 76,478 families in *Satkhira* and *Khulna* alone (the two worst affected districts) (IOM Assessment Mission, 2009).

### 5.2.4. Droughts in the North-western Region

Droughts have emerged as a new form of disaster causing tribulation to poor farmers and agricultural labourers. A UNDP study projects that Bangladesh experiences major droughts once in five years. Droughts at local scale are much more frequent and affect part of the crop life cycle. The western part of the country is vulnerable to drought during pre-monsoon period. During the last 50 years, Bangladesh suffered about 20 drought conditions. The drought condition in north-western Bangladesh in recent decades had led to a shortfall of rice production of 3.5 million tons in the 1990s (Rahman, A., et. al., 2009: 35-36). In this region *monga*-a seasonal hunger has become a problem for many landless poorer people. The EMT-DAT database has recorded four major drought periods in Bangladesh (cf. Matthijs, 2010: 45). Current Severe drought can affect yield in 30 per cent of the country, reducing national production by 10 per cent in 2030. Temperature increase of 0.5°C and annual rainfall reduction of 5 per cent could reduce runoff into the Ganges, Brahmaputra and Meghna Rivers by 14 per cent, 11 per cent and 8 per cent, respectively. With 12 per cent reduction in runoff, the population living in severe drought-prone areas increases from 4 per cent to 9 per cent under moderate climate change (Rahman, A., et. al., 2009: 36).

### *5.2.5. River Bank and Coastal Erosion*

In the recent decades, riverbank and coastal erosion with loss of habitat and available cultivable land has become one of the major natural hazards in Bangladesh. The issue now is a major policy concern in Bangladesh since this territorial loss poses significant threat to her national and human security. This erosion though does not directly affect people life but it terms of economy and property loss, it takes a huge tolls. It makes them asset poor. A study by the Centre for Environmental and Geographic Information Services (CEGIS) assumes that since 1973, over 158,780 hectares of land has been eroded and in 2010 alone, river-bank erosion was expected to displace 11,000 people living on the banks of the Jamuna, and more than 5,000 living alongside the Ganges and Padma rivers (CEGIS, 2010). This trend of erosion may be exacerbated due to an increase in monsoon rainfall increasing human impacts. A survey carried out in 1998 among 230 households in Serajganj, found that 5,500 of

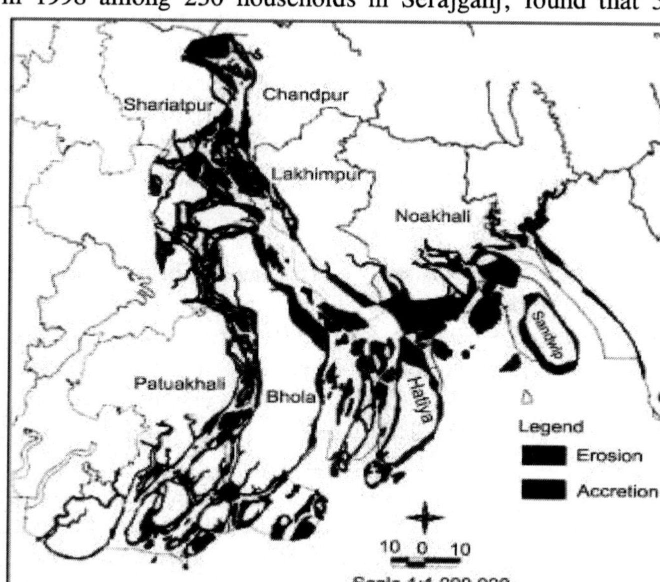


**Figure: Erosion and Accretion on Bangladesh**

*Source*: Matthijs van der Hoorn, 2010: 46.

30,000 slum dwellers were erosion-affected displaces (Hutton and Haque, 2004: 41-62).

Moreover, verified climatological and hydrological data suggests that climate change will exacerbate coastal erosion primarily through rising sea levels, changes to river flow and other hydrological dynamics. Outreach coastal islands like *Bhola, Sandwip, Hatia, Kubdia,* are on high risk of erosion. Along these belts, erosion is almost a regular phenomenon. One study identifies rates of erosion on *Bhola* Island as between 0.31 to 0.43 cm in a day (Krantz, 1999: 21). Shamsuddoha et. al. report that riverbank erosion has reduced *Hatiya* from 1000 sq. km. to only 21 sq. km. over 350 years and *Swandip* has lost 180 sq. km. in the last 100 years. According to their findings, over the last 40 years, *Bhola* Island has been squeezed to 3400 $km^2$ from 6400 $km^2$ in 1960, which means *Bhola* suffered net loss of 3000 $km^2$. An outreach island of 97 Square Miles (250 $km^2$), *Kutubdia*, under Cox'Bazar district, is under threat of continuous erosion, due to strong tidal action and cyclonic effect. According to a count within a 100-year period from 1880 to 1980, *Kutubdia* reduced to 23 square miles (Shamsuddoha et. al., 2007: 22-23).

### *5.2.6. Sea Level Rise*

Sea level rise has emerged as one of the most enduring threats of climate change to Bangladesh (Mahtab, 1989), (BCAS, 1994). IPCC assesses that globally average sea levels rose by a rate of 1.8 mm per year from 1961 to 2003 and the most severe of the IPCC six scenarios in the Fourth Assessment Report predicts a further total increase of between 26 and 59 cm by the end of the 21st century (cf. Walsham, 2010: 16). Based on changing trend SAARC Meteorological Research Council (SMRC) predicts increase of annual mean temperature 0.22°C until 2050 and 0.41°C until 2100.

Moreover, using 22 years historical tidal data of the three coastal stations of Bangladesh, a recent study carried out by SMRC reveals that the rate of sea level rise during the last 22 years is many fold higher than the mean rate of global sea level rise over 100 years (cf. Dr Atiq Rahman, 2009: 29). One study suggests that sea-level rise will lead to a potential loss of 15,668 $km^2$ land, which is expected to affect 11 per cent of the population or 5.5 million people. It predicts that if sea level rise goes up by one meter, the implications will include a 20.7 per cent land loss, affecting 14.8 million people (WEDO, 2008: 3). *Patuakhali, Pirojpur, Barisal, Jhalakati, Bagerhat* and *Narail* are the most vulnerable areas in terms of sea level rise. Due to increased rainfall in addition to 62 cm sea-level rise, the inundated area is projected to be about 16 per cent (551, 500 ha) and more area will be inundated in the year 2080 (Preliminary Assessment of ASPS-II). The following information reveals a grim reality in terms of sea level rise in Bangladesh (Shamsuddoha et. al., 2007: 13-15):

- Over the last 100-year Bangladesh has warmed up by about 0.5°C and 0.5 m rise of sea level in the Bay of Bengal (BUP, 1993).
- In the South western Khulna region 5.18-mm/year sea-level rise is recorded which might reach to 85 cm by 2050.
- World Bank's study on the impact of Sea-level rise in Bangladesh reveals that, 100 cm sea level rise within next 100 years will inundate 15 to 17 per cent of country's land area, i.e., 22135 to 26562 square kilometres, which will make 20 million people environmental refugee.
- A 2° temperatures and 45 cm sea-level rise would increase 29 per cent risks of flooding of country's low laying areas and may cause permanent inundation of 145 km long coastline stretches from Cox's bazar to Badar Mokam.
- Metrological data from 1977 to 1998 clearly shows annual sea level rise at the rate of 7.88 mm, 6 mm and 4 mm, respectively in Cox's bazar, Chardanga at Hatiya and Hiron Point in Sundarban.
- A study published in 1998 shows that sea level rise would inundate 58 thousand hectares of agricultural land.

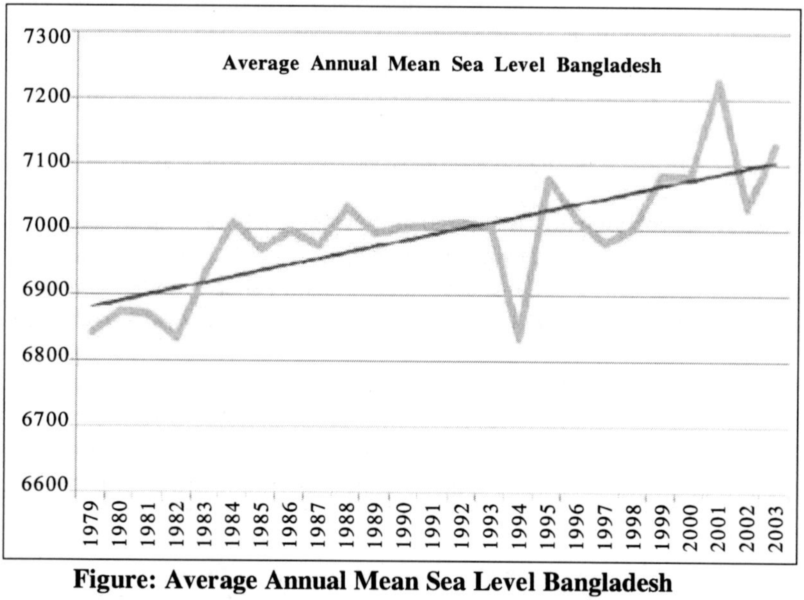


**Figure: Average Annual Mean Sea Level Bangladesh**

*Source:* Permanent Service Mean Level (PSMSL) database, Proudman Oceanographic Laboraton.

Bangladesh 'Climate Change Strategy and Action Plan' estimates an impending threat of displacement of more than 20 million people in the event of sea level rise. The settlement of this massive displacement will be a serious problem making the case the worst humanitarian crisis ever.

### *5.2.7. Salinity Ingress and Saline Water Intrusion*

Along the coastal line of Bangladesh the salinity intrusion has become a grave concern which is eating away the little arable land, hampering the agricultural productivity and disrupting the biodiversity and ecosystem. National Adaptation Programme of Action (NAPA) projects that out of 2.85 million hectares of the coastal and offshore areas about 1.2 million hectares of arable land are affected by varying degrees of soil salinity (NAPA, 2005:4).

A study projects that in dry season alone, when the flows of upstream water reduce drastically; the saline water goes up to 240 kilometres inside the country and reaches to Magura district. Presently around 31 upazillas of *Jessore, Satkhira, Khulna, Narail, Bagerhat and Gopalganj* districts are facing severe salinity problem. If some 16,000 sq km of coastal land is lost due to a 45 cm rise in sea level, the salinity front would be pushed further inland (EJWG, 2007). Mapping out the impacts salinity in the coastline of Bangladesh a study predicts:

- 10 per cent more land (relative to 1990) will be saline-affected and intensity will be increased by 10 per cent;
- decreases availability/productivity of agricultural land;
- ncreased food insecurity as naturally-growing species disappear;
- serious scarcity of safe drinking water;
- loss of biodiversity, e.g., decrease in tree species and freshwater fish;
- socioeconomic problems, generally women will be more vulnerable.

However, salinity intrusion in surface water has a seasonal variation making a dominant factor to coastal echo-system, fisheries and agriculture. In case of any unusual change on present spatial and temporal variation of salinity, the biophysical system of entire coastal area will be severely affected. It is found that 5 ppt isohaline (line of equal salinity level) intrude more than 70 km landward in the western part of *Sundarbans*.

The present interface between freshwater and saline water lies around 120 to 160 km inland in the southwest, and this could well be pushed northward as far as central *Jessore* region in the event of a sea level rise (BIDS/ADB, 1994). Moreover, salinity ingress also causes an increase in soil salinity. 'Soil Resource Development Institute' of the Government of Bangladesh reported that, soil salinity levels south of Khulna and Bagerhat towns ranged between 8 to 15 dS/m during the low flow season. The report reveals that several sub-districts such as *Kachua, Mollahat,* and *Fultali* south of the *Sundarbans* have begun to develop soil salinity during

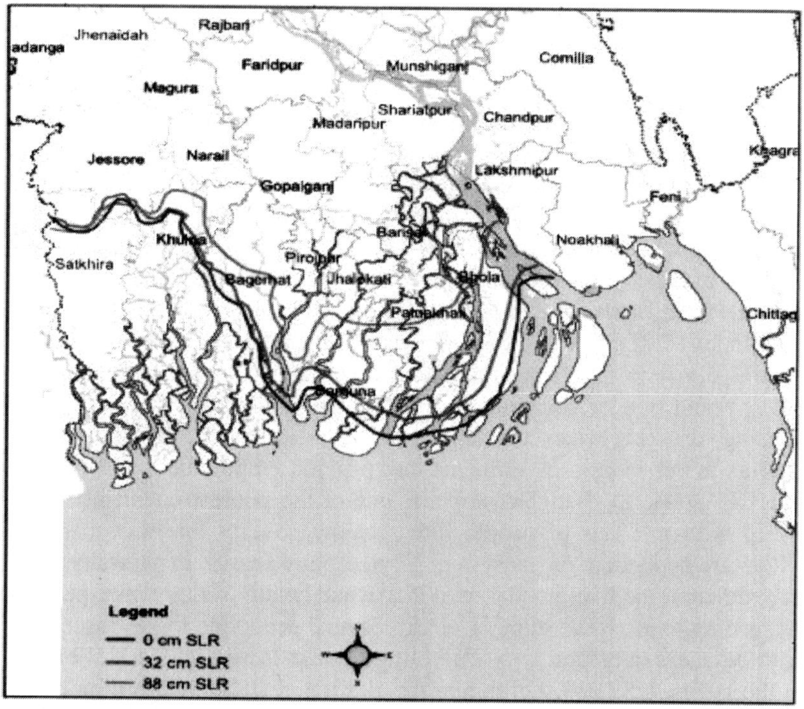


**Figure: Likely Salinity Intrusion in Southern Bangladesh Due to Different SLR**

(*Source*: Bangladesh Climate Change Strategy and Action Plan, 2009: 15).

the low flow seasons of 1980s. Information suggests climate change will increase the salinity ingression in the following ways:

- Increased sea level will cause water ingression in the rivers;
- decreasing trend of fresh water flow from the upstream will cause intrusion of saline water;
- upward pressure of the saline and fresh water interface in the level of underground Aquifer;
- downward seepage of saline water from surface and salinisation of underground water;
- the pace of evaporation in winter will increase soil salinity frequency and intensity of tidal surges will increase ingression of saline water.

### *5.2.8. Climate Change, Poverty and Development*

"Poverty and climate change are the two great challenges of the 21st century. Our responses to them will define our generation, and because they are linked to each other, if we fail on one, we will fail on the other" (Lord Nicholas Stern, 2010).

Climate change exerts a heavy toll on development in Bangladesh putting infrastructure to test and overall human development at risk. Climate change, indeed, produces a 'cascade effect' with its growing and complex interaction with other socio-economic factors. Many studies apprehend that such impacts have the potential to roll back many of the gains in development made to date. To demonstrate the magnitude of climate change the 'National Strategy and Action Plan' counts that once in every four to five years, there is a severe flood that covers almost 60 per cent of the country causing huge loss of life, livelihoods and substantial damage to infrastructure, housing and agriculture. The UNDP Disaster Risk Index (DRI) (UNDP, 2004) shows that the intensity and frequency of natural disasters may limit the social investments aiming to ameliorate poverty and hunger, provide access to education, health services, safe housing, drinking water and sanitation, or to protect the environment as well as the economic investments that provide employment and income (UNDP, 2004: 2). Bangladesh being one of the poorest countries in the world with millions of people living below poverty line and many of whom are dependent on ecosystem services, any change in natural system may threaten the livelihoods, food intake and health of these poor people. According to a UNDP study, globally annual economic losses associated with disasters averaged US$ 75.5 billion in the 1960s, US$ 138.4 billion in the 1970s, US$ 213.9 billion in the 1980s and US$ 659.9 billion in the 1990s (UNDP, 2004: 31). 'National Climate Change Strategy and Action Plan', asserts that over the last 35 years, national disasters alone diverted more than $10 billion which could be invested in other significant human development sectors (GoB, 2009: xvii). A UNDP study finds the following link between disaster and development:

**Table: Disaster-Development Nexus**

| | *Economic Development* | *Social Development* |
|---|---|---|
| Disaster limits development | Destruction of fixed assets. Loss of production capacity, market access or material inputs. Damage to transport, communications or energy infrastructure. Erosion of livelihoods, savings and physical capital. | Destruction of health or education infrastructure and personnel. Death, disablement or migration of key social actors leading to an erosion of social capital. |

(*Source*: UNDP, 2004: 20).

Moreover, climate change has serious negative impacts on achieving Millennium Development Goals (MDGs) and its targets. 'The National Strategy for Accelerated Poverty Reduction (2009-2011)' has recognised, therefore, climate change as one of the emerging issues in pro-poor growth. The following table shows how the climate change interacts with other socio-economic factors and limits the achievement of MDGs in Bangladesh:

### *5.2.9. Threats to Agricultural Productivity Loss*

Bangladesh's variable climate presents significant threats to its agricultural sector which accounts for about 20 per cent of the country's Gross Domestic Product (GDP) and which provides more than 60 per cent of the national employment. Erratic behaviour of present climate such as higher temperature, higher rates of evaporation and changing rainfall are expected to pose significant threats to impressive gains Bangladesh achieved in agricultural production to date. With increasing flood and water logging in the central region, flash flood in the northeast region and drought in the northwest and southwest region and salinity intrusion in coastal region, extreme weather events such as floods, droughts and heat stress are found to have an extra bearing on the agricultural sector and to put crop productivity at risk. About 1.32 m ha of cropland is highly flood-prone and about 5.05 m ha moderately flood-prone. According to a count, in two severe flood years in 1974 and 1987, the shortfalls in production from trend were about 0.8 and 1.0 Mmt of rice, respectively. During 1984, flood affected both *Aus* and *Aman* rice crop and the shortfall was about 0.4 Mmt (NAPA, 2005:10). In 2007, agricultural production losses were about at 1.3 million tons due to flooding across the country. The IPCC's Fourth Assessment Report predicts an overall decline in rice production in Asia of 3.8 per cent by the end of the century and a decrease in the production of rice and wheat in Bangladesh by as much as 8 per cent and 32 per cent, respectively by 2050 (Walsham, 2010: 20).

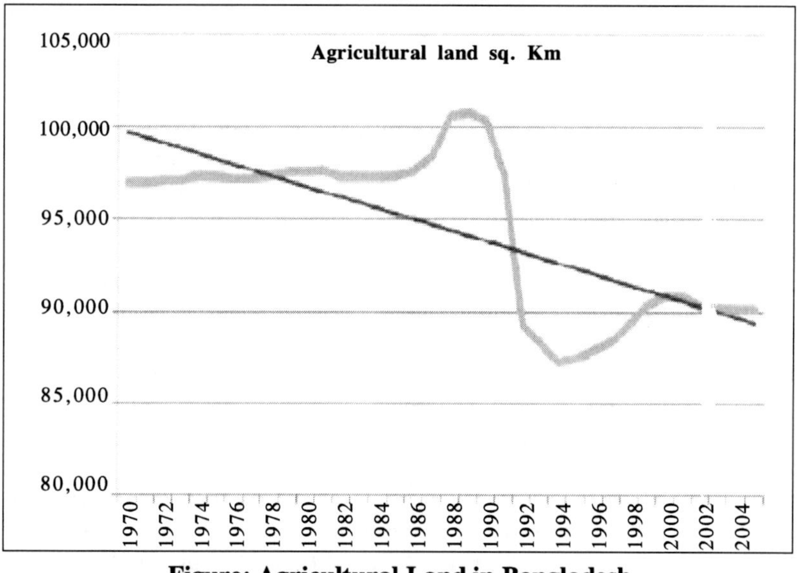


**Figure: Agricultural Land in Bangladesh**

(*Source*: World Bank, 2009: World Development Indicators, DPP Quick Query).

Given existing environmental trends coupled with the continuing loss of arable land to urban, industrial development, human settlement and unsustainable use of water, the country faces mammoth challenge in sustaining the impressive gains in agricultural sector. Now the annual loss of arable land stands at about 100 thousand ha per year. Moreover, salinity intrusion is changing the organic matter-physical, chemical and biological properties of soil. 45 per cent net cultivable area has even less than 1 per cent organic matter. Out of 2.85 million hectares of the coastal and offshore areas, about 1.2 million hectares of arable land are affected by varying degrees of soil salinity (NAPA, 2005: 4). In the Ganges flood plain, tube-well irrigation is not feasible because the upper aquifer has become saline severely limiting the dry season crop productivity. Moreover, the GFDL model predicts about 17 per cent decline in overall rice production for Bangladesh and as high as 61 per cent decline in wheat production under 4 degree changes in temperature (Rahman and Alam, 2003). Climate change is expected to exacerbate the situation further.

### *5.2.10. Chronic Hunger and Threats to Food Security*

With substantial damage to human habitat climate change threatens the food security in Bangladesh and leads to increasing hunger. NAPA reports that high temperature reduces yields of HYVs of *aus, aman* and *boro* rice. Food-grain production, due to increasing temperature, now is projected to decrease 28 per cent and 68 per cent for rice and wheat, respectively. A 60 per cent moisture stress on top of other effects might cause as high as 32 per cent decline in *boro* yield (NAPA, 2005:14). In addition, the soaring food prices coupled with shortage of food availability further worsen the situation. The country's population is projected, over the next 30 years, to grow to 200 million and feeding this burgeoning population will add an extra burden on the current growth rate of food production. With declining trend in regional food production, due to climate change, the possibility of agro-food imports will face significant setback. Identifying the food insecurity as a critical national issue in Bangladesh, Food and Agriculturs Organisation (FAO) of the United Nations estimates that currently, over 41 million people or 27 per cent of the total population are malnourished (FAO). Moreover, in recent years, the country's agricultural imports outnumber its exports twelve-fold. In 2006-07, agrofood imports represented approximately \$1.9 billion. Available statistics projects within the 20 years, climate change would reduce global food production by approximately 50 million tons, which in turn, could force up global food prices by 20 per cent. FAO estimates that 925 million people were undernourished in 2010. It is estimated that in the eastern part of Bangladesh—the most richly and productive region of the country—1400 tons of grain production would be lost due to sea level rise in 2030 and 252,000 tons by 2075.

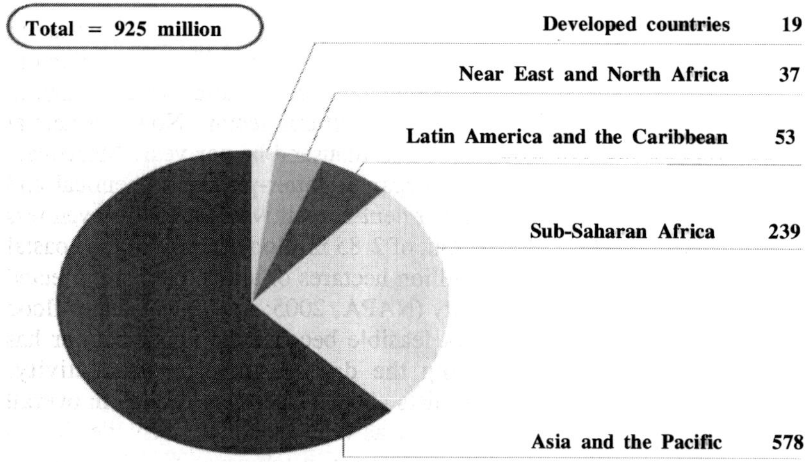


**Figure: Under-nourishment in 2010, by Region (millions)**

(*Source:* http://www.fao.org/hunger/en).

### *5.2.11. Threats to Economic Security*

"Climate change is the greatest market failure the world has ever seen, and it interacts with other market imperfections... with other market failures and economic dynamics, giving rise to many complex policy problems" (Nicholas Stern, 2006).

With the Stern's Review the implication of climate change to economics is now much established. Climate change is an "economics spoiler"- it impedes the growth and sustained development. This impediment, however, has a spatial distribution across the regions. Developing countries are the hardest hit. Bangladesh has a vulnerable economy which is not climate-neutral. The combined effects of low economic growth, inadequate infrastructure and low-level social development, lack of institution capacity, high frequency of natural disasters, high dependency on climate-sensitive and low resource base, have made the country vulnerable to climate stimuli. Prolonged floods and inundation, water shortage, sea level rise, widespread drought, extreme weather events all take significant toll on economic prosperity of the country and all these have significant negative impacts on sustained growth, investments and savings.

Though, Bangladesh has moved from a low human development category to a medium human development (UNDP, 2004) category with macro economic situation getting better but still 36 per cent of the people are living on a dollar per day or less (NAPA, 2005: 3). Beyond agricultural loss, climatic variability has large impacts on industry and infrastructure. Floods have severe impacts on small and medium handloom industry in

**Figure: Top 10 Natural Disasters in Bangladesh for the Period 1900-2001 Sorted by Economic Damage Costs**

| *Disaster* | *Date* | *Damage (000 US $)* |
|---|---|---|
| Flood | 5-July-1998 | 4,300,000 |
| Storm | 15-Nov-2007 | 2,300,000 |
| Flood | 20-Jun-2004 | 2,200,000 |
| Flood | Jun-1998 | 2,137,000 |
| Storm | 29-Apr-1991 | 1,780,000 |
| Storm | 15-May-1995 | 800,000 |
| Flood | Aug-1987 | 727,500 |
| Flood | July-1974 | 579,200 |
| Flood | Sep-2000 | 500,000 |
| Earthquake (Seismic activity) | 26-Dec-2004 | 500,000 |

(*Source*: http://www.emdat.be/result-country-profile).

Pabna, Sirajganj and Bogra. Cyclone and storm surge in coastal region destroy infrastructure such as industries, machineries, water supply, network system, power supply network, all of which cause huge loss in

**Table: Serious Floods in Bangladesh and Estimated Economic Damages**

| *Event* | *Impact* |
|---|---|
| 1984 flood | Inundated over 50,000 sq.km, estimated damage US $ 378 million |
| 1987 flood | Inundated over 50,000 sq. km, estimated damage US $ 1 billion, 2,055 deaths |
| 1988 flood | Inundated 61 per cent of the country estimated damage US$ 1.2 billion, more than 45 million homeless, between 2,000-6,500 deaths |
| 1998 flood | Inundated nearly 100,000 sq. km., rendered 30 million people homeless, damaged 500,000 homes, heavy loss to infrastructure, estimated damage US $ 2.8 billion, 1,100 deaths |
| 2004 flood | Inundation 38 per cent, damage US $ 6.6 billion, affected nearly 3.8 million people. Estimated damage over $ 2 billion, 700 deaths |
| 2007 flood | Inundated 32,00 sq. km, over 85,000 houses destroyed and almost 1 million damaged, approximately 1.2 million acres of crops destroyed or partially damaged, estimated damage over $1 billion 649 deaths. |

(*Source*: National Climate Change Strategy and Action Plan, 2009: 9).

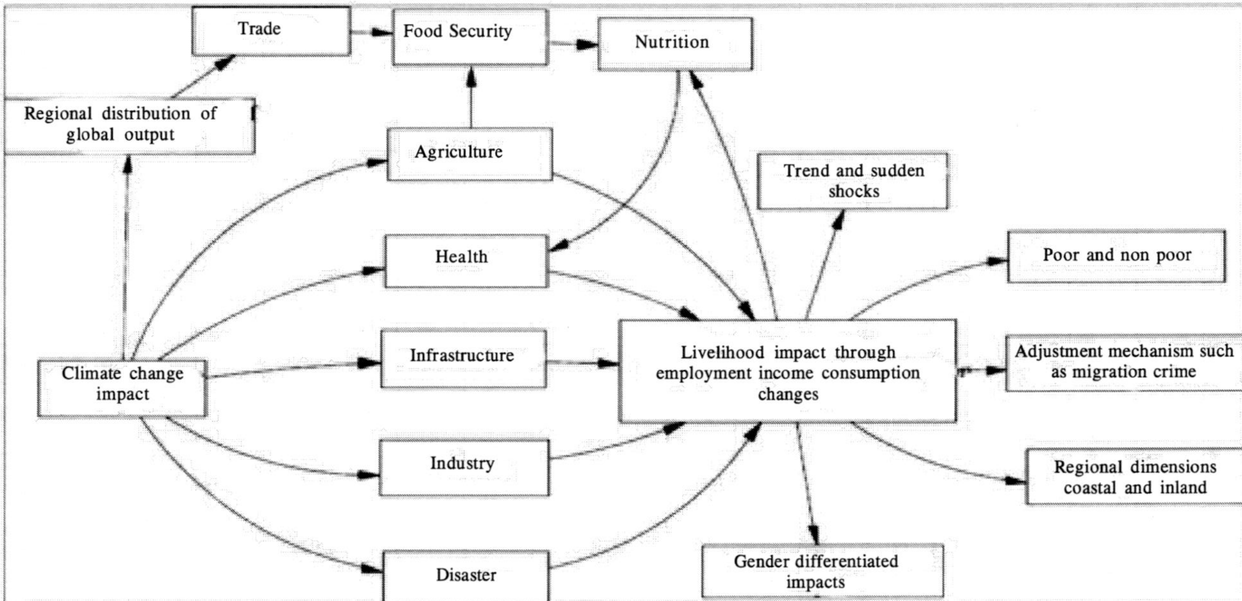


**Figure: Intensity of Impacts on Different Sectors Due to Climate Change**

(*Source*: NAPA, 2005: 16).

terms of economy. 'The Khulna Paper Mill' had been closed due to increase in water salinity of the river. Climate change significantly hampers costal economy with huge toll on tourism industry. The *Sundarban* mangrove area is big source of the country's economy. Climate change puts these economic bases at risk.

### *5.2.12. Fragility and Political Insecurity: Climate Change Erodes the 'Social Contract'*

'Climate change may significantly increase instability in weak or failing states by overstretching the already limited capacity of governments to respond effectively to the challenges they face. The inability of a government to meet the needs of its population as a whole or to provide protection in the face of climate change-induced hardship could trigger frustration, lead to tensions between different ethnic and religious groups within countries and to political radicalisation. This could destabilise countries and even entire regions' (European Council, 2008).

CNA report recognises climate change as a 'threat multiplier' (CNA Corporation, 2007: 6) which exacerbates existing vulnerabilities, tensions, and instability. A study by International Alert titled *A Climate of Conflict: The Links between Climate Change, Peace and War* recognised that while the physical consequences of climate change is determined by the nature, the consequences of these consequences are not. People already living in poverty, in underdeveloped and unstable states, under poor governance will be hardest hit of climate change; these by a complex interaction with a mix of social, economic and political factors will aggravate and worsen the adaptive capacity of many developing countries (Smith and Vivekananda, 2007: 8).

Bangladesh being in the 'warning zone' (Fund for Peace, 2011) of the 'Failed States Index 2011' with widespread corruption and poor governance is exposed to potential political threats emanating from climate change. Governance deficit is still an acute problem and climate change in terms of economic collapse, food insecurity and impacts on livelihoods will severely limit state's ability to meet the additional burden. Failure to render services to the already stressed communities to cope with additional burden on their life and livelihoods, existing frustration will intensify. This, unmet frustration coupled by weak institutional capacity will make people desperate, one of the possible consequences of which will be the violent breach of 'social contract' between people and states. In addition, the polemic political culture, nascent democracy, intra and inter-party feuds, radical Islamist terrorism and radicalisation put Bangladesh on high alert. Moreover, Worldwide Governance Indicators (WGI) of the World Bank shows a declining trend of political stability in Bangladesh since 1998 (World Bank, 2009). 'Where governance is poor, a social disaster

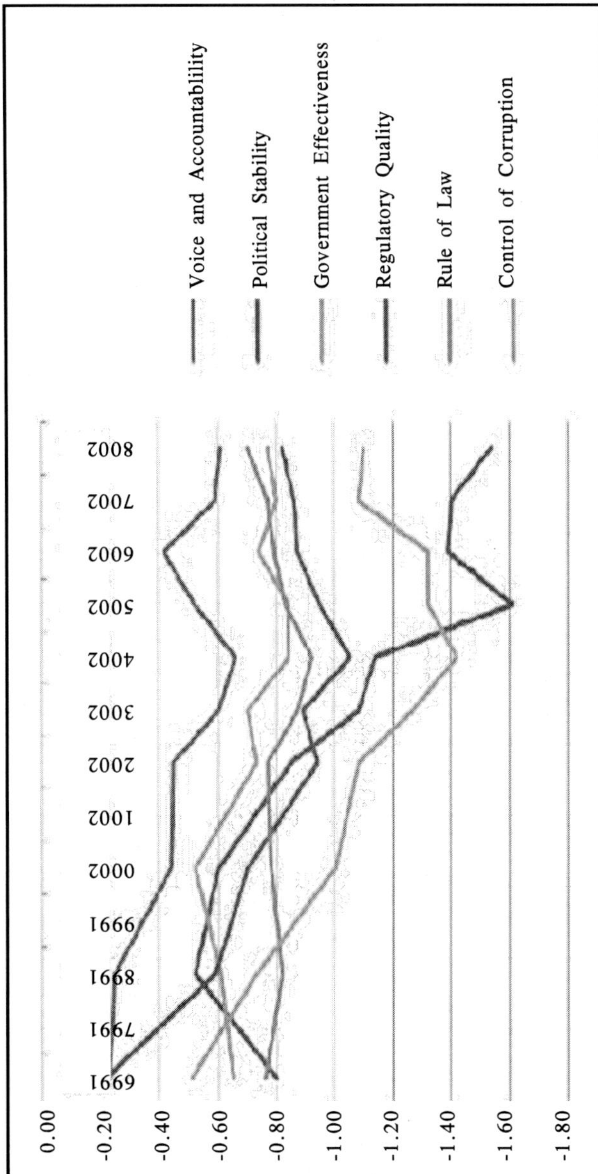


**Figure: Worldwide Governance Indicators (WGI) Bangladesh**

(*Source*: Worldwide Governance Indicators', World Bank, 2009).

seems close to inevitable' taking this hypothesis for granted it is assumed that in Bangladesh climate change will have a destabilising effect on the level of political stability.

### *5.2.13. Threats to Health Security*

With severe impacts on poor people's health, climate change poses significant risks to human health and threatens health security of Bangladesh. It reverses the progress made in the health sector and aggravates the existing health-related problems. Climate change impacts on human health in two ways—First, increased temperature, precipitation change, sea level rise, frequent natural disasters will lead to increased health-related casualties, putting pressure on already stressed medical facilities and resources. Second, a warmer and more variable climate leads to higher levels of some air pollutants and increases transmission of diseases through unclean water and contaminated food. Moreover, heat waves and water shortage will adversely impact safe drinking water and sanitation system.

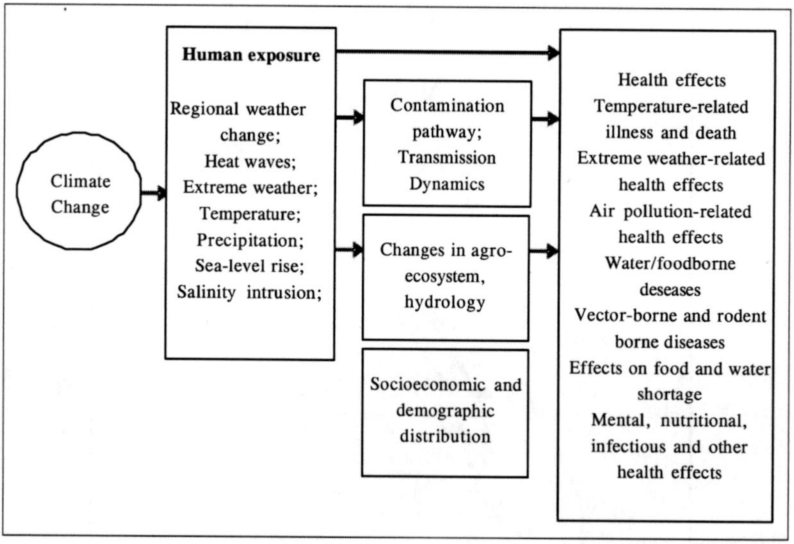


Bangladesh is vulnerable to outbreaks of infectious, waterborne and other types of diseases (World Bank, 2000). World Health Organisation (WHO) reports that the incidence of malaria increased from 1,556 cases in 1971 to 15,375 in 1981, and from 30,282 cases in 1991 to 42,012 in 2004 (WHO, 2006). Moreover, predicted precipitation change will compromise both the quality and supply of water increasing the risks of water–borne diseases. Prolonged inundation, water logging and floods will affect sanitation system increasing diarrhoea and cholera.

**National:**

- As of 2004, 16388 dengue cases were reported of which 210 were fatal.
- Kala-azar has remerged since the cessation of dichlorodiphenyl--trichloroethane (DDT) spraying operations. At least 20 million people in more than 27 districts are at risk.
- Eighteen million people in 12 districts are considered to be at risk of filariasis
- Acute respiratory infection (ARI) accounts about 145,000 (33 per cent) deaths annually among children less than five years of age.

Sound public health rests on sanitation, sufficient food, safe drinking water, secure shelter and good social condition, climate change threatens all of these dynamics.

**Table: Major Climate-sensitive Diseases Occurring during the Last Few Decades in Bangladesh**

| *Diseases* | *Total cases per period* | *Period* | *Average annual cases* |
|---|---|---|---|
| Diarrhoea | 48 302 636 | 1988-2005 | 2 842 273 |
| Skin diseases | 23 697 833 | 1988-1996 | 2 623 092 |
| Malaria | 1 018 671 | 1974-2004 | 33 956 |
| Mental disorders | 201 881 | 1988-1996 | 22 431 |
| Dengue | 19 830 | 1999-2005 | 3305 |

*Source*: Data modified from WHO, 2006, Director General, Health (Bangladesh); 1996, 1997, MoEF.

### *5.2.14. Climate Change and Forced Migration in Bangladesh*

As the climate change unfolds, environmental refugee or climate refugee has emerged as one of the phenomenal national problems in Bangladesh. Faced with an unconceivable scale of environmental change, migration for vulnerable people has been an adjustment mechanism of first resort, or a survival mechanism of last resort (UN University and IEHS, 2008). For millions of people living in the coastline, char lands temporary or permanent migration, due to floods, sea level rise, coastal salinity, shoreline erosion, food insecurity and livelihood loss, is not a new phenomenon. Environmental degradation shapes people migration decision-making process and migration has emerged as a cooping strategy. Indeed, the correlation between climate change and migration is complex. Increased number and intensity of disasters forced people to flee because it destroys their homes and habitats. Environmental degradation and rising sea level, coastal erosion doom their livelihood and force communities to abandon their traditional habitat for more accommodating environment elsewhere.

The IPCC and the Stern review estimates that by 2050, 150 million and 200 million people could permanently be displaced due to rising sea-levels, floods and droughts. An estimated 26 million of the 350 million displaced worldwide are considered climate displaced people. Of these, 1 million each year are estimated to be displaced by weather related disasters brought on by climate change. The World Bank estimates that sea level rising a single meter would displace 56 million people in 84 developing countries. In Bangladesh alone, 20 million would be affected. Media reports and surveys suggest that many of the internal migrant from environmentally vulnerable areas end up in the urban slums or (*bastes*) putting pressure on already stressed urban capacities.

### *5.2.15. Threats to Water Security in Bangladesh*

Humanity will face major water challenges in the next few decades in certain regions of the world related to the impacts of climate change and rapidly growing human demands for water (GHF, 2009: 40). Indeed, climate change poses gruesome risks to the water security in Bangladesh. It exacerbates water crisis across the South Asian region. Today, over 1.3 billion people worldwide are water stressed. Hundreds of millions more people are projected to become water stressed by the 2030s due to climate change (GHF, 2009:40). It is likely that increased temperature and erratic behaviour of rainfall will lead to increased water demand and drought while sea level rise and salinity intrusion will deteriorate water quality. Moreover, chronic shortages of freshwater are likely to threaten food production, reduce sanitation, hinder economic development and damage ecosystems. About 70 per cent of the world's runoff water withdrawals are used to irrigate farmlands—this total rises to almost 95 per cent in developing countries (GHF, 2009: 40). Climate change exacerbates water scarcity and adds new risks to this farming system. The coastal region of Bangladesh faces a double-edge problem. Excessive rainfall in the monsoon, causes flood and water logging which results in scarcity of drinking water, water borne diseases' outbreak, damage of crops, infrastructure and reduces income source and livelihood options while in the dry season, lack of rainfall causes scarcity of water for drinking, household activities and irrigation. Coastal cyclone damaged water infrastructure, increases contamination of fresh water sources with saline water and causes water pollution. Moreover, shrinking water resources may trigger distributional conflicts posing major security challenges to water management system across the South Asia. The rapid melting at the Asian Water Tower--the Himalayan Glaciers—which feeding 6 major river systems in Asia and supply water to 2 billion people - has posed significant risks of trans-boundary hydro-conflicts over the access and management of water resources.

**6. Conclusion**

The above discussions indicate that the human face of climate change has become a pressing human security concern currently dominating the climate policy choices in Bangladesh. Climate change poses ultimate threat to the survival of this delta. It endangers the geo-physical and hydrological features of Bangladesh. Hardly any single facets of our life on climate change has no latent effect. On many dimensions, the physical change of climate has a variety of dire consequences for man's livelihood and wellbeing, irreversibly and strongly linked to the questions of social security, health security, food security, political stability, and economic prosperity in Bangladesh. The convergence of environmental degradations with other socio-economic and political aspects seriously corrodes institutional capacity and thus limits state's core functions to render services. Besides, the adverse impacts of climate change are expected to act as a threat multiplier by adding additional strains to already vulnerable poor people highly exposed to climatic variability. It has the potential to cripple agro-based economy. River bank and coastal erosion, sea level rise and saline intrusion, food productivity loss have already become a grave concern across the coastal belt destroying habitat and putting overall coastal life at risk. Moreover, severe tropical cyclones, flooding, inundation and tidal surges often take a huge toll on vital infrastructure in rural area claiming huge loss of lives, property and livelihoods. Shrinking water resources aggravates agricultural productivity, thus, limiting basic access to food for many poor people. Climate change eats away development gains thereby jeopardise state's capacity to uplift the standard of life and livelihood. Climate change exacerbates hunger, limits social investments necessary to ameliorate poverty level and to provide access to food. It aggravates health-related problems, deteriorates sanitation system and jeopardizes environment and economic investments that generate income and employments.

These implications of climate change with human security concerns on a variety of issues suggest for a combination of adaptive and mitigation measures including reduction of underlying vulnerabilities and risks, community resilience building as well as better planning for management in order to ensure adequate assistance and protection of the affected people and communities. Indeed, the social impacts of climate change are diffuse and diverse; therefore, integrating strategies between adaptation, mitigation, development and disaster risk reduction must be mutually reinforcing; each of them should be supported by different sets of institutions, policy frameworks and funding mechanisms.

Climate change has sweeping impacts on many aspects of human life and it undermines many dimensions of human efforts. Therefore, the guiding principle that ensue the mitigation and adaptation measures must ensure mainstreaming climate change concerns within ongoing human development practices. Key sectors of economic planning — agriculture, tourism, land-use planning, public health, environmental management and basic infrastructure provision— will all need to consider climate change. This development must not be mere sustainable but climate-proof. Moreover, community-based social resilience building must be added as an additional pillar to the concept of sustainability. As climate change unfolds, climate justice has become a vital concern for people affected by it; therefore, the core strategy of any viable responsive climate policy must be to ensure empowerment of vulnerable people and communities to cope with these challenges. Climate-proof and wider drive for a dignified existence for all must rest on harmony with the environment as well as safety from it.

## References

Ali, Anwar. 1999.'Climate Change Impacts and Adaptation Assessment in Bangladesh', *Climate Research*, Vol. 12, Space Research and Remote Sensing Organisation (SPARRSO), Agargaon, Sher-e-Bangla Nagar, Dhaka 1207, Bangladesh.

Annan, Kofi. 2000. 'Secretary-General Salutes International Workshop on Human Security in Mongolia', Two-day Session in Ulaanbaatar, May 8-10, 2000. Press Release SG/SM/7382, Available on: (www.un.org/News/Press/docs/2000/20000508.sgsm7382.doc.htmkl).

Ashton, John. 2007. UK Foreign Secretary's Special Representative for Climate Change, in a Statement at the, 'Conference on the Climate Change: The Global Security Impact', Royal United Services Institute, available on: (http://www.e3g.org/programmes/climate-articles/climate-change-the-global-security-impact).

Bangladesh Bureau of Statistics (BBS), Population Census. 2001.

BCAS. 1994. *Vulnerability of Bangladesh to Climate Change and Sea Level Rise: Concepts and Tools for Calculating Risk in Integrated Coastal Zone Management*, Vols. 1 and 2. Technical Report, Bangladesh Centre for Advanced Studies, Dhaka.

Buzan, Barry. 1991. *People, States and Fear: An Agenda for International Security Studies in the Post-Cold War Era*, Hertfordshire: Harvester Wheatsheaf.

CCCM Cluster Bangladesh, 'Survey and Assessment, available on: (http://www.cccm.iom.org.bd/index.php?option=com_ content&view=article&id=3&Itemid=8).

Centre for Environmental and Geographic Information Services (CEGIS). 2010. 'Prediction of Riverbank Erosion along the Jamuna, Ganges and Padma Rivers in 2010', Dhaka.

Danish Embassy Bangladesh, 'Preliminary assessment of ASPS-II Components vulnerability to Climate Change', available on (http://www.docstoc.com/docs/22885528/Climate-Change-Scenario-for-Agriculture-in-Bangladesh).

Dankelman, Irene, et al. 2008. *Gender, Climate Change and Human Security: Lessons from Bangladesh, Ghana and Senegal*, a report prepared for ELIAMEP, Women's Environment and Development Organisation (WEDO), New York.

Disaster Management Bureau, Ministry of Food and Disaster Management and Comprehensive Disaster Management Programme. 2007. 'Consolidated Damage and Loss Assessment, Lessons Learnt from the Flood 2007 and Future Action Plan (Executive Summary).

Ecological Institute. 2011. 'CLICO Meeting in Berlin—Discussion on Human Security', available on: (http://ecologic.eu/4016).

Elhance, A. O. 1999. *Hydro Politics in the Third World: Conflict and Cooperation in International River Basins*, United Nations Institute of Peace Press, Washington, D.C.

EM-DAT: CRED International Disaster Database. 2011. 'Country Profile', available on: (http://www.emdat.be/result-country-profile)-Universite.

Equity and Justice Working Group (EJWG) and COAST Trust.2007. 'Increasing Salinity Threatens the Productivity of Bangladesh', A Campaign Brief, Dhaka, available on (http://www.equitybd.org/newsletter/issue/02/3.pdf).

European Council. 2008. '*Climate Change and International Security*', Paper from the High Representative and the European Commission to the European Council.

Fund for Peace. 2011. '*Failed States Index 2011*' available on: (http://www.fundfor peace.org/global/?q=fsi-grid2011).

Gary, K., and Murray, C. 2001. 'Rethinking Human Security', *Political Science Quarterly* 116(4): 585-610.

General Economics Division, Planning Commission, Government of the People's Republic of Bangladesh. 2008. 'Moving Ahead: National Strategy for Accelerated Poverty Reduction II (FY 2009 -11)', available on: (http://www.usaid.gov/bd/files/7c.PRSP.pdf).

German Advisory Council on Global Change (WBGU). 2008. '*Climate Change as a Security Risk*', Earthscan, UK and USA, available on: (http://www.wbgu.de).

Global Humanitarian Forum (GHF). 2009. *Climate Change: The Anatomy of a Silent Crisis,* Human Impact Report, Geneva.

Huq, S., and Pettengell, C. 2008. 'The Impact of Climate Change on Human Security', *The Bridge,* Issue 10, available on: (http://www.bridgemag.com/magazine/index.php?option=com_content&task= view&id=519& Itemid= 79).

Hutton, D. and Haque, C. E. 2004. *Human Vulnerability, Dislocation and Resettlement: Adaptation Processes of River-Bank Erosion-induced Displaces in Bangladesh*', Disasters, 28(1), 41-62.

IOM Assessment Mission, International Organisation for Migration.2009. '*Post Cyclone Aila*', available on: (http://www.cccm.iom.org.bd/file/pdf/18.pdf).

IPCC Fourth Assessment Report, 2007.

Islam, S.M.R, Huq, S, and Ali, A. 1999. 'Beach Erosion in the Eastern Coastline of Bangladesh', in: Huq S, Karim Z, Asaduzzaman M, Mahtab F (eds)

*Vulnerability and Adaption to Climate Change for Bangladesh*. Kluwer Academic Publishers, Dordrecht.

Kaplan, D. R. 1994. 'The Coming Anarchy', *Atlantic Magazine*, available on: (http://www.theatlantic.com/magazine/archive/1994/02/the-coming-anarchy/4670/3/).

Lippman, Walter, US Foreign Policy: Shield of the Republic, Pocket Book, New York. 1943. Cited in Maier, Charles S. *Peace and Security for the 1990s*. Unpublished paper for the MacArthur Fellowship Programme, Social Science Research Council, 12 Jun 1990, as quoted in Romm 1993, p. 5.

Lord Nicholas Stern. 2010. quoted in Nigel Thornton, '*Realising Development Effectiveness: Making the Most of Climate Change Finance in Asia and the Pacific*', a synthesis report from five country studies in Bangladesh, Cambodia, Indonesia, Philippines and Vietnam, Asia Pacific Climate Change Finance and Aid Effectiveness Dialogue 19th–20th October 2010, p. 3.

Magnus Krantz. 1999. *Coastal Erosion on the Island of Bhola, Bangladesh*, Department of Physical Geography GÖTEBORG.

Mahtab, F. 1989. 'Effect of Climate Change and Sea Level Rise on Bangladesh', *Expert Group on Climate Change and Sea Level Rise*, Commonwealth Secretariat, London.

Marianne Stone. 2009. 'Security According to Buzan: A Comprehensive Security Analysis', available on: (http://geest.msh-paris.fr/IMG/pdf/Security_for_Buzan. mp3.pdf).

Matthew, A. R. et al. (eds.) (2010), *Global Environmental Change and Human Security*, The MIT Press, Cambridge, Massachusetts.

Matthew, W. 2010. Assessing *the Evidence: Environment, Climate Change and Migration in Bangladesh*, International Organisation for Migration (IOM).

Matthijs van der Hoorn. 2010. '*Climate Change and Conflict in Bangladesh*', A Master's Thesis, Department of Human Geography, Radbound University Nijmgen, available on: (http://geography2.ruhosting.nl/masterthesis/scripties/Hoorn Matthijsvander.pdf).

McBride J. L. 1995. "Tropical Cyclone Formation," in Elsberry RL (ed.) *Global Perspectives on Tropical Cyclone*. WMO Tech Doc WMO/TD-No. 693, Report No. TCP-38, WMO, Geneva, p. 63–105 (reprinted 1996).

Ministry of Environment and Forest (MEF), Government of the People's Republic of Bangladesh. 2005. *National Adaptation Programme of Action (NAPA)*, final report.

Ministry of Environment and Forests (MEF), Government of the People's Republic of Bangladesh. 2009. *Bangladesh Climate Change Strategy and Action Plan*.

Neumann C. J..1993. Global Overview, Chap.1: Global Guide to Tropical Cyclone Forecasting. WMO, Geneva.

Ogata, Sadako and Sen, Amartya. 2003. '*Human Security Now: Protecting and Empowering People*', Commission on Human Security, New York.

Ogata, Sadako. 2003. 'Human Security Now', Fifth Ministerial Meeting of the Human Security Network Graz, Austria.

Prabhakaran Paleri. 2008. *National Security: Imperatives and Challenges,* Tata McGraw-Hill Publishing Company Ltd., New Delhi.

Rahman, A., and Alam, M. 2003. *'Mainstreaming Adaptation to Climate Change in Least Developed Countries (LDCS)',* Working Paper 2: Bangladesh Country Case Study, Available on: (http://pubs.iied.org/pdfs/10003IIED.pdf).

Rahman, Atiq et al. 2009. *'The Probable Impacts of Climate Change on Poverty and Economic Growth and the Options of Coping with Adverse Effect of Climate Change in Banglades',*, A Policy Study, Support to Monitoring PRS and MDGs in Bangladesh, General Economics Division, Planning Commission, Government of the People's Republic of Bangladesh and UNDP Bangladesh.

Shamsuddoha, M., and Chowdhury, R. K. 2007. *'Climate Change Impact and Disaster Vulnerabilities in the Coastal Areas of Bangladesh'*, COAST Trust and Equity and Justice Working Group (EJWG), available on: (http://www.equitybd.org/newsletter/english/Issue-5/Disaster_BD.pdf).

Smith, D., and Vivekananda, J. 2007. *A Climate of Conflict: The Links between Climate Change, Peace and War*, International Alert.

Stern Review: 'The Economics of Climate Change' 2006. available on: (http://siteresources.worldbank.org/INTINDONESIA/Resources/226271-11709110 56314/3428109-1174614780539/SternReviewEng.pdf).

Tadjbakhsh, S., and Chenoy, M. A. 2007. *Human Security: Concepts and Implications,* Routledge, New York, USA.

The CNA Corporation. 2007. *National Security and the Threat of Climate Change*, available on: (http://securityandclimate.cna.org/report/National%20Security%20and%20the%20Threat%20of%20Climate%20Change.pdf).

UNDP (1994), *Human Development Report 1994*, New York: Oxford University Press.

United Nations Commission on Human Security. 2003. *Human Security Now: Protecting and Empowering People,* United Nations.

United Nations Development Program (UNDP). 2004. *'Reducing Disaster Risk: A Challenge for Development'*, A global report, Bureau for Crisis Prevention and Recovery, available on: (www.undp.org/bcpr).

United Nations Food and Agricultural Organisation, 'Country Brief', available on: (http://www.fao.org/countries/55528/en/bgd/).

United Nations Food and Agricultural Organisation, 'Hunger', available on: (http://www.fao.org/hunger/en/).

United Nations University and Institute for Environment and Human Security Report. 2008. *Human Security, Climate Change and Environmentally Induced-Migration*, available on: (http://www.each-for.eu/documents/ELIAMEP.pdf).

Viles, H. and Spencer, T. 1995. *Coastal Problems: Geomorphology, Ecology and Society at the Coast.* London: Edward Arnold.

World Bank. 2009. 'Worldwide Governance Indicators', Country Data Report Bangladesh 1996-2009, available on: (http://info.worldbank.org/governance/wgi/sc_chart.asp).

World Bank. 2009. 'World Development Indicators', DPP Quick Query, available on: (http://data.worldbank.org/country/bangladesh).

# 5

# Climate Change, Human Security and Sri Lanka

*Erandathie Lokupitiya*

## Introduction

Sri Lanka is an island located in the Indian Ocean, with most of its land surface falling within 200 m above sea level. The country is located between the latitudes 5°54′ and 9°52′ north and longitudes 79°41′ and 81°54′ east. The country covers an area of 65,610 km$^2$ encompassing land under different land use and water bodies. There are three distinct climate zones based on the mean annual precipitation: dry zone (mean annual precipitation less than 1,750 mm), wet zone (mean annual precipitation greater than 2,500 mm), and intermediate zone (mean annual precipitation between 1,750 and 2,500 mm). Rainfall in Sri Lanka comes mostly through Southwest (May-September) and northeast (December-February) monsoons. The average temperature in the country is about 27°C, and only 3 per cent of the country has cooler temperatures of less than 20°C.

Climate change has been a known term for some people in Sri Lanka, mostly during the last two decades, because of certain devastating impacts they have been experiencing. The country has been vulnerable to the impacts associated with changes in temperature and precipitation and sea level rise.

Globally anthropogenic climate change has mostly occurred due to greenhouse gas (GHG) emissions from increased level of fossil fuel burning and deforestation. Due to increased demand for housing and other land use, Sri Lanka's forest cover had declined from 80 per cent at the beginning of the 20th century to about 32 per cent including forest plantations, by 2000 (Lokupitiya, 2008). Since 1990, a logging ban has been lifted for natural forests (Bandaratillake, 2001), and there is no good record of any illicit felling. Aforestation and reforestation activities are being promoted in the country, and the existing forests are managed and harvested sustainably. With the increased population density and increased GDP, overall GHG emissions from the major economic sectors in the country have increased. Agricultural practices with heavy fertiliser use, increased importation of motorised vehicles, and establishment of new industries have led to increased GHG emissions over time. However, per capita GHG emissions are low compared to certain other countries in the region.

Although the country was under a civil war which impacted human security since 1980 for nearly three decades, human security impacted by climate change could only be considered under personal and community safety, with regard to food, drinking water, and health, etc. This paper reviews the status of Sri Lanka with regard to such aspects of human security, focusing on the vulnerability and impacts related to climate change.

**Sri Lanka's Stand on Climate Change**

Recognising the need for high focus and serious concern on climate change, Sri Lanka ratified the United Nations Framework Convention on Climate Change (UNFCCC) in November 1993 and Kyoto Protocol in September 2002. Sri Lanka has actively participated in global and regional meetings related to the issue, and conducted awareness programs and other activities at national level. The country has developed the national climate change adaptation strategy (2010) and the national climate change policy of Sri Lanka (2012). A national policy of clean development mechanism (CDM) and a national carbon finance strategy have been drafted and a national carbon fund has been established. The Ministry of Environment serves as the national focal point for the UNFCCC, and the Climate Change Secretariat of the Ministry of Environment serves as the designated national authority (DNA) for clean development mechanism (CDM) related activities.

Given the relatively low amounts of GHG emissions by the country, Sri Lanka has shown greater concern over vulnerability and adaptation to climate change, even though the country has also adopted certain GHG mitigating, cleaner technologies. Ministry of Environment and certain other stakeholder institutions dealing with economic sectors relevant to vulnerability and adaptation are involved in addressing the issue, considering both long-term and short-term impacts. Major impacts on human security could be discussed under different categories of human security relevant to Sri Lanka.

## Human Security in Relation to Climate Change

Increased GHG emissions and direct and indirect impacts of climate change have influenced the livelihoods of humans, mostly the poor. Although the developed countries have made a bigger contribution towards climate change, they have also enacted better adaptation measures in addressing any impacts. Countries like Sri Lanka are still in the process of developing adaptation measures to climate change, given the high costs incurred in both facing and adapting to climate change. Most of the impacts the country is faced with are related to the increased temperatures, changes in precipitation, and sea level rise. The impacts and adaptation measures associated with the latter issues follow.

## Impacts due to Climate Change

Climate change impacts on human security in Sri Lanka have occurred mostly due to temperature and precipitation changes, and sea level rise. The direct impacts of temperature and precipitation changes involve droughts, flood events, landslides, and extreme wind events, etc. Sea level rise has also caused flooding and other impacts in coastal areas. In general, the number of extreme weather events involving temperature and precipitation fluctuations, of which floods and droughts being the largest, has increased over the past four decades. The overall highest frequency, magnitude, and impacts of these were reported during the last decade. Whether such instances are directly related to anthropogenic climate change might be still debatable; however, an overview of the situation with regard to these impacts and their effect on human security is a timely concern.

## Droughts and Flood Events

In Sri Lanka, the annual mean air temperature has increased between 1960 and 1990, at a rate 0.16°C per decade (Chandrapala, 1996a), and it has increased by 0.45°C during the last two decades (Central Bank of Sri Lanka, 2012). A temperature increase of up to 2°C has been observed for the minimum temperature of Nuwara Eliya, a location with an average annual temperature of 15.5°C. Most of the impacts of warming have been associated with more frequent and longer droughts, which involve much lower precipitation during monsoon and inter-monsoon times, compared to the average. Droughts occur mostly in August, and they are more common in Ampara, Anuradhapura, Hambantota, Kurunegala, Moneragala, Polonnaruwa, and Puttalam districts.

Besides droughts, the next major climatic events affecting the food security and overall livelihoods are floods. Floods mostly occur with the increased intensity of monsoonal rains. Parts of southern, western, and sabaragamuwa provinces are vulnerable to floods during southwest monsoons, and areas in northern, eastern, and north-central provinces are vulnerable during northeast monsoons. Due to changing weather patterns, the intensity of rains has increased, especially in lowland areas (Ranatunge et al., 2003) of Sri Lanka, although the overall annual rainfall has decreased. The average annual rainfall during the three decades extending from 1961 to 1990 is 7 per cent lower compared to that during the previous three decades (Chandrapala, 1996; Jayatillake, 2005). De Silva and Sonnadara (2009) observed a significant decrease in rainfall in the hill country, especially in southwest monsoons. On the average, the number of rainy days has also decreased in most of the meteorological stations in the country (Manawadu and Fernando, 2008).

Sri Lanka has 103 river basins, out of which 10 are considered as major rivers. The basins of major rivers Kalu, Kelani, Gin, Mahaweli, and Nilvala are often subject to frequent flooding. Colombo, Galle, Gampaha Kalutara, Kegalle, and Ratnapura districts are subject to floods during Southwest monsoon rains, while Ampara, Badulla, Batticaloa, Matale and Monaragala, Polonnaruwa, and Trincomalee are subject to heavy rains from northeast monsoons. High intensity (i.e., greater than 25 mm $hr^{-1}$) rainfall events due to climate change has led to more frequent occurrence of flash floods within the last decade; un-preparedness and defects in the canal and drainage system in the capital led to significant flooding in May and November 2010. Filling of wetlands in urban areas due to increased demand for housing and other infrastructure also has led to more frequent flooding with high intensity rainfall events in urban areas, as wetlands have served as natural sponges, which absorb and slowly release floodwaters from heavy rainfall. Northeast monsoons in 2010-2011 also led to intense rainfall and flooding, causing damage to more than 20 per cent of the rice area and affecting the livelihoods of people. Certain coastal areas also get flooded during storm events, leading to coastal erosion and inundation of lowland areas including coastal agricultural lands. Severely eroded coastal areas mostly lie between Kalpitiya (north-western province) and Matara (southern province).

**Landslides and Cyclones**

One impact of heavy rainfall is the increased frequency of landslides; forest clearance in steep slopes of mountainous areas and unplanned agricultural activities are also contributing factors. The frequency of landslide occurrence in the mountain slopes of the central and south-western parts of the country has increased during the last two decades, affecting the livelihoods of people by causing some deaths and making some families homeless. Landslides mostly occur during the November-January period, and mostly in Badulla, Moneragala, Nuwara Eliya, Ratnapura, and Kegalle districts. In Ratnapura District, the occurrence of landslides has increased with short duration, high intensity rainfall, and increasing temperatures, showing its correlation with climate change (Nawagamuwa et. al., 2011). Cyclones are relatively less common and felt occasionally in Sri Lanka, as the country is located outside the cyclone belt; so far Northern, Eastern, and North Central provinces have been affected mostly.

How the above climate change related direct impacts affect human security could be evaluated considering the security of livelihoods at community and individual level. The security to be considered here has been categorised below as food security, water security, secure power generation and energy supply, secure health, and Overall security of the Livelihoods of people.

***Food Security***

Sri Lanka is predominantly an agricultural country, with 30 per cent of the land area under permanent agriculture. Thus, the country inherits an agriculture-based economy, rice being the staple food for centuries. In addition, tea, rubber, and coconut stand as the main plantation and export crops, while coffee, cocoa, and cinnamon and other spices are exported. Agricultural crops contributed to about 8.5 per cent of the GDP in 2011. The paddy land extent is about 57 per cent of the total area under the principal plantation crops (i.e., tea, rubber, coconut) and paddy (Central Bank of Sri Lanka, 2012). Frequent droughts in rice areas in the country such as Anuradhapura, Polonnaruwa, Kurnegala, etc., have affected rice and other crops through affecting the irrigation water availability while causing threats to food security. According to United Nations Food and Agricultural Organisation (FAO), agriculture uses about 90 per cent of the available water in Sri Lanka.

Agriculture is also affected due to reduced yields and wind-driven soil erosion during droughts in the prominent agricultural areas; soil is dried quickly due to increased evaporation rates with hotter temperatures. Given the low amount of water in the major irrigation tanks, the area under cultivation becomes significantly low during the drought periods, affecting the livelihoods of the producers and food security for consumers of crops, especially rice.

The pest and disease occurrence have also increased due to hot temperatures, especially during drought periods. The plantation sector in Sri Lanka, most of which is found in wet and intermediate zones of Sri Lanka, is also prone to climate change impacts, as the increased temperatures have increased the frequency of diseases and pest outbreaks.

Compared to droughts, floods cause less impact on crops, mostly affecting the rain fed agricultural areas in the wet zone of Sri Lanka. The agricultural area in the wet zone is much less compared to the dry and intermediate zones; the paddy area in wet zone is 1/3 of that of the wet and intermediate zones. In coastal areas where salt water intrusion is a problem, certain agricultural land has been abandoned, and in some areas the crop yields are reduced due to salinity; for instance, in Hambantota district, rice yields have been reduced by 30-40 per cent, due to salinity (Patabendi, 2009).

Next to agriculture, fishery industry, which contributed to about 1.3 per cent of the GDP in 2011 (Central Bank of Sri Lanka, 2012), is also affected by the direct impacts of climate change. Fishery industry in Sri Lanka mostly comprises marine and coastal fishing, while inland fishery plays an important role in inland rural areas, especially in the north central province. Temperature increases due to climate change could impact fish stocks and their reproduction, breeding, and behavioural patterns, etc.

***Water Security (Safe Drinking Water)***

Increased demand for water consumption by domestic, industrial, agricultural, and other economic sectors of the country has exerted more pressure on the available sources of safe drinking water in Sri Lanka. Agriculture consumes the largest fraction of water in Sri Lanka. Currently water becomes available mostly through the monsoon rains and groundwater supply. Country has abundant water resources including rivers, lakes, tanks, lagoons, wetlands, and reservoirs, which are nourished by the rainfall. However, improper agricultural management practices and water disposal from various industries have affected the purity of the available water, especially for drinking purposes. In addition, changing climate, especially prolonged droughts, have affected the quality of drinking water in drought-stricken areas where drought conditions could lead to salt build up. Due to frequent droughts and water scarcity, the level of fluoride in the ground water in most of the paddy areas in the North central part of the country has become much higher than the critical level recommended by the World Health Organisation (WHO). Floods could also affect the drinking water; following floods there is a high possibility of spreading germs into unprotected, untreated drinking water, leading to various diseases including diarrhoea.

Currently only 14 per cent of the rural population and 75 per cent of the urban population are supplied with treated piped water from the National Water Supply and Drainage Board (NWSDB) (NWSDB, 2009). Unsustainable river sand mining in certain places like Ambatale and Kalutara, which serve as locations of water intake for national water supply, has led to saltwater intrusion during drought periods.

***Secure Power Generation and Energy Supply***

Drought impacts the water level in about twenty main reservoirs used for either irrigation purposes or production of generating hydroelectricity. Currently, about 40-50 per cent of electricity generation in Sri Lanka comes from hydropower. However, as the water levels in the major reservoirs drop with heavy droughts, the country's power supply is significantly affected, needing occasional power cuts, as the thermal power generation as the main alternative is much more costly due to higher oil prices. In June—July 2012, over 80 per cent of the power generation has been thermal power, due to reduced water level in the reservoirs caused by the existing drought in most part of the country. The water level has been extremely low due to low rainfall and increased evaporation rates caused by the hot weather.

Improper agricultural practices in mountain slopes in hill country, especially in Upper Mahaweli Catchment area, have led to heavy soil erosion, especially during heavy rains, leading to siltation of major hydropower reservoirs, affecting the power generation capacity (Hewawasam, 2010). High intensity rainfall due to climate change could further aggravate the situation in future.

***Secure health and hygiene***

Droughts in Sri Lanka could affect health and hygiene mostly through scarcity of clean water, leading to diseases such as diarrhoea and skin diseases. There is also a relatively high incidence of chronic renal failure (CRF) among the inhabitants of Anuradhapura and Polonnaruwa districts (Dissanayake, 2005), which are frequently affected by droughts and poor groundwater quality. This is especially a problem in the rural North-central Sri Lanka, as more than 80 per cent of the rural population in Sri Lanka depends on ground water and other water resources described above for drinking and other purposes.

Climate, mostly the rainfall and temperature, influences the incidence of vector borne diseases such as malaria. Malaria epidemics are more frequent under drought conditions, as the very low water level remaining in the rivers or tanks could serve as suitable breeding grounds for vector mosquitoes. However, due to proper disease management and campaign on disease prevention, the incidence of malaria in the country has reduced by 99 per cent, from 210,000 cases in 2000 to only 124 cases in 2011. Temperature increases due to global warming have caused a significant increase in the spread of *Aedes aegypti*, the vector of the dengue fever; the incidence of which peaks with monsoonal rains. The disease was first reported in the country in 1962, and the first outbreak occurred in 1965. However, since 2002, the dengue epidemics are more frequent and reported every other year (Ministry of Health, 2012). The Dengue cases have significantly increased in 2012, compared to 2011; in 2012, 15,900 dengue cases and 78 dengue deaths were reported within the country by July; the numbers in 2011 were 28,473 cases and 185 deaths, according to the data

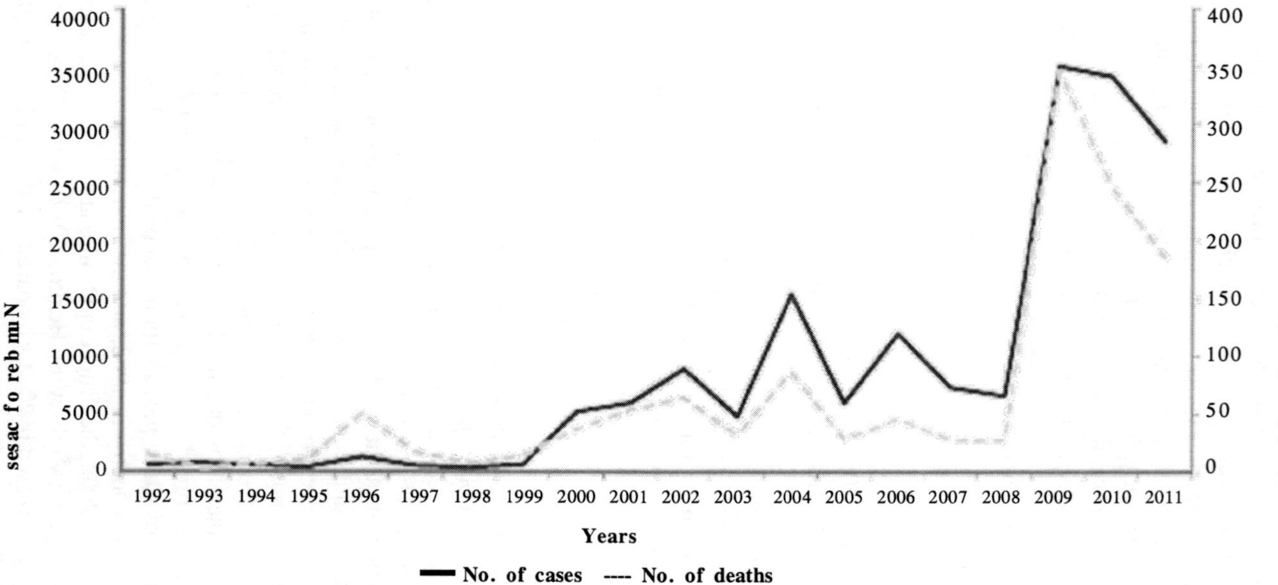


**Figure 1. Total Number of Dengue Cases and Death Reported during the Period 1992 to 2011**

(*Source*: Ministry of Health, 2012).

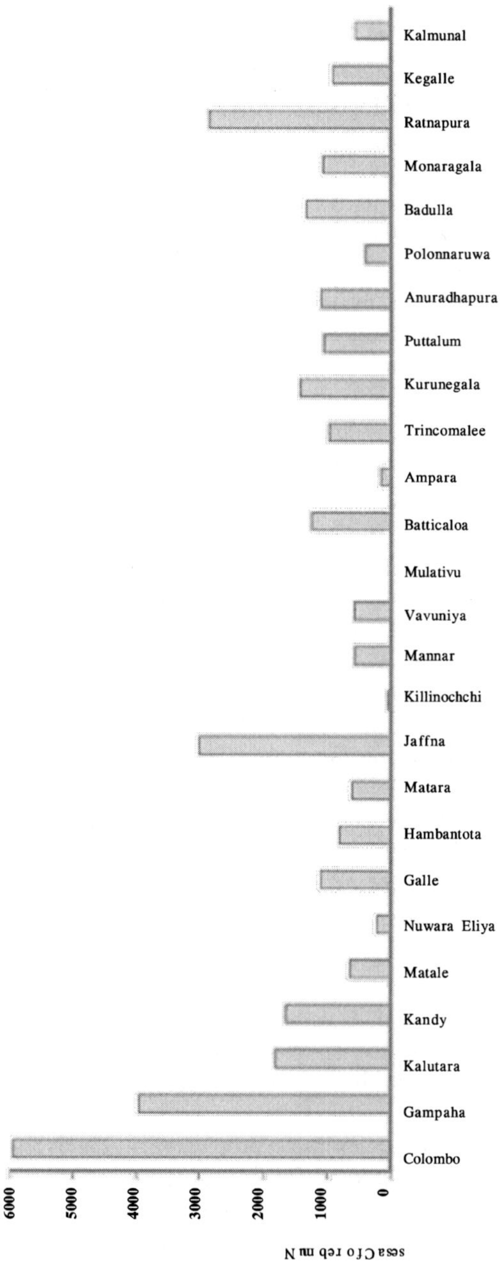


**Figure 2. Number of Dengue Cases Reported in 2010 by District**

(*Source*: Ministry of Health, 2011).

from the Ministry of Health. Although there is a downturn due to control of the vector and proper management of the disease, the incidence of dengue was very low in Sri Lanka about 10 years ago. And there was nearly a 10-fold increase and remarkable spread to newer areas from 2000 to 2010 (Figure 1), indicating some correlation with the changing climate and environmental conditions. The breakdown of incidence of dengue in 2010 by district wise is given in Figure 2.

Other than the mosquito-borne diseases such as dengue, there is a possibility for spread of diseases including hepatitis A, cholera, diarrhoea, typhoid, flu, Leptospirosis (rat fever), following flood events. Health security is also impacted by the increased air pollution, which mostly includes the emissions from fossil fuel burning in the energy sector including electricity generation, transport, and industries. Carbon dioxide, ozone, nitrous oxide, and methane emissions are some of the greenhouse gases among many other pollutant emissions. Some of these gases could cause various illnesses, mostly respiratory disorders and eye irritation.

*Overall Security of the Livelihoods of People*

During the last decade, the most severe droughts occurred in 2001 and 2004, with the total number of people affected amounting to about 3 million and 2.7 million, respectively. There was severe crop loss, water scarcity and various health impacts affecting the livelihoods of people. Livelihoods have been severely affected during these drought years with people living in Anuradhapura, Hambantota, Kurunegala, Moneragala and Puttalam districts being the most affected, while those in Colombo, Gampaha, Kalutara, Kegalle and Nuwara Eliya districts being the least affected. According to the Central Bank (2012), livelihoods of 1.8 million people (i.e., 8-9 per cent of the population in Sri Lanka) depending on agriculture were impacted by the changing weather patterns in Sri Lanka. Frequent crop losses due to extreme weather has also affected the mental stability of the people; there is a high incidence of committing suicide among the farmers in north central part of Sri Lanka, mostly due to financial losses or inability to cover debts. This has severely affected their livelihoods, as the main income of the people in this part of the country has been farming for generations, and it is hard for them to find alternative employments.

Although droughts cause more crop damage, the overall livelihoods of people are mostly affected by floods. Floods in Sri Lanka have led to damage to housing and infrastructure including roads and other constructions, requiring the evacuation of people from time to time. This has significantly affected the livelihoods of people in certain coastal and other flood prone areas in the country.

In living a healthy lifestyle, Sri Lankans have always interacted with various ecosystems and biodiversity in the country, which along with Western Ghats, has been named as one of the biodiversity hotspots in the world, for its rich fauna and flora. However, increasing temperatures and frequent weather fluctuations have led to the extinction of a significant number of species (Table 1), affecting proper ecosystem functioning. The length of the coastline of Sri Lanka is 1500 km, and currently about 30-35 per cent of this is subject to coastal erosion, at a rate of 0.3-0.5 meters per year (Patabendi, 2009). Sea level rise and high sea surface temperatures have affected coastal ecosystems including mangroves and coral reefs, threatening the fish and other life forms depending on these ecosystems; damage to coral reefs, mangroves, coastal wetlands, estuaries, lagoons affect feeding and breeding of various fish species.

**Table 1. Summary of Threatened Inland Indigenous Vertebrate and Invertebrate Fauna of Sri Lanka**

| *Taxon* | *Total Species* | *Total threatened*[*] |
|---|---|---|
| **Vertebrates** | | |
| Mammals | 91 (16) | 41 (14) |
| Birds | 227 (33) | 46 (16) |
| Reptiles | 171 (101) | 56 (37) |
| Amphibians | 106 (90) | 52 (51) |
| Freshwater fishes | 82 (44) | 28 (20) |
| Total | 677 (284) | 223 (138) |
| **Invertebrates** | | |
| Butterflies | 243 (20) | 66 (13) |
| Dragonflies | 120 (57) | 20 (20) |
| Freshwater crabs | 51 (51) | 37 (37) |
| Theraphosid spiders | 7(5) | 1 (1) |
| Land snails | 246 (204) | 33 (32) |

(*Source*: IUCN Sri Lanka and Ministry of Environment and Natural Resources 2007).
*- total of the species of three cateogories- extremely high risk of extinction; endangered, and vulnerable.

Sea level rise and coastal erosion could affect the fishing communities through impacts on their housing, territories, and damage to boat landing sites and infrastructure including roads and railways. Since fishing is possible only at certain times of the year, climate change impacts could seriously affect their annual income and livelihoods.

Tourism has been an attractive way of income for Sri Lanka, and recent extreme weather events related to climate change have affected some hotels and other infrastructure relevant to tourism in coastal areas, affecting the employment and livelihoods of the people concerned. Increased temperatures have also increased the demand for more electricity for cooling purposes inside buildings. High temperatures could lead to melting of bitumen on some roads, affecting safe road transport.

Thus, there are many direct and indirect impacts, which could be attributed to climate change. In addressing these, the country has taken some important adaptive measures, including policy, technical, technological, and other measures.

## Policy and Other Measures Adopted by Sri Lanka to Ensure Human Security in a Changing Climate

Since certain measures so far taken by the country could collectively address several categories of human security described above, the measures taken could be mainly categorised as policy- and other measures. There are certain important adaptive and mitigatory measures falling under these two categories, as described below.

### *Regulatory and Policy Related Action for Ensuring Human Security Against Climate Change*

Sri Lanka has taken further regulatory and policy measures in addition to those directly relevant to UNFCCC, as described earlier. Considering the recent higher occurrence of natural disasters, the Disaster Management Center (DMC) was established under the Disaster Management Act No. 13 of 2005. A ministry has been established for disaster management, and the Department of Meteorology, DMC, and National Building Research Organisation (NBRO) come under the purview of this ministry. Various projects are being implemented by the DMC in mitigating the impacts of extreme weather conditions, including droughts and floods. NBRO has played an active role in disaster mitigation; for instance in controlling landslides, landslide hazard zonation mapping in the relevant districts, risk assessment, and application of structural mitigation measures including trimming and benching, rock blasting and rock bolting, soil nailing, construction of drainage and turfing, etc., have been carried out. The Ministry of Disaster Management developed a road map for disaster risk management, considering seven thematic areas; under this, early warning systems for various hazards have been developed, and risk assessments for floods and other disasters are being carried out by the DCM in collaboration with other relevant agencies (Ministry of disaster management and human rights, 2006).

In order to control gaseous emissions from fossil fuel burning and other sources, emission standards have been set up through gazette notifications according to the National Environment Act. The National Environment Policy (Ministry of Environment and Natural Resources,

2003) does not seem have much concern about climate change, although it is an important policy in relation to overall environmental management in the country, while achieving sustainable development. However, the National Environmental Action Plan 2008-12 (Ministry of Environment, 2008) includes one chapter on climate change discussing the issue with relevant impacts and the action to be taken by the country, including the promotion of good agricultural practices and technologies and crop varieties suitable for climate change.

A national policy on rainwater harvesting was approved by the government in Sri Lanka in 2005. This policy was followed by amendments to certain existing by-laws under the Urban Development Authority, making rainwater-harvesting mandatory for certain categories of new buildings in areas coming under municipal and urban councils.

The National Policy on Agriculture (NAP) of 2007 was developed with the objectives of fulfilling the basic needs of farming communities, ensuring food security, income generation, and promoting the economically viable and environmental friendly agricultural production technologies, marketing and other relevant strategies. It also promotes proper land use planning, soil and water conservation and management, and use of traditional crops.

The Climate Change Secretariat of the Ministry of Environment of Sri Lanka plays a key role in relation to both adaptation and mitigation activities conducted by the country. As mentioned earlier, the National Climate Change Adaptation Strategy for Sri Lanka 2011to 2016 was developed in 2010 (Ministry of Environment, 2010). Climate change Technology Needs Assessment with regard to several economic sectors of the country is under way, and the National Climate Change Policy was published recently (Ministry of Environment, 2012).

Coastal erosion and other impacts associated with climate change have been addressed in Coastal Zone Management Plan (CZMP) and the Master Plan for Coast Erosion Management (MPCEM). These management plans have enforced no-build buffer zones along the coastline, to reduce disastrous impacts.

***Other Measures taken in Addressing the Impacts of Climate Change***

The other measures are taken by the country so far mostly aim at ensuring food security, cleaner drinking water and health safety. In developing resilience to frequent droughts, Sri Lanka needs to conserve available water, ensuring both food security and drinking water security. Restoration of the ancient tank system and rainwater harvesting are especially helpful in facing the future challenges of climate change (Eriyagama, 2010). Domestic rainwater harvesting has been promoted since 1995 through the Community Water Supply and Sanitation Project, and currently there are over 31,000 domestic rainwater systems in the country (Ariyananda, 2010). Rainwater harvesting and other safety measures are also promoted in the national agricultural policy (Ministry of Agricultural Development and Agrarian Services, 2007).

Reforestation and afforestation activities and restoration of wetlands and coastal mangrove ecosystems, etc., are being promoted, as these serve as natural buffers against extreme weather events and other disasters leading to coastal flooding. Considering such soft technologies/methods has been less costly, compared to hard, structural technologies. Certain structural measures also have been taken to control coastal erosion, such as breakwaters, revetments, and other constructions on southern and western coasts; however, coastal erosion is still a severe problem in certain parts of the country.

Flood- and drought-resistant rice varieties are being tested in Sri Lanka. Promoting high yielding rice varieties with a shorter growth period and high water use efficiency and traditional rice varieties that are tolerant of high salinity in soil and water has been done in some parts of the country (e.g., Hambantota), to prevent abandonment of agricultural land. This has already been practiced in certain areas of Hambantota. The addition of organic fertilizer has also helped reduce salinity problem in soil. Using traditional farming methods, especially pest control methods using conventional pesticides, could be considered in integrated pest management to reduce pest damage and outbreaks. Apart from rice, the plantation industry has also shown some climate change concerns; for instance, tea and rubber research institutes have been concerned in including the drought tolerance and/or pest. Disease resistance into their breeding programmes, while coconut research institutes have developed models to predict yields based on the consequences of climate change, including extreme weather events (IRI, 2005).

Climate change could cause extremely dry seasons in the future, aggravating salt-water intrusion and affecting drinking water intakes, if they are located close to the coastline. Therefore, the new intakes should be chosen to avoid further salinity intrusion. Coast Conservation Department has constructed Coast protection bunds as an adaptation to protect the coastal zone from coastal erosion; there are also regulations to prevent unsustainable sand mining.

In controlling possible flooding in the greater Colombo areas, dredging of the drainage canal system and construction of retention lakes in upper catchment areas have been done by the Land Reclamation and Development Corporation, which is under the purview of the Ministry of Defence. Proper land use planning is essential in avoiding the occurrence of landslides in hill country areas; this is especially important in choosing land for developing roads, buildings, and other infrastructure.

Since more energy is wasted for cooling purposes with the trend of temperature rice from climate change, buildings should be well-architected allowing better ventilation to keep up with hot temperatures. Less GHG emissions and energy saving are promoted through regulations and cleaner technology options for day to day life, which include the promotion of energy saving bulbs, solar and other renewable energy sources. Banning of the use of three-wheelers with 2-stroke engines since 2008 and increased tariff rates for high electricity consumption beyond a certain threshold number of units, etc.

With regard to health impacts, high occurrence of vector-borne diseases, especially dengue, has been a severe threat to human security. Given the increasing number of cases reported each year, the Ministry of Health have undertaken a country-wide survey to detect any probable mosquito breeding areas and taken measures to penalise those who have been ignorant and promoting such vector breeding grounds of the disease. The Health Ministry has declared the month extending from June 25 to July 25, 2012 as the National *Dengue* Eradication Month. Similarly, there has been more awareness creation among the public on dengue, rat fever, and other diseases which have been aggravated with more frequent weather fluctuations.

In summary, Sri Lanka has been impacted by climate change, mostly due to extreme weather fluctuations and sea level rise. Climate change has affected the human security in Sri Lanka through direct and indirect impacts on major economic sectors including agriculture, water, health, and power sector. The country in turn has taken several important policy-regulatory-, and other measures in addressing the issue. Improper land use and agricultural practices have further aggravated the climate change impacts on the country's major economic sectors. Therefore, in future, more awareness creation among people about different aspects of the climate change issue and its impacts are essential, to avoid any unplanned constructions and other environmentally damaging activities in vulnerable areas, in addition to any further adaptive and mitigatory measures the country will be undertaking.

## References

Ariyananda, T. 2010. Domestic Rainwater Harvesting as a Water Supply Option in Sri Lanka. Hydro Nepal: Journal of Water, Energy and Environment 6 : 27-30.

Central Banks of Sri Lanka. 2012. Annual Report 2011. Central Bank of Sri Lanka. Colombo, Sri Lanka.

Chandrapala, L. 1996a. Calculation of aerial precipitation of Sri Lanka on district basis using Voronoi Tessalation Method. Proceedings of National Symposium on Climate Change. Central Environmental Authority, Colombo, Sri Lanka, March 7-8, 1996.

Chandrapala, L. 1996b. Long term trends of Rainfall and Temperature in Sri Lanka. In: Climate variability and agriculture, eds., Abrol, Y.P.; Gadgil, S., Pant, G.B. (eds). New Delhi, India: Narosa Publishing House.

De Silva, G.J., Sonnadara, D.U.J. 2009. Climate Change in the Hill Country of Sri Lanka. Proceedings of the Technical Sessions, 25 (2009) 7-12. Institute of Physics- Sri Lanka.

Dissanayake, C.B. 2005. Water Quality of Dry Zone of Sri Lanka- Some Interesting Health Aspects. Journal of National Science Foundation Sri Lanka 33(3): 161-68.

Eriyagama, N. 2010. Impacts of Climate Change on Water Resources and Agriculture in Sri Lanka: Vulnerability Hotspots and Options for Adaptation. Water matters. News of IWMI Research of Sri Lanka. Issue 5: 6-7.

Hewawasam, T. 2010. Effect of Land Use in the Upper Mahaweli Catchment Area on Erosion, Landslides and Siltation in Hydropower Reservoirs of Sri Lanka. Journal of the National Science Foundation of Sri Lanka 38(1): 3-14

IRI. 2005. IRI Technical Report 10-06. Impact Assessment and Adaptation to Climate Change of Plantations in Sri Lanka. International Research Institute for Climate and Society (IRI). Earth Institute at Columbia University, USA.

IUCN Sri Lanka and Ministry of Environment and Natural Resources. 2007. The 2007 Red List of Threatened Fauna and Flora of Sri Lanka, Colombo, Sri Lanka. XIII + 148 pp.

Jayatillake, H.M., Chandrapala, L., Basnayake, B.R.S.B., Dharmaratne, G.H.P. 2005. Water Resources and Climate Change. In: Proceedings of Workshop on Sri Lanka National Water Development, eds, Wijesekera, N.T.S., Imbulana, K.A.U.S., Neupane, B. World Water Assesment Programme (WWAP). Paris. France.

Lokupitiya, E. 2008. Estimation and Analysis of Greenhouse Gas Emissions from Agriculture, Forestry, and Other Land Use Sectors for the National Greenhouse Gas Inventory of Sri Lanka. Ministry of Environment. Battaramulla, Sri Lanka.

Mandawadu, L., Fernando, N. 2008. Climate Changes in Sri Lanka. Colombo Review, 1(2).

Ministry of Agricultural Development and Agrarian Services. 2007. National Agticulural Policy. Ministry of Agricultural Development and Agrarian Services, Sri Lanka.

Ministry of Disaster Management and Human Rights, 2006. Towards a Safer Sri Lanka Road Map for Disaster Risk Management. Disaster management center, Ministry of Disaster Management and Human Rights, Sri Lanka.

Ministry of Environment and Natural Resources. 2003. National Environmental Policy for Sri Lanka. Ministry of Environment and Natural Resources, Sri Lanka.

Ministry of Environment, 2010. National Climate Change Adaptation Strategy for Sri Lanka 2011 to 2016. Ministry of Environment, Sri Lanka.

Ministry of Environment, 2012. The National Climate Change Policy of Sri Lanka. Ministry of Environment, Sri Lanka.

Ministry of Health. 2011. Surveillance of Dengue Fever. Epidemiological Bulletin Sri Lanka 52: 6. Third Quarter 2011. Epidemiology Unit, Ministry of Health, Colombo 10, Sri Lanka.

NWSDB, 2009. National Policy on Drinking Water. National Water Supply and Drainage Board, Sri Lanka.

Nawagamuwa, U.P., Rathnaweera, T.D, Rangana. H.L.L. and Palihawadana, 2011. M.P. Effects of Global Warming on Landslide Frequencies in Ratnapura District, Sri Lanka. IESLSSMS Joint International Symposium on Social Management Systems 2011. 14–16 September, 2011 in Colombo, Sri Lanka.

Patabendi, P. 2009. 'Global Climate Changes-—Impact and Adaptation:A Case Study from Southern Sri Lanka'. Third United Nations International UN-SPIDER Bonn Workshop 21-23 October 2009, Bonn, Germany.

Ranatunge, E., Hayashi, Y., Mikami, T., Morishima, W., Yokozawa, M. and Nishimori, M. 2003. Changes in the Southwest Monsoon Mean Daily Rainfall Intensity in Sri Lanka: relationship to the El Niño–Southern Oscillation. Palaeogeography, Palaeoclimatology, Palaeoecology 197(1s-2): 1-14.

# 6

# Climate Change and Threats to Human Security

*Dasarathi Bhuyan*

The prospect of humankind is shaped by issues that are beyond any individual nation's ability to solve, i.e., global issues. Among these issues climate change is the most debated topics on global issue, and the National Environmental Satellite, Data and Information Service (NESDIS) data centres are central to answering some of the most pressing global change questions that remain unresolved.

Climate change poses several threats to human security. Hydro-climatic hazards such as droughts and floods have the potential to trigger or exacerbate social tensions, intra- and interstate conflict. This study examines the relationships between hydro-climatic hazards, human security and conflict. The project will also develop recommendations for the best types of policies and institutions to avoid or better prepare for water conflicts related to climate change. The project's second policy brief is now available for download.

Potential links between climate change impacts and violent conflict have received wide public attention. International resolutions such as the Rio Declaration on Environment and Development, the European Security Strategy, and the UN High Level Panel on Threats, Challenges, and Change address these potential links between climate change and violent conflict. However, the alleged causal relationship between water shortages and violent conflict is as a general rule based on single case studies and is not been confirmed by large comparative studies. The vulnerability, capacity to adapt and resilience of different regions and communities to climate change differs greatly, and so may the propensity for violent conflict.

## Main Objectives

In this context, this study pursues the following objectives:

- To understand and model the relationships between hydro-climatic hazards, climate change vulnerability, human security and conflict, on the basis of theoretically-informed, comparative empirical research.
- To map international and national policies for security and adaptation in water resources and hazard management, and develop a policy model for security against hydro-climatic hazards ("hydro-security") in the Indian region.

Climate change is a significant and lasting change in the statistical distribution of weather patterns over periods ranging from decades to millions of years. It may be a change in average weather conditions or the distribution of events around that average (e.g., more or fewer extreme weather events). Climate change may be limited to a specific region or may occur across the whole Earth.

## Terminology

The most general definition of *climate change* is a change in the statistical properties of the climate system when considered over long periods of time, regardless of cause (IPCC, 2001 and Meteorological Services, Canada, 2002). Accordingly, fluctuations over periods shorter than a few decades, such as El Niño, do not represent climate change.

The term sometimes is used to refer specifically to climate change caused by human activity, as opposed to changes in climate that may have resulted as part of Earth's natural processes (Challenor P. et. al, 2005). In this latter sense, used especially in the context of environmental policy, climate change is synonymous with anthropogenic global warming. Within scientific journals, however, *global warming* refers to surface temperature increases, while *climate change* includes global warming and everything else that increasing greenhouse gas amounts will affect (Julio Godoy, 2006).

## Causes

Climate changes in response to changes in the global energy balance. On the broadest scale, the rate at which energy is received from the sun and the rate at which it is lost to space determine the equilibrium temperature and climate of Earth. This energy is then distributed around the globe by winds, ocean currents, and other mechanisms to affect the climates of different regions.

Factors that can shape climate are called climate forcing or "forcing mechanisms"(NASA, 2006). These include such processes as variations in solar radiation, deviations in the Earth's orbit, mountain-building and continental drift, and changes in greenhouse gas concentrations. There are varieties of climate change feedbacks that can either amplify or diminish the initial forcing. Some parts of the climate system, such as the oceans and ice caps, respond slowly in reaction to climate forcing, while others respond more quickly.

Forcing mechanisms can be either 'internal' or 'external'. Internal forcing mechanisms are natural processes within the climate system itself, e.g., the meridional turnover. External forcing mechanisms can be either natural (e.g., changes in solar output) or anthropogenic (e.g., increased emissions of greenhouse gases).

Whether the initial forcing mechanism is internal or external, the response of the climate system might be fast (e.g., a sudden cooling due to airborne volcanic ash reflecting sunlight), slow (e.g., thermal expansion of warming ocean water), or a combination (e.g., sudden loss of albedo in the arctic ocean as sea ice melts, followed by more gradual thermal expansion of the water). Therefore, the climate system can respond abruptly, but the full response to forcing mechanisms might not be fully developed for centuries or even longer.

**Internal Forcing Mechanisms**

Natural changes in the components of earth's climate system and their interactions are the cause of internal climate variability, or 'internal forcing'. Scientists generally define the five components of earth's climate system to include atmosphere, hydrosphere, cryosphere, lithosphere (restricted to the surface soils, rocks, and sediments), and biosphere (National Oceanic and Atmospheric Administration, 2006). The ocean is a fundamental part of the climate system, some changes in it occurring at longer timescales than in the atmosphere, massing hundreds of times more and having very high thermal inertia (such as the ocean depths still lagging today in temperature adjustment from the Little Ice Age (NASA, 2006).

Short-term fluctuations (years to a few decades) such as the El Niño--Southern Oscillation, the Pacific decadal oscillation, the North Atlantic oscillation, and the Arctic oscillation, represent climate variability rather than climate change. On longer time scales, alterations to ocean processes such as thermohaline circulation play a key role in redistributing heat by carrying out a very slow and extremely deep movement of water, and the long-term redistribution of heat in the world's oceans.

**Expected Effects of Climate Change**

Rise in global temperatures may in turn cause bigger changes, including glacial retreat, Arctic shrinkage, and worldwide sea level rise. Changes in the amount and pattern of precipitation may result in flooding and drought. Other effects may include changes in agricultural yields, addition of new trade routes, reduced summer stream flows, species extinctions, and increases in the range of disease vectors. Other expected effects include water scarcity in some regions and increased precipitation in others, changes in mountain snow pack, and adverse health effects from warmer temperatures. According to United Nations Environment Programme (UNEP), economic sectors likely to face difficulties related to climate change including banks, agriculture, transport and others.

**Climate Change: India's Perception**

Indians should be concerned about climate change since this phenomenon might have substantial adverse impacts on them. Not all possible consequences of climate change are yet fully understood, but the three main 'categories' of impacts are those on agriculture, sea level rise leading to submergence of coastal areas, as well as increased frequency of extreme events. Each of these poses serious threats to India. However, these are long-term issues. The overriding immediate concern for India should be the fast pace at which negotiations are taking place on the climate front. India's main energy resource is coal. With the threat of climate change, India is called upon to change its energy strategy based on coal, its most abundant resource, and to use other energy sources (e.g., oil, gas, renewable and nuclear energy) instead, which may turn out to be expensive. Thus, an immediate issue is to come up with a better negotiation strategy such that we have more freedom to decide which type of energy we use, how we generate power, how to reduce methane emissions by agricultural practices or forestry and so on. Negotiations are important for us as a means to reduce or postpone future vulnerability by getting the developed countries to reduce their emissions.

**Risk of Lower Agricultural Production**

The FCCC objective states that GHG concentrations should be stabilised at levels where food production is not threatened (UN, 1992). Thus, by examining the impact on agriculture of different climate change scenarios, one can get an idea of what is tolerable. Rosenzweig and Parry (1994) have estimated significant adverse impact on the agriculture of many developing countries. In a more detailed study of India, Kumar and Parikh (2001a and 2001b) examined the impact of climate change on agricultural crop yields, GDP and welfare. Considering a range of equilibrium climate change scenarios which project a temperature rise of 2.5°C to 4.9°C for India, it was estimated that—(a) without considering the carbon dioxide fertilisation effects yield losses for rice and wheat vary between 32 and 40 per cent, and 41 and 52 per cent, respectively; (b) GDP would drop by between 1.8 to 3.4 per cent. Their study also showed that even with carbon fertilisation effects, losses would be in the same direction but somewhat smaller. Using an alternative methodology Kumar and Parikh (2001a) showed that even with farm-level adaptations, the impacts of climate change on Indian agriculture would remain significant. They estimated that with a temperature change of +2°C and an accompanying precipitation change of +7 per cent, farm level total net-revenue would fall by 9 per cemt, whereas with a temperature increase of +3.5°C and precipitation change of +15 per cent, the fall in farm level total net-revenue would be nearly 25 per cent. For a developing country, these very large changes can cause much human misery. From India's point of view, a 2°C increase would be clearly intolerable. Other developing countries may be even more vulnerable (possibly Bangladesh or small island states).

**Risk of Sea Level Rise**

Large-scale emigration from coastal zones is expected due to submergence of coastlines after sea levels have risen. This will create large numbers of environmental refugees especially from low-lying delta regions in poor countries. Furthermore, intrusion of seawater in the ground water and changes in temperature can reduce agricultural and fishing incomes. Countries dependent on coastal fishery and agriculture, which most often include developing countries, are likely to be adversely affected. If a one-meter sea level rise were to take place today, it would displace 7 million persons in India (ADB, 1995). In the future, many more may be displaced. 35 per cent of the land in Bangladesh would be submerged by a one-meter rise. The estimates for costs to build walls along the zones vulnerable to sea level rise for the USA is $107 billion in 1989 prices (Yohe, 1990). That may be a small share of the GDP of developed countries, but such measures, even scaling for their coast lines, for say, Bangladesh, could require a very large share of its GDP. Who shall pay Bangladesh or India for such a wall? Given that these countries are unlikely to be able to pay for protective measures, tens of millions of people will be displaced in Bangladesh and many of them could spill over into India.

**Risk of Extreme Events**

Increased occurrence of extreme events due to climate change will also affect the poor most. In the cyclone in Andhra Pradesh in India in 1996, more than 1,000 people died and there was huge property loss. Cyclones of similar intensity in advanced countries like the US may not lead to any deaths and much hardship, due to stable and durable housing and other infrastructure and extended safety net available to the people in distress.

**The Failures of States and International Bodies**

The world's primary international agreement on combating global warming is the Kyoto Protocol, an amendment to the UNFCC negotiated in 1977. The Protocol now covers more than 160 countries globally and over 55 per cent of global greenhouse gas emissions. Only the United States and Kazakhstan have not ratified the treaty, with the United States historically being the world's largest emitter of greenhouse gas. This treaty expires in 1012, and international talks began in May 2007 on a future treaty to succeed the current one. China and India, though exempt from its provisions as developing countries, have ratified the Kyoto Protocol. China may have passed the U.S. in total annual greenhouse gas emissions according to some recent studies. Poor regions, particularly Africa and South Asia, appear at greatest risk from the projected effects of global warming, while their emissions have been small compared to the developed world. Today, when the globe is threatened by the very technology that we have created in our thirst for wealth, we need modernisation more than ever before. The desire for unlimited growth must cease, for in the long-run, simple living and high thinking is better.

**The Climate Change Challenges**

Climate change is one of the greatest global challenges of the 21st century, and increasing evidence of present and anticipated impacts of climate change highlight the need for action. The IPCC's Third Assessment Report noted that the Earth's average surface temperature increased 0.6 ± 0.2°C in the 20th century. This trend is expected to persist, with an increase of 1.4 to 5.8°C by 2100. Even with 'best case' mitigation efforts, some climate change cannot be avoided due to the inertia of the global climate system. Warming will vary by region and be accompanied by significant changes in precipitation patterns as well as changes in the frequency and intensity of some extreme events. Average global sea levels are projected to rise between nine and 88 cm by 2100, with implications for the 50 to 70 per cent of the world's population currently living in low-lying coastal areas. The probability of large-scale and irreversible impacts, such as the collapse of the Greenland Ice Sheet and the slowing (or shutting down) of the Atlantic Ocean conveyor belt, is expected to increase with the rate, magnitude and duration of climate change.

A number of climate records have been set recently, with most climatologists pointing to climate change as a significant factor:

- record temperatures were recorded in the July 2006 heat wave in Europe;
- five of the last 10 years have set records as the warmest ever (1998, 2002, 2003, 2004 and 2005). 2005 was the warmest year on record;
- there were 15 hurricanes in 2005 surpassing the previous record of 12, the most Category Five storms and the most hurricane damage ever; and

- a record low for Arctic sea ice was recorded in June 2004—six per cent below average.

Changes in temperature and precipitation patterns have already impacted natural and human systems. Observed changes include *inter alia* shrinking glaciers; thawing permafrost; later freezing and earlier breakup of ice on rivers and lakes; rising sea levels; extended mid-to-high-latitude growing seasons; pole ward and altitudinal shifts of plant and animal ranges; and declines of some plant and animal populations (IPCC, 2001). Documented impacts also include the spread of disease vectors including malaria; the destruction of coral reefs from warmer seas and extreme weather events; and threats to low-lying island states (Kelly Levin et. al, 2006).

The provision of ecosystem goods and services will be disrupted with coastlines and mountaintops more susceptible to irreversible losses. Projected climate change is expected to increase agricultural production in industrialised countries, while developing countries will face a decrease in agricultural land, potential cereal production and food security. More people will become water-stressed as hydrologic variability affects water quality and supply (IPCC, 2001).

In addition to altering biophysical systems, climate change will affect human health and socio-economic well-being. Recent increases in floods and droughts have already led to corresponding increases in damages and insurance impacts. In the future, even modest levels of warming are expected to increase the risks of hunger and disease (Parry et al., 2001: 181—83).

Although climate change impacts will affect all countries, the poor, primarily but by no means exclusively in developing countries, will be disproportionately affected. Their reliance on local ecological resources, coupled with existing stresses on health and well-being (e.g., HIV/AIDS, illiteracy) and limited financial, institutional and human resources leave the poor most vulnerable and least able to adapt to the impacts of climate change. Consequently, there is growing recognition that climate change may undermine the ability of developing countries to meet the targets put forth in the Millennium Development Goals (MDGs). Most ironically, the vast majority of those most vulnerable to the impacts of climate change are also the least responsible for contributing to it in the form of GHG emissions—whether the indigenous in Northern communities, the subsistence farmer in Sub-Saharan Africa or the small island dweller in the South Pacific.

Nor should we be oblivious to the possible opportunities that may present themselves under climate change scenarios. Changes in temperature and precipitation regimes might make it possible to grow food crops in new locations, potentially contributing to increased food security. In the

Arctic regions, warmer winters and ocean waters could lengthen the summer ice-free season, creating economic opportunities for Northern communities (while simultaneously raising concerns about potential ramifications for Northern ecosystems and traditional ways of life). As the process of global warming continues, it will be important to identify, assess and take advantage of new opportunities as they emerge. Countries are only beginning to grapple with the reality of rising GHG emissions, and a number of processes at the multilateral level offer opportunity to implement strategies to allow countries to engage in dialogue and take action on climate change. (Prominent multilateral initiatives are outlined in the Appendix.) Yet, the threat posed by climate change indicates that decisions and actions will need to be beyond environmental policy and be addressed on a number of fronts.

**International Peace and Security**

In a major London address in early 2006, British Defence Secretary John Reid warned that global climate change and dwindling natural resources are combining to increase the likelihood of violent conflict over land, water and energy. Climate change, he argued, 'Will make scarce resources, clean water, viable agricultural land even scarcer'—and this will 'make the emergence of violent conflict more, rather than less, likely' (Klare Michael, 2006). This speech reflects a shift in strategic thinking, where military analysts are increasingly recognising that environmental degradation and resource scarcity—stressors that will be exacerbated by climate change—could be potent sources of instability and armed conflict in the years to come. The European Security Strategy notes that climate change will aggravate competition for natural resources, and likely increase conflict and migratory movements in various regions (European Union, 2003).

**Critical Issues: The Impact of International Peace and Security on Climate Change Efforts**

The impact of climate change in fragile states around the world may not be so much a case of entirely new security threats, but more of enhancing existing instabilities and threats. There is no clear, mono-causal link between climate change and conflict. Environmental factors are rarely, if ever, the sole cause of violent conflict; however, it is clear that environmental stress can increase the severity, duration and collateral impacts of a conflict. The security dimensions of climate change are significant and is to be affected both the developed and the developing worlds. Four dimensions of this scenario are of particular relevance to Danish foreign and security policy: environmental degradation; resource scarcity; movements of environmental refugees; and contests over access to newly accessible resources.

## Climate Change May Aggravate Environmental Degradation

Climate change is a powerful force that will affect our environment in complex and interrelated ways. Early climate change scenarios predicted that the most serious impacts of climate change would fall on northerly and southerly latitudes. However, models that are more recent suggest that climate change could result in less rainfall around the equator and lead to a 'drying out' of the tropics. This would entail serious security implications. Coupled with rapid population growth, there is a very real concern that climate change could undermine the 'carrying capacity' of many developing countries—that is the capacity of countries to provide adequate food and water for its population. In particular, countries suffering environmental stress but with weak institutions and limited coping capacity are likely to experience increased crime and social unrest as well as the rise of radical social movements.

Developed countries are also not immune to the impacts of climate change and the expected increase in the severity and frequency of catastrophic weather events (Hans Gunter Brauch, 2002). Hurricane Katrina on the Gulf Coast of the United States in 2005 was a powerful demonstration of the ability of such events to overwhelm rapidly the resources and the adaptive capacity of even the wealthiest countries.

## Climate Change May Accelerate Resource Scarcity

Linked to accelerating environmental degradation is the prospect that climate change will further limit the supply of already scarce resources such as agricultural land and water. How countries and communities in resource-poor regions respond and adapt to reduced access to key resources is central to the possible security implications of a changing climate. It is predicted that Egypt, for example, will be affected by both temperature increases and sea-level rise. The first will increase evapotranspiration and the water needs of agriculture resulting in declining yields for agriculture. The second will inundate some of the most fertile land and densely populated regions. In Egypt, a sea-level rise of 37 cm by 2060 would cause food self-sufficiency to decline from 60 per cent in 1990 to 10 per cent by 2060 (BBC News, 2005). Former UN Secretary General Boutros Ghali warned in a 2005 interview that war between the countries of the Nile Basin over water is almost inevitable (Barnett, Jon, 2003).

## Climate Change may Drive 'Distress Migration'

The combination of environmental degradation, more frequent extreme weather events and resource scarcity can be expected to displace large numbers of 'environmental refugees' within countries and across borders. A sea-level rise of 45 cm in Bangladesh would result in a loss of nearly 11 per cent of Bangladesh's territory, forcing the relocation of an estimated 5.5 million people. Increasing desertification is expected to cause the movement of some 60 million people from Sub-Saharan Africa towards Northern Africa and Europe between 1997 and 2020 (UN, 2004). Large population movements are already recognised by the UN Security Council as constituting a potential threat to international peace and security, particularly if there are pre-existing ethnic and social tensions (Sindico Francesco, 2005). Immigration is, of course, already a very contentious issue within Europe.

Current projections of sea-level rise coupled with a likely increase in the frequency and severity of storms may make many small island states, such as Kiribati and Tuvalu, uninhabitable (Mc Leman Robert et. al., 2004). In this sense, climate change presents the most serious security problem that any country can face—the entire loss of its territory. The President of the Federated States of Micronesia put this bluntly, 'sea-level rise and other related consequences of climate change are grave security threats to our very existence as homelands and nation states' (Barnett Jon, 2003). The social, economic and political ramifications of large numbers of environmental refugees could be profound, and are increasingly becoming a reality with South Pacific island nations pursuing relocation options. Tokelauans have access to New Zealand, the Marshallese can settle in the United States under the Compact of Free Association, and Canada is funding the relocation of residents inland from parts of Vanuatu affected by climate change (Pacific News Service 2006).

**Conflict over Strategic Trade Routes and Newly Accessible Resources**

Climate change's impact on polar ice cover is redrawing global trade routes and opening up new areas to oil and gas exploration. Already ice cover in the Northwest Passage is thinning (Huebert Rob, 2001). By September 2005, Arctic sea ice had dropped to its lowest level on record. In fact, the same year, the Northern Sea route along the Siberian coast—once not navigable in its entirety—was free of ice for a whole month. Moreover, in August 2005 a Russian ship named the Akademik Fyodorov was the first ship to be able to reach the North Pole without the help of an icebreaker. The retreat of the sea ice has serious implications for control over these strategic and potentially valuable trade routes. Reliable transit through the Northwest or Siberian passages would dramatically reduce freight transport time. For example, ships travelling between Copenhagen and Yokohama have to travel only half the distance if they go along the north Siberian coast as through the Suez Canal. In terms of carbon dioxide emissions, shorter transport routes would have a positive environmental effect.

The receding sea ice is also opening up new possibilities for oil and gas exploration in the increasingly accessible Barents Sea. The Arctic as a whole is believed to contain as much as a quarter of the world's unextracted reserves of oil and gas. Thirty-seven rising temperatures may also change the distribution of other resources such as cold-water fish stocks, which could move north. At stake are sovereign rights to enormous quantities of natural resources as well as control over potentially valuable trade routes. Norwegian foreign minister Jonas Gahr Stoere noted in October 2005: 'In the years to come, it may be from this high North that both continental Europe and the United States will be looking for additional supplies of oil and gas' (Duval Smith Alex, 2005). Somewhat ironically, climate change might be revealing the very fossil fuels that could add even more carbon dioxide to the atmosphere.

## The Security Link

Fundamentally, 'climate change threatens to undermine governments' ability to ensure security and stability. While the majority of the scientific data is not disputed, climate change still appears low on a list of foreign policy priorities, in part because the threat feels abstract and response measures are too often poorly defined and communicated. The 'security link' conveys added, and arguably necessary, gravitas to the debate on climate change; and an appreciation of the security implications of climate change could give new impetus to the climate change agenda.

Stringent commitments to reduce carbon dioxide emissions may generate their own security implications. If renewable energy sources became more valuable because of an ambitious post-2012 climate agreement, the political economy of energy could shift dramatically. This could generate new tensions over access to different sources of energy or resources (such as water downstream of large dams). In addition, there are questions as to whether a widespread move towards nuclear power as a less carbon dioxide intensive form of energy production would have an impact on nuclear proliferation.

## Economic Wisdom Requires a Total Re-orientation of Science and Technology

*The idea of unlimited growth by almost all the countries of the globe must be questioned on two grounds; the availability of basic resources and also the capacity of the environment to cope with abuse by human beings. Both are strictly limited.* So much for the physical-material-aspects of the matter, let us now turn to certain non-material aspects. Modern economic systems are driven by limitless greed and envy. But can these forces be effective for long, or do they carry within themselves the seeds of destruction? (Schumacher, 2000).

Economic progress is possible only through the powerful impulses of greed and selfishness, which religion and traditional wisdom ask us to resist. Modern economic systems are driven by limitless greed and envy. But can these forces be effective for long, or do they carry within themselves of destruction?

If human vices such as greed and envy are systematically cultivated, the result is nothing less than a collapse of intelligence. A man driven by greed and envy cannot see things as they really are, in their roundness and wholeness; his very successes become his failures.

*The foundations of peace can not be laid by universal prosperity because such prosperity, if feasible at all, can be attained only by cultivating the forces of greed and envy, which destroy intelligence, happiness and serenity and thereby the peaceful instincts of man.* The rich can become richer only by continuing to make intolerable demands on limited world resources.

The economic wisdom requires a total re-orientation of science and technology. Scientific or technological 'solutions' which poison the environment or degrade society or the individual are of no benefit. Ever bigger machines, ever-bigger concentrations of economic power, causing ever-greater violence against the environment, do not represent progress: they are a denial of wisdom. Wisdom demands a new orientation of science towards the organic, the gentle, the simple, the elegant and beautiful, instead of using our technology to create machines, which churn out more and more goods to satisfy our ever-increasing wants and destroy our beautiful earth. We must try to bring about a revolution in technology, which will reverse the present trend towards destruction (*ibid*).

The wise men of the globe, such as Buddha, Christ and Gandhi, have long preached the virtue of moderation in all things, or renunciation and sacrifice. Today, when the globe is threatened by the very technology that we have created in our thirst for wealth, we need moderation more than ever before. The desire for unlimited growth must cease, for in the long run, simple living and high thinking is better.

## References

Alex Duval-Smith, 2005. 'Arctic Booms as Climate Change Melts Polar Ice Cap'. The Guardian, November 27.

BBC News, 2005. Ex-UN Chief Warns of Water Wars. February 2. Available at: http://news.bbc.co.uk/2/hi/africa/4227869.stm.

Challenor, P., R. Hankin and B. Marsh. 2005. 'The Probability of Rapid Climate Change'. A presentation made at the conference: Avoiding Dangerous Climate Change, February 1–3, 2005, Exeter, UK, A study by the Southampton Oceanography Centre and the Tyndall Centre for Climate Change Research focused on improving estimates of the probable collapse of the thermohaline circulation system. Preliminary results suggest that the probability of this event happening is 10 times higher than originally predicted.

Climate Change and Foreign Policy: An exploration of options for greater integration 37

EU. 2003. A Secure Europe in a Better World: European Security Strategy. Available at: http://ue.eu.int/uedocs/cmsUpload/78367.pdf .

Francesco Sindico. 2005. 'Ex-Post and Ex-Ante [Legal] Approaches to Climate Change Threats to the International Community'. New Zealand Journal of Environmental Law 9: 209–238.

Hans Günter Brauch. 2002. Climate Change, Environmental Stress and Conflict. AFESPRESS Report for the Federal Ministry for the Environment, Nature Conservation and Nuclear Safety. p. 87.

Intergovernmental Panel on Climate Change (IPCC), 2001. Third Assessment Report, Synthesis: Summary for Policy Makers. United Nations.

IPCC. 2001, and Arctic Climate Impact Assessment. 2004. Impacts of a Warming Climate. Cambridge, UK: Cambridge University Press.

Jon Barnett. 2003. 'Security and Climate Change'. Global Environmental Change, Vol. 13, Pergamon, p. 9.

Julio Godoy. 2006. Global Warming, Not Just Heat Wave. Inter Press Service, July 21, 2006.

Kelly Levin and Jonathan Pershing. 2006. Issue Brief: Climate Science 2005: Major New Discoveries.Washington, DC:World Resources Institute.

Meteorological Services Canada. 2002. Frequently Asked Questions about the Science of Climate Change. Available at: http://www.msc-smc.ec.gc.ca/education/scienceofclimatechange/ understanding/FAQ/index_e.html.

Michael Klare. 2006. "The Coming Resource Wars." The Energy Bulletin, March 6. Available at: http://www.energybulletin.net/13605.html>.

NASA Earth Observatory. 2006. Record Low for June Arctic Sea Ice. Available at: http://www.earthobservatory.nasa.gov/Newsroom/NewImages/images.php3?img_id=16978

NASA Goddard Institute for Space Studies Surface Temperature Analysis. 2006. Global Temperature Trends: 2005 Summation. Available at: data.giss.nasa.gov/gistemp.

National Oceanic and Atmospheric Administration. 2006. NOAA Reviews Record-Setting 2005 Atlantic Hurricane Season: Active Hurricane Era Likely To Continue. Available at: http://www.noaanews.noaa.gov/stories2005/s2540.htm

Pacific News Service. 2006. 'Niue: No Response Yet To Tuvalu's Resettlement Proposal'. Pacific Magazine. Available at: http://www. tuvaluislands.com/news/archives/2006/2006-01-12.htm; and Alan Dupont and Graeme Pearman. 2006. Climate Change and Security. Lowry Institute Paper 12. Double Bay, Australia: Lowry Institute for International Policy.

Parry et. al., predict that even at modest levels of global warming, by 2080, tens of millions of people worldwide will be put at additional risk of experiencing hunger and coastal flooding, hundreds of millions at risk of experiencing malaria, and billions of people at risk of experiencing water shortages. See Martin Parry, Nigel Arnell, Tony McMichael, Robert Nicholls, Pim Martens, Sari Kovats, Matthew Livermore, Cynthia Rosenzweig, Ana Iglesias and Gunther Fischer. 2001. 'Millions at Risk: Defining critical climate change threats and targets'. Global Environmental Change, (11): 181–83.

Rob Huebert. 2001. 'Climate Change and Canadian Sovereignty in the North-west Passage'. Isuma, 2(4),Winter.

Robert McLeman and Barry Smit. 2004. Climate Change, Migration and Security. Commentary No. 88, Canadian Security Intelligence Service.

Schumacher, E. F. Small is Beautiful, Orissa State Bureau of Text Book Preparation and Production, Bhubaneswar, 2000, pp. 226-52.

United Nations (UN). 2004. 'Facts and Figures: Desertification and Drought'. 2003: International Year of Fresh Water. Available at: http://www.wateryear2003.org/en/ev.php URL_ID=5137&URL_ DO=DO_ TOPIC&URL_SECTION=201.html.

# 7

# India's Perceptions, Positions and Policies on Climate Change and Human Security

*Amulya Kumar Tripathy*

## Introduction

The threat of climate change that led to the Framework Convention on Climate Change (FCCC) at Rio, is perceived differently by different countries. This fact has delayed any effective international agreement on how to deal with the problem. In the case of the Montreal Protocol covering ozone-depleting substances, there was a wide consensus and effective action was mobilised quickly. Thus, an understanding of perceptions and positions of different countries makes it easier to explore possibilities of effective action. In this paper, we present India's perceptions on the problem of climate change and sustainable development; the kind of negotiating positions that follow from these perceptions; the policies India has undertaken so far and finally, India's possibilities for action that can help contain the threat of climate change.

## Climate Change Perception

### *Unsustainable Consumption Patterns*

Unsustainable consumption patterns of the rich industrialised nations are responsible for the threat of climate change. Only 25 per cent of the global population lives in these countries, but they emit more than 70 per cent of the total global $CO_2$ emissions and consume 75 to 80 per cent of many of the other resources of the world (Parikh et al., 1991). In per capita terms, the disparities are also large: an Indian citizen emits less than 0.25 tonnes of Carbon per year whereas a citizen of the USA, for example, emits more than 5.5 tonnes. Parikh et al. (1991) emphasise the need for an equitable and efficient solution to climate change and suggest that efficiency can be obtained through a system of tradable emission quotas and equity through equal allocation of global environmental space to all human beings. These findings were well received by developing countries and are echoed in the UNFCCC, which recognised the rights of developing countries to economic development and also the 'common but differentiated responsibilities' of different countries.

**Methane Emissions and Subsistence Emissions**

Another theme of Indian analysts has been the lack of reliability of GHG emission estimates, particularly of methane. According to initial estimates, large emissions of methane from paddy fields were ascribed to developing countries. However, the empirical basis of these estimates was questioned; subsequently experimental measurements by Indian researchers showed these doubts to be well-founded (Mitra, 1992, 1996). Moreover, Agarwal and Narain (1991) argue that emissions by poor who live on the margin of subsistence should be considered a basic human right and should not be counted when ascribing responsibilities for emission reduction.

**Why should India be Concerned about Climate Change?**

Indians should be concerned about climate change since this phenomenon might have substantial adverse impacts on them. Not all possible consequences of climate change are yet fully understood, but the three main 'categories' of impacts are those on agriculture, sea level rise leading to submergence of coastal areas, as well as increased frequency of extreme events. Each of these pose is a serious threat to India. However, these are long-term issues. The overriding immediate concern for India should be the fast pace at which negotiations are taking place on the climate front. India's main energy resource is coal. With the threat of climate change, India is called upon to change its energy strategy based on coal, its most abundant resource, and to use other energy sources (e.g., oil, gas, renewables and nuclear energy) instead, which may turn out to be expensive. Thus, an immediate issue is to come up with a better negotiation strategy such that we have more freedom to decide which type of energy we use, how we generate power, how to reduce methane emissions by agricultural practices or forestry and so on. Negotiations are important for us as a means to reduce or postpone future vulnerability by getting the developed countries to reduce their emissions.

**Risk of Lower Agricultural Production**

The FCCC objective states that GHG concentrations should be stabilised at levels where food production is not threatened (UN, 1992).

Thus, by examining the impact on agriculture of different climate change scenarios, one can get an idea of what is tolerable. Rosenzweig and Parry (1994) have estimated significant adverse impact on the agriculture of many developing countries. In a more detailed study of India, Kumar and Parikh (2001a and 2001b) examined the impact of climate change on agricultural crop yields, GDP and welfare. Considering a range of equilibrium climate change scenarios which project a temperature rise of 2.5°C to 4.9°C for India, Kumar and Parikh (2001a) estimated that: (a) without considering the carbon dioxide fertilisation effects yield losses for rice and wheat vary between 32 and 40 per cent, and 41 and 52 per cent, respectively; (b) GDP would drop by between 1.8 to 3.4 per cent. Their study also showed that even with carbon fertilisation effects, losses would be in the same direction but somewhat smaller. Using an alternative methodology Kumar and Parikh (2001a) showed that even with farm-level adaptations, the impacts of climate change on Indian agriculture would remain significant. They estimated that with a temperature change of +2°C and an accompanying precipitation change of +7 per cent, farm level total net-revenue would fall by 9 per cent, whereas with a temperature increase of +3.5°C and precipitation change of +15 per cent, the fall in farm level total net-revenue would be nearly 25 per cent. For a developing country, these are very large changes, which can cause much human misery. From India's point of view, a 2°C increase would be clearly intolerable. Other developing countries may be even more vulnerable (possibly Bangladesh or Small Island States).

**Risk of Sea Level Rise**

Large-scale emigration from coastal zones is expected due to submergence of coastlines after sea levels have risen. This will create large numbers of environmental refugees especially from low-lying delta regions in poor countries. Furthermore, intrusion of seawater in the ground water and changes in temperature can reduce agricultural and fishing incomes. Countries dependent on coastal fishery and agriculture, which most often include developing countries, are likely to be adversely affected. If a one-meter sea level rise were to take place today, it would displace 7 million persons in India (ADB, 1995). In the future, many more may be displaced. 35 per cent of the land in Bangladesh would be submerged by a one-meter rise. The estimates for costs to build walls along the zones vulnerable to sea level rise for the USA is $107 billion in 1989 prices (Yohe, 1990). That may be a small share of the GDP of developed countries, but such measures, even scaling for their coast lines, for say, Bangladesh, could require a very large share of its GDP. Who shall pay Bangladesh or India for such a wall? Given that these countries are unlikely to be able to pay for protective measures, tens of millions of people will be displaced in Bangladesh and many of them could spill over into India.

**Risk of Extreme Events**

Increased occurrence of extreme events due to climate change will also affect the poor most. In the cyclone in Andhra Pradesh in India in 1996, more than 1,000 people died and there was huge property loss. Cyclones of similar intensity in advanced countries like the US may not lead to any deaths and much hardship, due to stable and durable housing and other infrastructure and extended safety net available to the people in distress.

**The Costs of Changing Energy Strategy**

If India has to reduce its carbon emissions, it would mean a major reorientation of her energy strategy, especially if that warranted a shift from its current coal-based to an oil and gas based energy system. Murthy, Panda and Parikh (2000) estimated costs associated with a low GHG energy strategy in terms of foregone income and welfare of the poor. They examined the consequences of alternative $CO_2$ emission reduction strategies on economic development and, in particular, the implications for the poor by empirically implementing an economy-wide model across India over a 35-year time horizon. A multisectoral, inter-temporal model is used for this purpose. The model has specific technological alternatives and endogenous income distribution with dynamic behaviour; it covers the whole economy in an integrated top-down-bottom-up model. The results show that $CO_2$ emission reduction imposes costs in terms of lower GDP and higher poverty. A 30 per cent $CO_2$ reduction over a period of 30 years using annual emissions reduction targets leads to a fall in GDP of 4 per cent and raises the number of poor by 17.5 per cent in the 30th year (that is, if 2000 were taken as the baseline, these changes would occur by 2030). Cumulative emission reduction targets are, however, preferable to annual reduction targets and an optimum strategy where the country is free to decide when to reduce emissions as long as over 30 years the same amount of cumulative emissions is reduced. It reduces the hardship of emission reductions with a cumulative reduction target of 30 per cent (i.e., not specifying annual targets), the fall in GDP is 1.4 per cent and increase in the number of poor is 6 per cent in the 30th year both of which are less than in the case of annual reduction. The scenarios involving compensation for the loss in welfare are not very encouraging, as they require large capital inflows. The required minimum compensation to maintain welfare levels is $278 billion for annual reduction and $87 billion for cumulative reduction respectively. The payments are large, and are concentrated in the early years of the period concerned. A more realistic strategy would call for compensation payments, which are spread out over the years even though it may require larger total inflow over the years. Contrasted with these, scenarios involving tradable emission quotas give India an incentive to be carbon efficient. It becomes a net seller for the first 25 years, say up to 2020, if the quota is in the range of 1 ton per person as per 1995 population. Because of the reduction in carbon intensity due to the incentives provided by tradable quota, less is emitted in later years when India becomes a net buyer. The results suggest that for India, and other developing countries, the window of opportunity to sell carbon quotas is only in the next two decades or so. Thus, precious time is lost during prolonged negotiations.

**Profile of India: Emissions from an Input-Output Model**

Parikh and Gokarn (1993) put in the first effort to identify India's emissions profile by using an input-output model. They aggregated the 1979-80 tables into 40 sectors. The Planning Commission updates these input-output coefficients at least every 10 years. Subsequently, the 1989-90 tables was available some time in the early nineties and was used by Murthy et. al., (1997a, 1997b) to study the structure of $CO_2$ emissions where one looks at the coal, oil, gas and electricity rows that cut across different sectoral columns multiplied with the level of activities. We studied emissions

- From different sectors: power, steel, transport, and so on.
- By different expenditure groups, rural, urban, low, middle and high income.
- For different purposes: different fuels (coal, oil and gas).

**Mitigating Climate Change and Pursuing Sustainable Development in India:**

India has for quite some time pursued GHG friendly policies in her own interest. India's obligation to minimise energy consumption - particularly oil consumption - and to deal with its environmental problems prompt it to follow many such policies. Directly or indirectly these efforts are made by Government as well as by people to reduce energy consumption. These include:

(a) Emphasis on energy conservation.
(b) Promotion of renewable energy sources.
(c) Abatement of air pollution.
(d) Afforestation and wasteland development.
(e) Economic reforms, subsidy removal and joint ventures in capital goods.
(f) Fuel substitution policies.

Some of these efforts are ongoing for several decades and are institutionalised in a number of ways through policies, programmes and the creation of specific institutions. These are government efforts; in addition, there are a number of measures taken by people themselves.

Some because of resource minimising cultural traditions as well as good practices that exist in India and some due to sheer poverty and deprivation. We discuss each of the above separately. While some of the energy savings are due to conscious resource utilisation practices in a positive sense, the dark side has to do with human drudgery and deprivation. These include 'compulsory (or perverse?) forced energy savings' by the poor due to deprivation. These range from:

(a) Lack of electricity connections and if connected, then a lack of electric appliances and even of adequate light bulbs in rural households.
(b) Lack of piped water or pumps that require long trips by women and children on foot to obtain surface water.
(c) Lack of even cooking fuels due to which the poor depend on biomass rather than clean and convenient fossil fuels.
(d) Lack of fans and heating devices for a large percentage of households that are necessary for comfort and productivity.
(e) Lack of basic infrastructure such as schools, hospitals, and roads that are essential elements for human development.

All of the above save energy at the cost of human welfare. Clearly, it is not recommended to continue the existing state of affairs. If India is committed to human development, poverty eradication should take place. This may result in an increased energy use. This may be considered a due right of the poor, even if it increases India's GHG emissions.

**Energy Efficiency and Conservation**

Energy conservation and increased efficiency is gradually taking an important place in the energy and industry sector plan in India. As was noted earlier, the power sector is responsible for the highest direct $CO_2$ emissions in India (42 per cent), followed by iron and steel, road, railways and air transport, and coal. (Parikh et al. 1993). Improving the efficiency of coal and electricity use could significantly reduce emissions from these sectors. Tangible results of efforts to increase efficiency can be seen (Nag, B. and J. Parikh, 2000) in declining energy intensity for industry and transport sectors where many factors jointly contribute to such a result. Unfortunately, there is a marginal economy-wide increase in commercial energy intensity due to the substitution of non-commercial energy by commercial energy Noncommercial energy is not accounted for in the conventional energy statistics which are used to derive carbon intensities. Non-commercial energy is largely used for cooking and unorganised industries (viz., fuelwood, crop residues and animal dung). Its share is now reduced by increased use of commercial energy, viz., kerosene and liquefied petroleum gas.

**Promotion of Renewable Energy**

Apart from energy conservation and efficiency improvements, the need to find, develop and exploit non-conventional energy sources, many of them clean and renewable, has long been recognised by the Government of India. Due to the importance given to the subject, the activity started in the fifties under the Ministry of Science and Technology, grew into a separate department under the ministry of energy and then became a full-fledged Ministry of Non-Conventional energy sources. An aggregate installed capacity of 2302 MW through various renewable energy sources, namely, wind farms, micro-hydroelectric plants, biomass and cogeneration power plants, biomass based gasifiers systems and solar photovoltaic systems was expected by the end of the eighth plan - 1992-1997. This is likely to reach to 6500 MW in the ninth plan - 1997-2000. By the end of the eighth plan about 2.6 million family size biogas plants (3M3 each) 25 million improved cooking stoves that use less fuel wood, about 390,000 [M2] solar thermal collector area and 430,000 solar cookers had been cumulatively installed. Expected energy generation and energy savings from these renewable energy systems come to about 26.3 million tons of wood equivalent, which amounts to a significant 12 to 15 per cent savings in total wood consumption.

**Abatement of Air Pollution**

Efficiency-enhancing measures in fossil fuel use may lead to the dual benefits of both local air pollution abatement and GHG emissions abatement. Pollution levels are prescribed for vehicles and have recently been made mandatory to reduce urban pollution. Therefore, the transport sector is gearing up to face the challenges of urban air pollution by providing for more energy efficiency in vehicles. In addition, air pollution regulations result in higher fuel quality, which in turn lead to reduced GHG emissions. For example, improved gasoline can make vehicles run more efficiently. It is difficult to say how much GHG can be saved due to such pollution control measures. Estimates indicate it to be in the range of 10 to 15 per cent up to some point: say 2015 by which time the current stock would be largely replaced by the new vehicles. Then future reduction will be significant only if still stricter norms are followed at a later stage.

**Afforestation and Wasteland Development**

It is estimated that land-use change in developing countries could contribute to global emissions to the extent of 1.6 billion tonnes of carbon. However, indicators from India show that India's share of this contribution is minimal. Biomass is widely used even today in India for a variety of purposes including fuel, timber and feedstock. Concern for the consequences — degradation of woods and forests and consequent degradation of soils — was expressed as far back as 1974 (Fuel Policy Committee, 1974). More recently, programmes for afforestation have found support from both governmental as well as non-governmental organisations.

These programmes aim not only to halt deforestation but to increase green cover. Various agencies spend as much as Rs.50 billion (US$1.25 billion) per year on wasteland development. If all the efforts at afforestation were to succeed, India's net emissions of $CO_2$ could come down significantly. In the nineties, the rate of deforestation has slowed considerably, if not marginally reversed as shown by recent satellite imagery (MOEF, 1995).

**Issues in Climate Change**

India should be an active and decisive partner, along with other developing countries, in climate change negotiations. We need to ask:

- What concentration levels, along with the associated risks, are acceptable to developing countries?
- How could it be ensured that the risks to the developing countries, and not just the costs to the developed countries, are minimised?
- If Annex-I countries postpone their commitments to reduce emissions, they use up limited carbon emission budgets available for future. How will it be available to the developing countries when they need it for their growth?
- How do we ensure that we have a fair share of the global environmental space?

The assumptions about greenhouse gas (GHG) concentration levels for stabilisation of atmospheric concentrations range from 450,550 and even 1,000 ppmv (for example, Wigley et al., 1996). It should be noted that according to the IPCC third assessment report, increase of $CO_2$ concentrations in this range could lead to an equilibrium warming of between 2.0°C to 4.8°C (IPCC, 2001). The IPCC report probably specified a specific $CO_2$ concentration increase with which the warming of 2 – 4.8 degrees was associated.

Each of these concentration levels permits different reduction strategies. Freedom to choose options is more limited in the case of 450 ppmv as carbon budgets are very low and therefore greater mitigation action is required in the nearer term. Integrated assessment models are currently being developed to consider these issues. Unfortunately, assumptions, premises and paradigms dictate the results of these models. In the view of the authors, often the developed countries' perspectives are hardwired into the models in such a manner that even if many scenarios are generated, the basic theme and results do not change (J. Parikh, 1992). For example, these models focus on minimising costs to the developed countries and not the risks to the developing countries. Yet, the decision about what is an acceptable level of climate change should centre around the risks to the developing countries. In fact, the developing countries should have a greater say as they are more vulnerable to the impact of climate change, they have a very small share in the cumulated emissions and thus have less responsibility for the problem of global warming; they are also poor and their emission trajectories are likely to rise due to development. In no other environmental issue are large polluters given opportunities to decide what cost and efforts are acceptable to them without full consideration of the vulnerability of the others. The level of effort needed to address an environmental issue is decided based on what is good for society. For climate change, this will depend upon what risks of climate change impacts are associated with different levels of emissions. We have already discussed in the previous section some of the types of risks and cost implications of climate change for India. India is a large country with wide ranging soil climate and other natural conditions. The results for India are thus likely to represent developing countries as a whole. Such risks to poor countries should be the primary focus of the climate change analysis, rather than costs to the developed countries. To this extent, a paradigm shift is necessary from the cost-minimisation to risk minimisation in the future analyses of IPCC.

Of course, the nature and extent of climate change and its impacts are uncertain. That, however, should not be grounds for inaction. We should find a way to deal with differing perspectives on uncertainty and risk. We offer a suggestion for this later.

**Emission Reduction Responsibilities**

What are the implications of the differentiated responsibilities accepted under FCCC? If we look at the factors driving emissions, we get an idea. Parikh J (1994) modification of the Kaya identity states:

**Discounting the Future**

Behind the suggested reduction time frame, there is an implicit discount rate. It is often argued that the future generations would be richer and hence we can pass on the burden of emission mitigation on them. Some even suggested that they have to be similar to those used in any other investment strategy. It is often argued by some Annex I countries that we should use a high discount rate in designing climate change mitigation strategies. In our view, this is not quite correct. Even after 50 years, Indian nationals are likely to be poorer than those of the OECD are today. Thus, by not taking actions now the burden is transferred not just to rich citizens of the OECD of the next generation but also to poorer Indians of tomorrow who would be poorer than today's citizens of OECD. A low discount rate is more appropriate when assessing optimal mitigation strategies. Emission mitigation by Annexure I countries is needed now.

**Delay is Free Riding**

Despite the commitments made at Rio, countries have taken little action to meet their commitments. At the Conference of Parties (COP) in Kyoto in 1997, too, and again delayed their commitments. Through delays, rich OECD countries are occupying global environmental space. During 1990 to 2020 (during which period they were supposed to act, have not acted and are not likely to act) OECD countries would have emitted more than India would emit in the next 30 years, assuming a 5 per cent increase in India's GHG emission every year. One can recognise that this delay is costing India and other developing countries opportunities to develop in the future. Through delay OECD, countries are further occupying global environmental space, and since Kyoto, they have asked for even more of a delay. Delaying Kyoto is really reneging on Rio.

**Delayed Action Occupies Global Environmental Space**

To discourage free riding during the negotiation period and beyond, we suggest that countries are accountable for their own emissions for a specific period, say after 1990 or 2000. That is, whatever decisions are arrived at, will be applicable retroactively from, say, 2000. That is, the clock starts ticking and all emissions are cumulated for each country even during negotiations. This way, negotiations will conclude faster and policy actions to reduce emissions will begin soon. Regardless of the outcome of the negotiations, these emissions will be shown against each country and that much less will be available to them in future. Thus, the countries taking actions in advance get their rewards and procrastinating countries will have to do more lately.

**Mitigation Costs and Benefits**

Reluctance to take action now implies a faith in technical progress to effectively deal with climate change later. This is a risky strategy which the poor and vulnerable will find hard to accept. Technical progress sometimes brings with it unanticipated consequences. When CFCs where introduced, they were hailed as a great technical innovation - but, as it known now, they had the side effect of leading to ozone depletion.

Thus, relying on technical solution alone can be risky. If dramatic technical progress does not take place, life style changes are inevitable if we want sustainable development.

**The CDM – A Step towards Equity?**

The Clean Development Mechanism (CDM) proposed in the Kyoto Protocol offers developing countries finance and technology by allowing countries to offset emissions through investing in emissions reduction in non-countries. Apart from the generally recognised problems of appropriate determination of the base line, India's concerns relate to getting fair compensation for sink projects, ensuring real transfer of technology and an uneasiness about selling 'low hanging fruits', i.e. the exploitation of cheap emissions reduction early on in the process by developed countries.

***Look Before you Leap – Sink Projects through CDM***

It is generally believed that the sink projects such as growing trees for afforestation and so on are some of the most attractive options. India has some 100 million hectares of wasteland and degraded forests on which such projects may be started. However, several major considerations may be important:

(a) The trees fix carbon only during the growing periods. After reaching maturity, they are carbon neutral. Thus, the carbon sink projects can create liabilities for the host country through committed land use.

(b) If at the end of maturity forests are removed, it may appear in the statistics of land use change of India. If the wood is burnt, will the $CO_2$ generated be the liability of India?

(c) If the forests are left intact, it may have implications if the opportunity costs of land become high in the meantime. However, only if the country had taken a careful long-term decision to create green cover on a permanent basis, may such projects be considered. In the cases where the forest is removed or burnt, the global environment does not benefit as it would have merely postponed the problem. The liabilities are reduced only if energy crops are grown that will replace fossil fuels, for example, for wood based methanol, or wood-based power generation. However, who claims the credit, the one who supported the plantation or the one who uses it to replace fossil fuels?

(d) Another difficulty relates to the measurement of carbon sequestered. This is not an easy task. All kinds of fudging are possible and there would be incentives to do so. One may also note that afforestation projects involve very little technology and hence very little technology transfer. Technology transfer is claimed to be a major advantage of the CDM.

***Technology Transfer (TT) and CDM***

TT and CDM should be linked to ensure wider adoption of environmentally beneficial technologies beyond the CDM project. India would like to see that a "CDM project" leads to real technology transfer giving the country the ability not only to operate the technology but also to replicate and innovate. Another concern of India is pricing of technology. There should be competition here. In a bilateral deal, the supplier of technology has monopoly power and the price charged for technology may be too high. In addition, projects such as sequestration projects do not involve technology transfer. One way to ensure that CDM projects involve technology transfer at competitive prices is to require that every CDM project, including sequestration projects, make a specific contribution to a technology acquisition fund with which the developing country is free to buy technology not necessarily related to the CDM project, from anywhere in the world. This can moderate excessively high charges for technology from a monopolist supplier.

### *Low Hanging Fruits and Pricing of Carbon*

When carbon is traded, what developing countries gain would depend on whether the market is competitive, whether futures markets exist, or whether the carbon is bilaterally traded in a project-by project basis, as is envisaged under CDM. Figure 10 illustrates who gains what from trading. India is concerned that in bilateral negotiations between project parties, Indian entrepreneurs might only look at their private gain and sell carbon at throwaway prices. Developing countries should resist such trade. A well functioning market along with a futures market is the best way to ensure a good price. Development of such a market will take time. Meanwhile, a global carbon price floor should be announced for emissions trading and not all developing countries should trade below this price. India may want to do so unilaterally for its own projects.

### Towards a Comprehensive Early Agreement

Disappointing delays in the implementation of the Kyoto Protocol lead us to suggest a system where all countries should be accountable for their cumulated emissions, say after 1990. When final negotiations are concluded, those countries that have taken early action will be rewarded and the others will have to do a lot more lately. We suggest the following:

- Despite the uncertainties surrounding climate change, the risks of potentially adverse impacts on the food system, coastal zones and increased occurrence of extreme events should be avoided by early action.
- Even during the negotiation period, an immediate decision to work from cumulative emissions for each country from a given year, say 1990 or 2000, should be taken. That is, whatever the final negotiated strategy; it will be applicable from the agreed reference year in the past to reward early actions by any country and perhaps to conclude negotiations faster. Delay to implement such a system only rewards current high emitters who do not take action to reduce emissions.
- Suppose we agree to limit climate change to 2°C of temperature increase. To give countries some leeway to deal with uncertainties

involved and their differing perspectives of risk, countries must be held accountable for the damages caused due to their cumulative emissions over the most pessimistic scenario (i.e., one that restricts the atmospheric $CO_2$ level to the lower value for such a temperature increase). However, a country may be permitted to emit up to their quotas as per the scenarios they consider reasonable. *Over time with research and better understanding of the global climate, system the uncertainty will reduce and estimates of* the range of emissions *required* to restrict warming to 2°C will narrow. Participating countries would be held responsible for the emissions that correspond to narrowed range. For greater flexibility, quotas should be leaseable. There are many desirable consequences of such a system. It will optimise response and reduce free-riding through delay. We have observed that the cost of delay in emission reduction (by the North) in terms of the South's foregone opportunities to development is substantial. This will impose many constraints on the way the South decides on policy options regarding issues such as how to generate power, how to use land, and what crops to grow and so on. Moreover, the South is highly vulnerable to the impact of climate change. Hence, unless the North acts now, North-South transfers of large amounts will be needed to compensate the South for the development opportunities foregone or for direct economic losses stemming from climate change.

- The risks to poor countries should be the primary focus of the climate change analysis, rather than costs to the developed countries. To this extent, a paradigm shift is necessary from the cost minimisation in the future analyses of IPCC.

## Conclusion

Our main arguments are as follows:

- India and other developing countries feel strongly that they are not responsible for the threat of climate change that has been created. Unsustainable consumption patterns of the rich industrialised nations in the world are responsible for it.
- Yet, India and other developing country economies may be highly vulnerable to climate change. India's food production would be adversely affected. Sea level rise would displace a large number of people. The developing countries are particularly vulnerable to the likely increase in the incidence of extreme events. The impacts of climate change could hinder development and delay progress in eradicating poverty, potentially aggravating social and environmental conditions in these countries

- An analysis of India's emissions shows that its per capita emission of carbon is one fourth of the global average. Even the top 10 per cent of urban population emits well below the global average per capita emission.
- India, and other similar types of developing countries, are making significant progress in limiting GHG emissions through normal policy developments such as those aiming to improve energy and economic efficiency of the energy and industrial production capacity, as well as energy development, both conventional and renewable, which target improved environmental quality and limit human health hazards from air pollution.
- India's energy intensity in industry and transport sector has come down. It has installed 2,300 MW of generating capacity based on various renewable. Deforestation is arrested and the vast potential of afforestation on wasteland is increasingly utilised.
- India and many developing countries have carried out price reforms and removed subsidies. These have resulted in substantial energy savings and reduction in emissions through greater use efficiency and fuel substitution.
- An equitable climate regime will focus on limiting the risks from climate change impacts to poor developing countries rather than on limiting the costs of mitigation per se. Options that improve economic efficiency of mitigation also need to address the distribution of economic costs associated with climate change. Such a system needs to be guided by a better understanding of the potential economic impacts and other risk to developing countries, which emanate from the climate change problem. One must also recognise the need for economic growth of developing countries.
- With differentiated responsibility, Annexure I countries have to take the lead. For a smooth transition, they should first stabilise their carbon emissions as soon as possible and then reduce them to sustainable levels over the coming decades. Emission of developing countries will need to grow even at increasing rates for some time. They would have to stabilise them somewhat later in the future and then reduce them.
- Unfortunately, Annexure I countries are delaying action. By their delay, they are occupying global environmental space and are free riding on developing countries. Compared to the carbon emission that OECD countries would have made had they followed the FCCC agreed on at Rio, they have emitted much more. In fact, even if they were to meet the Kyoto targets by 2012, the additional emission of OECD countries between 1992 and 2012, exceeds the emission

India is likely to make over 40 years assuming a 5 per cent growth rate of emission.

- If countries recognise the environmental, societal and ecosystem benefits of mitigation and value them properly, it would justify incurring large mitigation costs. We need to increase awareness of citizens.
- One promising option for organising mitigation over the long term is to hold countries accountable for all emissions from some fixed date in time, say 1990 or 2000. This provides incentives for early conclusion of negotiation.
- The CDM could be risky for developing countries because of perverse incentives to exaggerate valid credits (e.g., through exaggerated baselines) and because of likely imbalances in the power among the investor and host parties who will need to negotiate about important variables (e.g., type of project, baseline, credit sharing and possibly price or other terms of reference for the investor). Equitably to share the gains from CDM projects, we may start with fixing a global carbon price floor.
- A major attraction of the CDM for developing counties is technology transfer. However, carbon
- Sequestration projects do not involve any significant technology. In addition, the price at which monopolist suppliers provide technology may vary. We suggest a technology acquisition fund in which every CDM project, including sequestration project, is required to contribute to technology funds with which technology catering to specific needs of developing countries can be developed (form example, two wheelers transport with four stroke engines or certain cheap cooling equipment). Moreover, the developing country should be free to choose technology, not necessarily from the country that brings the CDM project, but from anywhere in the world.
- A more interesting option over the longer term could be to go to a fully-fledged emission trading system, which would increase the economic efficiency of long-term mitigation and, if emission quotas are allocated in an equitable way, begin to compensate developing countries for any costs that significant mitigation might impose on their developing economies.
- Many persons in India and other developing countries are concerned about selling off their cheap mitigation options (the 'low-hanging fruits'). One should weigh the price one gets today, the worth of such fruits in the future and the possibilities of their 'rotting' if unused.
- The need for an approach to mitigating that is the threat of climate change that is equitable and one that can accommodate differing

perspectives on risk need to be elaborated. To initiate action now even with differing perspectives of uncertainties and risks those different countries have, we suggest a scheme where a global trading system of carbon emissions with futures market is introduced. The allocations of quotas are made on an equitable basis. However, the total quota will depend upon each country's subjective trajectory that restricts global temperature change to a desired limit; say 2°C. Countries, however, are responsible for their cumulative emissions in carbon-ton-years that they have made and the range of permissible trajectories narrows as our knowledge and understanding improve.

In the context of the current debate about climate change, it is necessary to show that far from being inactive, the developing countries, especially India, are taking considerable actions in terms of policies, programmes and projects. Technology transfer can speed up the modernisation process and additional funds can accelerate Government initiatives in energy conservation. However, policies for poverty alleviation must take priority. It is shown that savings in GHG emissions by the poor should not be expected at the expense of development. Yet, other savings by developing countries can be increased by technology transfer, investment in better infrastructure, and efforts for modernisation, all of which require financial support. Encouragement to conservation and good practices would result in lower emissions. Far from free riding, low GHG emissions in developing countries have made it possible to sustain the high pattern of energy consumption by the industrialised countries for decades in the past, at present and in the future too.

## References

Asian Development Bank.1995.

Climate Change in Asia, Article by V. Asthana. Bolin B., (editor). 1995. *IPCC Guidelines for Estimating National Greenhouse Gas Inventories*, Vol. I, II and III, published by UNEP, OECD, IEA and IPCC, 1995.

Carbon Dioxide Information Analysis Center. 1993. Trends '91 *A Compendium of Data on Global Changes*, Oak Ridge National Laboratory, Oak Ridge, Tennessee.

Cline W.R. 1992. *Economics of Global Warming*, Institute of International Economics, Washington, DC.

Forest Survey of India. 1988. The state of forest report 1995. Forest Survey of India, Dehradun.

IPCC, (Intergovernmental Panel for Climate Change). 1992. *The First Assessment Report*, Cambridge University Press, Cambridge. IPCC, (Intergovernmental Panel on Climate Change), (1996). *The Science of Climate Change,*

*Vol. 1 of Climate Change. 1995*. IPCC second assessment report, Cambridge University Press, Cambridge.

Kumar, K.S. Kavi and Jyoti Parikh, (1997) "Potential impacts of global climate change on Indian agriculture", presented at the workshop, Measuring the impacts of Climate Change on Indian and Brazilian agriculture, held at the World Bank, Washington D.C., 5-7 May.

Kumar, K.S. Kavi and Jyoti Parikh, (1998) Climate change impacts on Indian agriculture: The Ricardian approach". in Dinar et.al. Measuring the impacts of Climate Change on Indian Agriculture, World Bank Technical Paper No. 402, 1998.

Kumar, K.S.Kavi, Parikh, J., 2001a. 'Socio-economic Impacts of Climate Change on Indian Agriculture', *International Review for Environmental Strategies*, 2(2).

Kumar, K.S.Kavi, Parikh, J., 2001b. 'Indian Agriculture and Climate Sensitivity', *Global Environmental Change*, 11, pp. 147-154.

Manne, A.S. and Richels, R.G., (1993) *Buying Greenhouse Insurance*, Cambridge, Mass: M.I.T. Press.

Mitra A.P. (Ed.), (1992) *Global Change: Greenhouse Gas Emissions in India - 1991 Methane Campaign*, Scientific Report No. 2, NPL, Publication and Information Directorate, CSIR, New Delhi, India, June 1992.

Mitra A.P., (Ed.). 1996. *Global Change Greenhouse Gas Emission in India - Methane Budget Estimates from Rice Fields based on Data available upto 1995*, Scientific Report No.10, Publication and Information Directorate, CSIR, New Delhi, India, July 1996.

Murthy, N. S., Manoj Panda and Jyoti Parikh. 1997a. 'Economic Development, Poverty Reduction and Carbon Emissions in India', *Energy Economics*, Vol.19, No. 3. Murthy, N. S., Manoj Panda and Jyoti Parikh. 1997b. 'Economic Growth, Energy Demand and Carbon Dioxide Emissions in India: 1990-2020', *Environment and Development Economics*, Vol. 2, No. 2 (forthcoming).

Murthy N.S., Manoj Panda and Kirit Parikh. 2000. 'CO2 Emissions Reduction Strategies and Economic Development of India', IGIDR Discussion paper. Nag, Barnali and Jyoti Parikh (2000) Indicators of Carbon Emission Intensity from Commercial Energy Use in India, *Energy Economics*, Vol. 22, pp. 441-61.

Painuly J.P. (2000) "Kyoto Protocol and Voluntary Commitments: What are the Risks to Developing Countries" Invited Paper, International Symposium on Development Policies for the New Millennium in Honour of Professor Kirit S. Parikh, IGIDR, July 12-14, 2000.

Parikh Jyoti, and Gokarn S. (1992)"Climate Change and India's Energy Policy Options:New Perspectives on Sectoral CO2 Emissionsand Incremental Costs", Global Environmental Change, Sept. 1993.

Parikh Jyoti and Gokarn S., (1993) Climate Change and India's Energy Policy Options, *Global Environmental Change*, Vol. 3(3), 276-291. Parikh Jyoti

(1992) "IPCC Response Strategies Unfair to the South", *Nature*, Vol. 360, pp. 507-508, 10th December.

Parikh Jyoti, (1993) "Joint Implementation and Sharing Commitments: A Southern Perspective". *Proceedings of a workshop held on Oct. 16-18, 1994 on Integrated Assessment for Climate Change* at International Institute for Applied Systems Analysis (IIASA), Laxenburg, Austria.

Parikh Jyoti, (1994) "North-South Issues for Climate Change", *Economic and Political Weekly,* pp. 2940-2943, November 5-12.

Parikh Jyoti, (1998) "Linking Technology Transfer with Clean Development Mechanism (CDM): A Developing Country Perspective", Presented at Columbia University.

Parikh Jyoti, Roy Culpeper and Davis Runnalls, J.P. Painuly (Eds.), (1997) *Climate Change and North-South Cooperation,* Tata McGraw Hills Publishing Co. Ltd. New Delhi.

PCRA Report (1994), A Pursuit with a Purpose, Petroleum Conservation and Research Association, New Delhi.

Reddy B.S. and Jyoti K. Parikh, (1997) "Economic and Environmental Impacts of Demand Side Management Programmes", *Energy Policy*, Vol. 25, No. 3.

Report of Task Force on Wastelands Development in the IX Five Year Plan (May 1996)

Wigley, T., R. Richels and J.Edmonds, (1996) "Economic and environmental choices in the stabilisation of atmospheric $CO_2$ concentrations", Nature, 379, pp. 240-243.

Yohe, G., (1990) "The Cost of Not Holding Back the Sea", Coastal Management, 18, pp. 403-431.

# 8

# India's Role in South-South Cooperation: A Case of Climate Change Negotiations

*Dhanshree Jayaram*

## Introduction

The human-induced process of environmental change that was set off way back during the era of civilisation building took a perilous turn in the twentieth century due to the industrialisation drive of the Western countries; it is now gradually worsening as the international community awaits a consensual deal. Climate change is one such threat and among all environmental threats, this is one phenomenon, which has gathered maximum amount of attention. One could say that it is not the definition of climate change that needs to be looked into; it is the definition of 'response to climate change' that needs to be transformed so that the 'measures' bear fruits. Geopolitically, climate change is an issue that encompasses all facets of international politics that steer a nation's march towards achieving national comprehensive power. Ironically, the international negotiations themselves have become synonymous with 'conflict' rather than 'cooperation'. The first step towards resolving the conflicts would be re-standardisation of international relations that would create political willingness based on 'complex interdependence'. South-South cooperation could be one of the pillars of such an order as attempts to bridge the differences between the South and the North have failed repeatedly.

South-South cooperation is nothing but the exchange of resources, technology and knowledge between the developing countries. A clear understanding of the evolution of both South-South as well as

North-South relations at the global climate change negotiations since the inception of the United Nations Framework Convention on Climate Change (UNFCCC) in 1992, especially at Copenhagen (2009), Cancun (2010) and Durban (2011) becomes crucial. The major fault-lines guided by multitude of factors including 'common but differentiated responsibility', bottom-up approach versus top-down approach and climate justice among others need to be dissected. These factors have played a decisive role in the formulation of India's perceptions and policies of climate change as well. It also seeks to probe the shift in global climate change in politics with the rise of the BASIC group (Brazil, South Africa, India and China) particularly the Asian giants and the emergence of G-2 (China and the United States (US); the subsequent assertion of the Least Developed Countries (LDCs), the Small Island Developing States (SIDS) and the Alliance of Small Island States (AOSIS) and the split in G-77 (a coalition of developing countries) are some of the other developments that require analysis. In the end, the effort would be to get out of the vicious web formed by the negotiations and look at the issue from a fresh perspective. India could be the pioneer of change in terms of environmental management.

**UNFCCC: A Divided World**

It was at the United Nations Conference on Human Environment (Stockholm Summit) in 1972 that the first UN Summit on environment was held. At this multilateral forum, the link between environment and development was established stressing on the importance of sustainable development; international agreements on ocean dumping, pollution from ships and endangered species trade were also finalised. At this conference, Indira Gandhi, the then Prime Minister of India famously proclaimed, 'Every palace one sees in India is a demonstration, not of her riches, but of the insolence of power that riches give to the few, who owe them to the miserably requited labours of the millions of paupers of India' (Ramakrishnan, 2011). At the First Climate Conference at Geneva, scientists from varied disciplines finally concurred that a wide range of human activities could indeed contribute to global warming, which eventually led to the establishment of the World Climate Programme and World Climate Research Programme in the form of the Intergovernmental Panel on Climate Change (IPCC). The scientific consensus gained momentum during the 1980s and 1990s at the Villiach Conference (1985), the Toronto Conference (1988) and the Second World Climate Conference (1990) that called for policy response including a 20 per cent cut in carbon emissions by 2005.

All these developments led to the agreement in the form of the UNFCCC at the 1992 Rio Summit (United Nations Conference on Environment and Development). Held from 3 to 14 June, 1992, the Earth

Summit discussed the stabilisation of greenhouse gas (GHG) concentrations in the atmosphere, but the treaty was legally non-binding. In addition, the parties to the UNFCCC were classified as Annex I countries (industrialised countries that have an obligation to limit anthropogenic emissions of GHGs), Annexure II countries (developed countries that have agreed to pay for the costs of developing countries) and Developing countries (that can volunteer to become Annexure I countries under the precondition of achievement of significant levels of development). In conjunction with this classification, the developed countries were assigned a 'voluntary goal of returning GHG emissions to 1990 levels' while the developing countries under the auspices of 'common but differentiated responsibilities' were entitled to receive financial and technological resources from developed countries to adapt to climate change (Carter, 2007: 251). Another major highlight of the Summit was the creation of a fund under the aegis of the Global Environment Facility (GEF). This Convention was initially signed by 155 countries and the European Union (EU) and entered into force on 21 March, 1994. Currently it has 194 parties—193 countries and one regional organisation. The classification of countries sowed the seeds of divisions within the UNFCCC, with some countries obligated to do much more than the rest owing to their greater share of responsibility for the problem. The objective of the conference was to bring the world together to address a global threat; instead, it initiated an eternal clash of interests between nation states.

Before coming to South-South cooperation, it is important to throw light on the various interrelated fault-lines that have had detrimental impact on the intergovernmental negotiations in general and South-South cooperation in specific. The first and the most important conflict is the one that exists between what the postcolonial experts call, the North and the South, or the industrialised and the non-industrialised countries. India is one of the prominent members of the South bloc. The two groups differ mainly on the issues of sovereignty and the 'common but differentiated responsibility.' 'The developing countries' sovereign right to use their own natural resources was diluted to some extent by including this principle in the Preamble and not in the Principles section. As Mathew Paterson points out, the North felt that 'putting clauses in the Principles section would create precedents and potentially create new general principles of international law' (Paterson, 1996: 74). The origin of the dispute over 'climate justice' lies in this fundamental divergence from the Declaration of the Stockholm Conference on the Human Environment which included this principle as an article but since it was just a declaration, it has no legal repercussions. Climate justice has always been widely debated by the developed and the developing countries and during one of the summits (at Copenhagen) one of India's senior negotiators argued that 'every human being has an equal right to resources of the atmosphere; therefore you have to take the per capita approach taking also into account the historical emissions' (Press Trust of India, 2011). On the second issue, while the developed countries endorse sharing of burden of reducing emissions, the developing countries accuse the developed countries of using climate change as a plain excuse to prevent the former from achieving developmental levels of the latter after causing enough havoc through uninhibited emissions. Though China is currently the largest emitter of GHGs, its per capita emissions are still far lower than most of the developed countries due to its high population. Therefore, questions over per capita emissions and population have always created further divisions between the industrialised and the emerging countries.

Another point of difference between the North and the South has been the debate between the top-down approach and the bottom-up approach in terms of emissions reductions strategies; while the North largely favours the former, the South has been an advocate of the latter. However, there are differences within the two blocs as well. In 'top-down approach,' 'actions are determined by the U.N. itself and nations do not have the freedom or flexibility to define their emissions reductions commitments' (although they can decide how to implement those). In 'bottom-up approach,' 'the focus is on nationally-defined measures, and nations are allowed to define commitments and actions, and make pledges unilaterally' (Caldas, 2011). Among the countries of the North, Japan supports the sectoral approach (bottom-up) to tackle climate change and the US espouses another kind of bottom-up approach by proposing National Strategies instead of the 'targets and timetables' approach backed by the EU. Among the countries of the South, the consensus has been by and large to adopt the portfolio approach in which the Kyoto-based approach for the existing parties, a hybrid of top-down and bottom-up approaches for the non-Annex I parties and pure bottom-up approach based on only national actions for the LDCs would be applied (Dubash and Rajamani, 2011: 596). The countries in the North, mostly the EU, argue that top-down approach would be required due to the urgency of the problem, the need for transparency at the international level and cost-effectiveness. The developing countries have in turn been contending that top-down approach could not only take away the attention of the policy-makers from adaptation, which is a grave concern for majority of the countries in the world currently; it could also lead to compromises and the result would always be the lowest common denominator. Generally, the Kyoto Protocol is regarded as a paradigm of top-down approach whereas the Copenhagen and Cancun agreements are heavily tilted towards the bottom-up approach.

However, the most pressing issue between the two groups of countries has been financial resources and technology transfer. The developed countries argue that the developing countries should not commit the same mistake that they did while industrialising themselves. This argument stands ground only if the developed countries agree to share their knowledge resource with the developing countries since they have more financial and technological resources and knowhow than the latter. It is true that the developing countries are obliged only to "establish national inventories of their GHG emissions sources and sinks;" at the same time one also has to keep in mind that they have "little incentive from the convention to limit their emissions growth" (Paterson, 1996: 76). One of the mediums of financial transfer, the GEF itself has been surrounded by controversy since its inception as the developing countries object to the spending being dictated by the voting procedures of the World Bank and not the Conference of Parties (COP). While the developing countries have overtly termed the climate change negotiations as an 'economic conference' instead of an 'environment conference,' developed countries particularly the US has time and again insisted on using the term 'technology cooperation' instead of 'technology transfer' (Paterson, 1996: 83-84).

Some of the other fault-lines, which have dominated the discourse of the climate change negotiations and have jeopardised South-South cooperation along with India's leadership of the South bloc are worth mentioning. One among them is the split between the major energy producers and others. Among others are the most vulnerable countries such as the LDCs and the SIDS. Countries, which are dependent on energy exports over the years, fear the negative consequences of emissions cuts on their economies. Organisation of Petroleum Exporting Countries (OPEC), especially the ones in the Middle East is a classic example. The OPEC, SIDS and LDCs are part of the G-77, yet they are divided on their positions on climate change; this has further divided the South. The division between those countries which are more vulnerable to the direct outcomes of climate change such as sea-level rise or erratic rainfall patterns and those which are relatively less vulnerable based on their degree of exposure to climate change as well as their economic resources to cope with the disasters – characterised by the labels of 'climate-ready' and 'climate-not-ready' respectively – is the next most prominent fault-line. For these reasons, Maldives, Vanuatu and Bangladesh have been more strident in their demand for a stronger global convention. Among the developing countries, the AOSIS and some of the LDCs would like to securitise the issue in order to force the rest of the world to adopt immediate and substantial cuts in the emissions of GHGs instead of mere stabiliation targets; while the Newly Industrialised Countries of East and Southeast Asia as well as the emerging economies such as India and Brazil would like to restrict the ambit of discussion to development and access to technology (Paterson, 1996: 85). The securitisation debate has been mostly driven by the developed countries with firm backing of the AOSIS and the LDCs even in the United Nations Security Council (UNSC).

**India's Role in the Creation of the Global Climate Regime**

Even before the negotiations started in 1992, the former Prime Minister, Rajiv Gandhi had openly tried to galvanise the support of the South to block international schema guided by the West to protect the environment. He argued that unless and until the developing countries are provided environmentally friendly technologies at affordable costs, they should not stop growing economically. Interestingly, Dr Rajendra Pachauri presented one of Rajiv Gandhi's ideas – that of a Planet Protection Fund (PPF) – at an international meeting to facilitate this (Rajan, 1997: 97). Similarly, India was one of the first countries that recommended the international community to bring the GEF under the Conference of Parties (COP). Another Indian proposal was to introduce the concept of per capita equity by which 'an equitable solution' can be found on the basis of 'significant reductions in levels of per capita emissions in developed countries, so that over a period of years these converge with rising per capita emissions in developing countries' (Rajan, 1997: 123) as enunciated by Chandrashekhar Dasgupta, one of the most seasoned climate negotiators for India for nearly 15 years. The efforts made by India to lead the developing world and acquire a favourable framework convention before the Earth Summit in 1992 were thwarted by the rest of the developing countries including China. The developed countries and even other developing countries struck down many ideas put forth by India due to lack of consensus among the latter. The developed countries had no intentions to assist the developing countries even long before countries such as China and India reached a position in which they could offer some amount of competition to the former. However, concerted efforts on the part of the latter to put pressure on the former would have produced some positive results for the latter. Instead, the developing countries were divided among themselves on the basis of their interests, whether it was in the form of founding of the AOSIS on the basis of the impact of sea-level rise on small island states or the BASIC on the basis of the developmental agenda of the countries that constitute it. The AOSIS came together primarily because their interests lie in addressing the effects of climate change and not the costs of strategies, which were of a bigger concern for India and China (Rajan, 1997: 98). Therefore, they supported the idea of international negotiations through which they would receive more visibility; it would also provide them a platform to cooperate with the industrialised countries.

Navroz K. Dubash's terminologies – 'Growth First Stonewallers', 'Progressive Realists', and 'Progressive Internationalists' – are fundamentally driven by their mottos – 'It's our turn!,' 'It's an unfair world!,' and 'Seize the moment!' respectively (Dubash, 2009: 9). These mottos are the propellers of our strategy at the climate change negotiations. On the grounds of sovereignty, equity and development, India has asserted its position that it would not take on the burden of international obligations and that the developed countries should be ready to provide financial and technological assistance to the developing countries. India has also strongly criticised the domination of bodies such as the IPCC by the developed countries (Rajan, 1997: 109); the inability of the developing countries to participate in the functioning of the IPCC in a holistic manner arose due to paucity of resources of all types – technological, financial, institutional and human – that are the prerequisites for participation in the IPCC. Ironically, it is an Indian (Dr Pachauri) who leads the IPCC team presently.

**The Chronicles of the Kyoto Protocol and the Bali Action Plan**

The Earth Summit was followed by COP-1 in Berlin in 1995. The first five years of the climate change negotiations saw active participation of and cooperation between the South bloc countries. This is one of the reasons why the 'Berlin mandate' recognise the necessity for the implementation of stronger commitments through a feasible protocol. COP-2 held in Geneva in 1996 that called for legally binding mid-term targets generated a momentum for the enforcement of a protocol, which materialised at COP-3 held in Kyoto, Japan in the form of Kyoto Protocol after ten days of rigorous negotiations. The Kyoto Protocol was one of the biggest achievements of South-South cooperation. India played a major role in not only galvanising the voices of the developing countries but also drafting the agreement that adheres to the 'common but differentiated responsibility.' It must be remembered that during the initial years of the negotiations, both India and China claimed to be the leader of the South bloc. However, always India led from the front verbally while China did it in a covert fashion. The Kyoto Protocol imposed legally binding overall reduction in GHG emissions of 5.2 per cent below the 1990 levels throughout the period 2008-12 (Carter, 2007: 253). The Kyoto Protocol also paved the way for an international emissions trading regime allowing industrialied countries to buy and sell emission credits amongst themselves; a Joint Implementation (JI) procedure enabling industrialised countries to implement projects that reduce emissions or remove carbon in another Annexure I country in exchange for emission reduction credits. And the Clean Development Mechanism (CDM) permitting developed countries to finance emissions-reduction projects in developing countries and receive credit for doing so (United Nations Framework Convention on Climate

Change). One hundred seventy-eight countries reached a binding agreement, though the US has still not ratified the protocol despite agreeing to a 7 per cent reduction target at Kyoto after it successfully bargained for a tradable permit system.

***The Post-Kyoto Picture***

Until 2001, the negotiations focussed mainly on the industrialised countries' strategies to reduce emissions. The Marrakesh Accords (2001) that came into force in 2005 were rejected by the US and even countries like Australia, which were rooting for carbon sinks in forest and agriculture as well as unrestrained emissions trading. In fact, in 2000, the split between the energy-rich and the imported energy-dependent countries had led to the suspension of talks. Not only did the Middle-Eastern/West Asian oil giants resist cuts, the parties also failed to reach consensus on the financial mechanism as well as consequences of non-compliance by countries. The differences between countries such as the US, Australia, Japan, Canada etc on the one side Europe on the other took the backseat thereafter. Post-Marrakesh talks have been largely dominated by the developed versus developing country positions debate. At Montreal (COP-11), the dialogue on the two-track approach for future was initiated – first, to negotiate the second period commitments for the developed countries under the Kyoto Protocol implemented by the Ad Hoc Working Group-Kyoto Protocol (AWG-KP); second, to discuss the adoption of new commitments by the Annex I parties along with Nationally Appropriate Mitigation Actions (NAMAs) by the developing countries under the Convention implemented by the Ad Hoc Working Group-Long-Term Cooperative Action (AWG-LCA). The two-track process fructified at COP13 after the dialogue ended. Held in 2007 at Bali (Indonesia), COP13 laid down the Bali Action Plan for a post-2012 climate change regime, that is, the Convention track. This track was supposed to conclude in Copenhagen with an 'agreed outcome'; instead, it could only generate an outcome, which could be dubbed a third kind of instrument – a high-level political instrument. The post-Kyoto period was marked by intra-South conflict. Except in the case of the Bali Action Plan South failed to speak with a common voice with tensions and conflicting interests between the emerging economies and the LDCs becoming sharper year after year.

**The Ultimate Political Showdown at Copenhagen and Cancun**

COP-15 took place in Copenhagen in 2009, in which the negotiations embodied an international system that some would say was tilted equally to the developing countries and their demands. This is the summit, which reinvigorated the concept of South-South cooperation after the drafting of the Kyoto Protocol in an unprecedented manner. India's role also came to the fore as it managed to bring at least the three emerging countries – China, South Africa and Brazil – together to thwart the attempts of the North to impose a legally binding agreement on them. However, the other countries in the South bloc constituted by G-77 were alienated. Therefore, South-South cooperation was compartmentalised wherein the emerging countries formed one group, BASIC and cooperated fruitfully, while the rest joined hands to pressurize both the developed and the emerging countries. The purpose of the Summit was to agree upon a mechanism for the period from 2012 when the first commitment period under the Kyoto Protocol expires. The voices of the developing countries were heard as much as that of the industrialised countries; this could well be the reason why instead of coming up with a "post-Kyoto" mechanism, it could only produce a 'political agreement' that had no legal subtexts Moreover, the 13-paragraph Copenhagen Accord is considered an external document as it was not negotiated within the UNFCCC process. The Copenhagen Accord was drafted by the BASIC countries as well as the US that recognised climate change as one of the greatest challenges of the present day and recommended actions to keep temperature increase below "2°C". It confronted enough criticism for not being legally binding, for not setting real targets to achieve in emissions reductions and for the scepticism over the mobilisation of US$100 billion every year by 2020 by the developed countries.

***Reactions and Shift in Relations***

The compartmentalisation of South-South cooperation could be seen in the way different groupings reacted to the developments. Several delegations including that of Tuvalu, Venezuela (who were represented by their respective Heads of State) and Sudan (Chair of G 77) threatened to walk out if the agreement did not mention legally binding emission cuts and temperature rise to well below 1.5°C. The summit was in reality reduced to a '25-member mini summit' with around 100 world leaders mostly belonging to African, Latin American and Asian under-developed and developing countries sitting in the plenary chamber next door believing that 25 rational parties would produce some sort of an encompassing document – plainly highlighting the almost trifling role played by those countries in the negotiations whose stakes are the highest in the evolving environmental and geopolitical scenario. Lumumba Stanislaus Di-Aping, the Sudanese negotiator said, "This is asking Africa to sign a suicide pact, an incineration pact in order to maintain the economic dependence of a few countries. It is a solution based on values that funnelled six million people in Europe into furnaces (Batty, 2009).

The failure of the Copenhagen Summit was largely attributed to global recession and conservative domestic pressure in the US and China.

The redundancy of Kyoto Protocol took the centre stage of discussions as the developed countries no longer wished to acknowledge the legal distinction between the developed and developing countries especially at a time when the developing countries were growing at a fast pace, largely unaffected by the economic recession of 2008 while the developed countries were reeling under the hardships caused by it. In such a situation, in which China and even India has emerged as competitors economically, the possibility of partnership between them was grim, especially since the developed countries would not be ready to provide any form of financial assistance to India and China. The industrialised countries did not approve of the two-track process and instead stressed on the need for a single binding agreement that would replace Kyoto under which both the developed and the developing countries would have to share the responsibility for the emissions cuts. Developing countries still believed in the developed countries' supreme responsibility for both causing climate change and tackling it.

China's international environment policy has been formulated on two grounds – firstly, non-compromise on economic development and secondly, avoidance of high-cost commitments (Jeon and Yoon, 2006: 847). China's role in the climate change negotiations is paramount as the US and China account for over one-third of the GHG emissions of the world. With the Copenhagen Summit, the concept of G-2 evolved that places the US and China at the forefront of global decision-making. One could say that this is a sign of the emergence of a truly multi-polar world as China is exceedingly making its presence felt all across the globe not just economically but also politically. The other stakeholders are Europe, Russia and India. In climate negotiations, the AOSIS and G -77 are equally important players as their pressure drives the industrialised countries as well as the emerging economies into taking at least the lowest common denominator. A section of the strategic thinkers saw the developments at the Copenhagen Summit as something which visibly displayed the emergence of a new political order, 'one shaped by the new self-confidence of the Asians and the powerlessness of the West' (Paskal and Savitt, 2009). India sent out a strong message to the West that it could no longer be dictated by the financial mechanisms designed by the West aimed at granting the responsibility of the administration of the Green Fund to institutions like the World Bank or International Monetary Fund (IMF) which continue to be governed by the financial contribution of each country and thus biased towards the West for obvious reasons. It also raised the issue of 'climate justice' in response to the demands of the developed countries to make the developing countries' domestic commitments legally binding and verifiable. China, on the other hand had multiple agendas – to engage in the 'climate market' negotiations as it is regarded as the biggest beneficiary of the CDM, as well as to reiterate its indisputable role of leadership in the developing world. Both the Indian and the Chinese contingents also asserted that the primacy of the UN process centred on the two-track dialogue should not be bypassed in order to include a third track based on the Copenhagen Accord that had not been endorsed by majority of the countries.

At the outset, one could haul out two major shifts in geopolitics and international relations – the ascendancy of the BASIC and the marginalisation of the West, especially Europe. The BASIC countries exhibited their disapproval a number of times during the Summit due to the refusal of the industrialised countries to dedicate enough attention to the second commitment period of the Kyoto Protocol. They also broke ranks with G-77 in order to broker an agreement. In effect, these countries not only demonstrated that since they are making a transition from a developing to a developed country, they are required to be in both camps to strike a compromise formula. G-77 itself got divided into various fragments as the Green Fund clearly draws a distinction between the emerging economies and the LDCs and the SIDS. At Copenhagen, the US had hardly any role to play initially as Shyam Saran says. With mere voluntary targets of the scale of 17 per cent emissions cuts by 2020 that are not subject to international compliance despite huge technological, technical and financial resources, US has indeed set a very low standard despite continuing to be one of the largest emitters in terms of both aggregate and per capita emissions. Ideally, the EU and the developing countries irrespective of their levels of development should have joined hands, but the EU squandered the excellent opportunity in response to the emerging economies' reservations about the Measuring, Reporting and Verification (MRV) and possible emissions cuts (Saran, 2011). The rigidity of the EU was counteracted by the flexibility of the BASIC, which made a few compromises that have already been reflected upon earlier in the paper. The leadership of the BASIC was the biggest highlight of the Summit. While talking about cooperation between the BASIC countries, one should also keep in mind that China is far ahead of the rest, especially India, in terms of 'total GDP, annual rate of growth in GDP, carbon intensity of economy, aggregate and per capita emissions, rate of growth in emissions and the contribution its emissions make towards the current rate of build-up of $CO_2$ in the atmosphere' (Jacob, 2010: 439). Therefore, close engagement with China on this issue could prove to be counterproductive for India.

There was no significant change in India's position and perception with respect to climate change until the Copenhagen Summit. At Copenhagen, India had to modify its stand and state its global climate change policy according to changing geopolitical scenario. Shyam Saran, who was Prime Minister's Envoy of Climate Change at the summit, outlines India's red lines when it comes to its stand on climate change at the negotiations level. First, the global climate regime should not be constrained and that it should take into consideration growth prospects, eradication of poverty, current reliance on fossil fuels, the costs involved in the transition to cleaner fuels and the need for technology transfer and financing. Second, mitigation of climate change has to be carried out in India's interests but it is contingent largely on a supportive global regime. Therefore, it is in India's interests to negotiate. Third, the principle of equity would be supplemented by three mandates – implementation of voluntary measures to reduce carbon intensity of growth whereas absolute reductions by the developed countries; mainstreaming of poverty eradication; and a two-track dialogue (Long-Term Cooperative Action and Kyoto Protocol) so that the Kyoto Protocol would not be abandoned. Fourth, the question of where is the money going to come from has to be addressed with seriousness. India had to opt out of the major recipients of fast-track finances so that the LDCs and the SIDS would get the requisite resources for both adaptation and mitigation. Fifth, the per capita principle should be applied by which India would not have to adopt any legal obligation. India is ready to take concrete actions and is ready to register them with the UNFCCC. International MRV for internationally aided actions is acceptable since India's actions are transparent. Last, India should focus on adaptation more because the impacts of climate change would not be curtailed even if emissions become nil owing to the accumulation impact. Development would contribute to the capacity to adapt (Saran, 2011).

### *As the Ice Floated to Cancun*

After a much-hyped Copenhagen Summit that produced hardly any tangible results, the COP16 at Cancun was deemed a success by majority of the analysts. Many steps envisioned in the Copenhagen Accord were reformed and formalised through an official UN Agreement. At the same time, the Accord has its shortcomings. Majority of analysts consider it yet another lowest common denominator that showcases compromises and confusions regarding the nitty-gritty and implementation; while others perceive it as a good foundation for moving forward. The Agreement provides emissions cuts targets for approximately 80 countries including China and India. This could clearly be earmarked as a country of India.

### *Yet another Game of World Politics*

At the Cancun Summit, South-South cooperation broke down completely. Former Minister of Environment and Forests, Jairam Ramesh, raised a few eyebrows when he said in his speech at the high level segment that 'all countries must take on binding commitments in an appropriate legal forum', which was contrary to India's earlier position. However, later he clarified that his attempt was 'to walk the thin line between safeguarding our position while showing a level of sensitivity to the view shared by the majority of countries at Cancun, including many of our developing country partners.' He also made a distinction between an 'appropriate legal forum' and a legally binding agreement (The Hindu, 2010). The BASIC partners, Brazil and South Africa switched their sides to the bloc of the AOSIS, LDCs, Africa and four of the SAARC partners (Bangladesh, Maldives, Nepal and Bhutan) that insisted on a legally binding agreement. At the same time, the Indian establishment claims that it has safeguarded its interests by successfully convincing the other parties to delete the clause, which wanted global GHG emissions to be reduced by half by 2050, which would have put the burden of fixed emissions cuts on developing countries like India as well. This would have curtailed India's right to equitable sustainable development. A document prepared by the Indian Government allegedly reveals that its push for easing Intellectual Property Rights (IPR) control on green technology was ignored. Somewhere down the line, the Indian Government feels that turning the Cancun Accords into firm actions by the end of 2011 might hurt the interests of the country. Earlier, Prime Minister Manmohan Singh had formally signed a declaration at Major Economies Forum in 2009 along with G-8 countries, which purportedly violated India's earlier position on 'equal per capita rights for all countries to the global common carbon space in the atmosphere' (Sethi, 2011). Therefore, India is said to have made a series of compromises in order to be part of the 'coalition of the willing', which was essentially a handiwork of the US meant to rope in as many countries as possible on its side to endanger the future of the Kyoto Protocol.

Clearly, there was a shift in the geopolitical scenario that was reflected at the Cancun Summit. The US and the EU, along with Japan, Australia and Canada had managed to keep the Kyoto Protocol being uncertain, while persuading the rest of the world to sign an agreement after the previous year's debacle. The marginalisation of the BASIC was imminent since the LDCs and the SIDS (including the AOSIS) vigorously demanded a formidable agreement that entails actions from both developed countries and the emerging economies. Countries like Brazil have always had a flexible stand on the issue of legally binding emissions reductions as long as no serious constraints are imposed on them. South Africa decided to go with the rest of the African continent that called for stronger commitments from the entire world in the wake of recurrent droughts. China has embarked on the path of a green growth in a full-fledged manner and it is one of the major hubs of the global carbon market. India was perhaps left with no choice to make heavy concessions on its earlier stand so that it is not isolated on an issue that is of concern for the whole globe. It is also guided by the domestic factors, which in turn are driven by internal contradictions. India's quality of environmental governance is yet to reach an appreciable state and considering that it is next to impossible that its emissions could peak before 2020 due to increasing energy demands, India was caught in a sticky situation. As far as the LDCs and the SIDS are concerned, they are exceedingly dependent on other countries – both developed and developing – for their adaptation and mitigation measures. It was at the Cancun Summit that these countries played the be-all-and-end-all role in ensuring that the BASIC did not prevail upon the rest of the international community. By creating fissures within the BASIC and G-77, they also ensured that the roles of EU and the US were enhanced.

## The Road to Durban (COP-17) and Future Negotiations: Recommendations and Growing Uncertainties

### *Technology is the Buzzword*

Though it was at Bali that the need for technology transfer was given the final push, it was the Cancun Agreement that made formal provisions for a Technology Executive Committee, a Climate Technology Centre and a Climate Technology Network. Movement of technologies from the developed to the developing countries has been obstructed by the former for a long time for various reasons including fear of losing hegemony in the technological market because of reproduction of technologies by the recipient countries. Instead, they have shown more willingness to sell or provide in aid-outdated technologies to the developing countries. The proposed network should not restrict itself to technology transfer in the context of North-South dialogue; it should be extended to South-South cooperation since countries such as China and Brazil have already taken huge leaps in this direction. Along with strengthening and redefining the roles of the exiting technological institutions, new centres have to be set up to meet demands especially in the case of the LDCs and the SIDS. The funds for technology R&D and transfer have to be secured through contributions from the developed countries as well as creation of a sustainable development fund in the developing countries. Autonomous scientific appraisal, public-private partnerships in technology development, involvement of the NGOs and the civil society in the implementation of the projects as well as establishment of a feedback system are essential.

Among the countries of the South, the two countries that have had the most significant amount of say are India and China. However, technology cooperation between the two is dismally low-key. China has the world's largest installed renewable capacity; it has the largest number of hydroelectric generators in the world; and it tops the world in solar power generation. China ($54.4 billion) was ahead of Germany ($41.2 billion) and the US ($34 billion) in terms of green technology investments in 2010. It has the largest wind resources in the world and according to a Global World Energy Council study if it continues to advance its renewable energy projects, the wind farms would be able to meet 50 per cent of China's power requirement by 2030. Another report released by the Breakthrough Institute and Information Technology and Innovation Foundation has also found that 'China is now poised to surpass the West in virtually all areas of clean energy, including transportation technologies, high speed rails, advanced batteries, solar water heaters and even next generation nuclear and 'clean coal' technologies' (Climate Avenue, 2011). Geopolitically, ties between China and India have been tense. Sino-Indian cooperation in Renewable Energy Technology can be boosted only by private players as the Government of India would be highly reluctant to engage with a country, which has menacing regional as well as global aspirations; added to that is the historical baggage of war and trust deficit. According to Shyam Saran, the West is still ahead of China in terms of cutting edge technology. In addition, the familiarity between the Indian entrepreneurs and the Western ones is more. Bilateral dealings are different from multilateral dealings. At the end of the day, India has to build its own capacity (Saran, 2011). Others would suggest that India should look forward to cooperating with every country including China despite political hostility as South-South cooperation on environment and development is the key to future discourse on climate change and sustainable development. Mr Niu Qingbao, Consul General of China in Mumbai said that the level of cooperation between India and China is very low currently even though the potential exists and that there are no obstructions whatsoever from China's side (Qingbao, 2011).

IBSA (India, Brazil and South Africa) is the best platform for South-South cooperation in terms of development of technology. India could collaborate with Brazil on the issue of biofuels in which the latter is the world leader. It has developed flex-fuel car in which both gasoline and ethanol can be used. Blending of ethanol with gasoline is being not only encouraged but also authorised in India as evidenced by the Government's decision in 2003 to authorise a five per cent blending of ethanol with gasoline in nine states and four Union Territories in 2003, which was extended to 20 States and eight Union Territories in 2006 (Rediff Business, 2010). Use of bio-fuels not only saves petrol but also results in lower emission of pollutants. On the flip side, there are reports which indicate a certain rollback on the part of the Indian Government on the front of bio-fuels, especially jatropha due to lower production of jatropha seeds, rising food prices, fear of diversion of land for bio-fuels and its impact on food production, and the purported negative consequences of first generation biofuels on the environment. Based on these reasons, the erstwhile Minister of Environment and Forests, Mr Jairam Ramesh rejected Executive Secretary, Christiana Figueres' suggestion that India should 'explore business possibilities of jatropha plantations for bio-fuels' with private bio-fuel companies backed by the International Air Transport Association (IATA) in 2011 (Sethi, 2011).

***Beyond Copenhagen: The Beginning of a New Journey at Durban***

The post-Copenhagen scenario had left several questions unanswered despite a so-called 'successful' Cancun Agreement. On the one hand, countries such as the US and Canada have failed on the mitigation front completely in the last one decade. On the other, the BASIC countries have expressed their readiness to adopt legally binding emissions reductions targets at a legally appropriate forum but also to be transparent with respect to their domestic actions. The EU has made considerable efforts to address the issue and with its historical tradition of propagating universalisation of the European ideals, it would be delighted to take the lead on this issue too.

The Durban Summit was the longest UNFCCC meeting ever as two more days were added to the original schedule to salvage the talks. The Summit came up with the 'Durban Platform for Enhanced Action', which entails commitments by every country to cut GHG emissions; it envisages the signing of a deal by 2015 and the coming into effect of the same deal by 2020. Some would call it a success while others would label it 'too little too late.' This document implies that all countries including the US, India and China would have to adopt legally binding targets; this is being viewed by the EU as a great success. The uncertainty about the Kyoto Protocol also got resolved as the EU and a few other Annexure I countries decided to adopt the second commitment period of the Kyoto Protocol, the first commitment period of which ends in 2012. However, countries such as Canada have refused to sign it and it has already withdrawn from it. Canada would have surely not been able to meet its target of the first commitment period and if it signs on to the second one, it would have to pay penalties under the 'flexible' compliance mechanism of the Kyoto Protocol. The US would still be at par with the developing countries and would only be fulfilling voluntary targets during the time between 2012 and 2020. In addition, it cannot be confirmed that countries would eventually reach a legally binding agreement by 2015, the negotiations for which would begin in 2012. This meeting has proven to be a big letdown for India as a leader of the South-South cooperation movement. Jairam Ramesh's concessions at Cancun had not paid off since the developed countries still stuck to the lowest common denominator although the G-77's demand of adoption of the second commitment period was accepted. India's demands to include unilateral trade measures, Intellectual Property Rights and equitable access to sustainable development (backed by the BASIC, Arab League and the G-77) were not met. The biggest setback was the non-inclusion of the term 'equity' in the document. The entire debate came down to the wording of the document on the question of whether to include the phrase 'legal outcome' (as India and China wanted) or a 'protocol or legal instrument' (as the EU wanted) and eventually the phrase – 'agreed outcome with legal force' was endorsed (Gray, 2011).

Take for example the BASIC group of countries. Though these countries came to the negotiations with a common agenda, the fissures within the group that had started to develop at Cancun widened during the Durban Summit. This time, even China did not show as much reluctance to signing a legally-binding agreement by 2020 as much as it had been over the years probably because the realisation had dawned upon the country's leadership that its aggregate emissions are the highest in the world; its per capita emissions could double by 2020. Mr Prodipto Ghosh points out, 'China has far greater resources to use strategically than India. China has invested heavily in green technologies while India does not have such resources. China could become the biggest economy by 2020 and by the same year, its emissions will be at par with the European levels in terms of per capita emissions. At the same time it does not want to break ranks with the G-77' (Prodipto Ghosh, 2012). Besides, China sees itself as one of the technology vendors by 2020 and therefore, it has not backed India's proposals on the relaxation of the IPR regime. South Africa being the host wanted some sort of a deal by the end of the negotiations. Even otherwise, Brazil and South Africa were flexible when it came to the signing of a legally binding agreement. With the decision to fully operationalise the Green Fund by 2012, majority of the countries in the South bloc gravitated towards the EU except India and China. The uncertainty surrounding the sources of finance for the Fund was overlooked by these countries. The AOSIS and the LDCs wanted a stronger deal but they were content with some movement in the direction of a legally binding agreement in future. China gave in to the pressure, as it did not have much of a choice especially when even the US also vaguely supported the deal. In the end, the Summit clearly became an epitome of a European diplomatic victory. India was perceived as the biggest villain in the whole process. Some would even argue that India was forced to sign the deal; if it did not, it would have been blamed entirely for the collapse of the Summit. On the other side, it could also be argued that South-South cooperation came of age during the Summit with India joining hands with the other developing countries to strike a deal, which it did at Cancun too. The onus now lies on the BASIC countries to apply pressure on the industrialised countries to push for stronger mitigation efforts on their part during the interim period. It is here that South-South cooperation could create the platform for India to turn the tables on the North.

***The Strategy-building Exercise in Tandem with the Rethinking Exercise***

By now, it is very clear that the climate change negotiations have virtually reached the crossroads. Majority of the decisions made at the Cancun Summit have not been implemented and no country has the political inclination to lead the negotiations in its current form or even refined form as the state of affairs might change in future. The pessimism, which has been a hallmark of the negotiations since the coming into effect of the Kyoto Protocol, is expected to continue at least for the next two-three years. The obsession with 'making money' among certain countries has made the negotiations seem like trade negotiations, which fundamentally follow the model of the World Trade Organisation negotiations. This has led to the entire focus of discussion being shifted to the 'cost' factor that overshadows several benefits of being environment-friendly. This is the exact reason why the negotiations have lost their credibility and the only way in which some life could be breathed into the negotiations is to take a few steps backward and rethink over the international community's approach towards climate change. India has to deal with the climate change negotiations and the major countries in the world from a position that lays emphasis on the 'right' issues and does not allow the attention to be shifted from the true 'national interests' as well as 'international peace and stability.' In the process, India also has to streamline the perspectives and positions of the entire South bloc. It is high time that countries such as India begin to facilitate the transition of talks from trade-related issues to environment-related issues rather than portraying themselves as obstacles in reaching a global climate change agreement.

This however requires countries such as India to come forward with fixed standpoints on the issue and bring to the attention of the international community about the lopsided nature of the regime. It has to convince the entire South bloc and the amicable developed countries that the way forward would be to concentrate on environmental change rather than mere climate change so that the threat of human-induced environmental change to the developed countries could be underscored. Due to the over-emphasis on carbon emissions because of business interests attached to the cap-and-trade mechanisms, the real issues that face countries such as India are being ignored. The need of the hour is to expand the international community's understanding beyond the carbon emissions and look at other GHGs as well besides other concerns such as groundwater depletion, land degradation, over fishing and deforestation among others.

The developed countries must get rid of their 'developed country complacency syndrome' and address the issue holistically. The West has been adding fuel to the issue of climate change only to put pressure on the emerging economies to halt their carbon-based growth. However, they themselves have been hardly paying attention to the threats posed by environmental change to them. For example, the havoc caused by Hurricane Katrina was exacerbated by a number of human-induced factors such as faulty levee design and implementation, lack of town planning and large-scale subsidence caused by draining of wetlands, extraction of groundwater and inappropriately designed waterways (Paskal, 2011). In fact, even in tackling climate change, the other GHGs such as methane, nitrous oxide and the most important factor, water vapour have been discounted. If one can break the debate down to issues such as methane emissions, an area that a country such as India could easily capitalise on, countries such as the US would be forced to come to the table. For example, one could argue that India should get credit for the overwhelming number of vegetarians in the country in comparison to the predominantly meat-eating West. The cattle industry is subsidised in the US. If one reflects on the impacts of cattle industry, one could easily say that the West needs to clean its backyard before pointing fingers at the developing countries. The water utilised to produce one pound of meat amounts to 2,500 gallons. More than half of the US agricultural land is used to produce meat and this leads to top soil erosion and depletion of forested areas (Gardella, 1999). Moreover, methane and nitrous oxide emissions which are considered twenty times and three hundred times (respectively) more potent than carbon emissions are mainly produced by the livestock and meat industries (PETA). The Indian livestock and meat industry is a dormant one in comparison to the ones that thrive in the West.

It is very clear that the global climate regime is highly biased due to its inherent economic and trade undertones. Energy security debate runs parallel to the climate change debate. The question that India has to ask is that how much should the country take the climate change-energy security nexus seriously and how much should it concentrate on the science. The negotiations have well bypassed the science of climate change debate and jumped directly into the energy security debate. Therefore, it is imperative for countries in the South to keep an eye on the science of climate change while being active at the negotiations. The word 'developed' does not indicate immunity from the vagaries of climate change. It does not imply absolute readiness to improve resilience either. Therefore, the attempts by the developed countries to portray the climate change problem, as a problem of the developing countries has to be countered. This could be done in two ways. First, pressure has to be put on the developed countries to broaden the debate to environmental change. Second, the notion that reduction in the consumption of fossil fuels, which in turn would lead to reduction in the GHG emissions, which in turn would lead to reduction in the effects of climate change, which in turn would reduce the extreme weather events, has to be discarded so that there is 'consistency' in terms of identification of the causes of and response to climate change. One of them is adaptation. This could easily bring some of the SIDS-LDCs and India together.

India, should therefore, participate in the negotiations and attain whatever they can be hold of in terms of financial and technological assistance. It is very clear that there can be no global climate deal without India's participation. Hence, India should use its influence at the international level to acquire a deal in its favour. It should continue to underscore the per capita emissions principle to ensure that unfair regime is imposed on it. On the negotiating table, it has to learn a few lessons from China, which agrees to comply with all the international norms, but is seen to be in contravention of those very norms as evidenced by its nuclear cooperation with Pakistan despite being a signatory to the Nuclear Non-proliferation Treaty (NPT) and Pakistan's dubious history of proliferation. India should not approach the negotiations with a black band, confront isolation, and pressure to toe the line of the West. It should try to engage with the rest of the international community including the West in such a manner that it is not labelled an 'encumbrance' on the one hand and 'moral' and 'stupid' on the other.

India has a number of small-scale traditional technologies that it could promote elsewhere in the world especially among the developing countries. Technology cooperation among the developing countries and even between the developed and the developing countries requires political will. Interestingly, environmental diplomacy has become an integral part of the struggle for supremacy, so much so that a country's expertise in environment and climate in various sectors could be a potential soft power tool just as public diplomacy is. China's entry into the scramble for resources in Africa after the exploitation of the continent for more than two centuries by the Western powers has raised serious concerns in some of the countries even though moves are being made by China to address environmental challenges through the Forum for China-Africa Cooperation. India, on the other hand, has a better international image in terms of environmental footprint abroad. Its knowledge pool in this field is considered one of the largest and thus, it should pursue a more vigorous environmental diplomacy to enhance its global reach. Environmental diplomacy, if utilised in an equitable fashion, could improve relations between India and the other developing countries, especially in Africa and Latin America. The need of the hour is to exhibit India's potential to exercise its soft power and develop new technologies to be self-sufficient.

For this, the coordination between the Ministry of Environment and Forests, the Ministry of Earth Sciences, the Ministry of New and Renewable Energy and the Ministry of External Affairs has to be wholesome. History has shown that the differences between the MoEF and the MEA have derailed India's negotiating positions on various occasions including the Copenhagen Summit in which reports insinuated that Shyam Saran (Prime Minister's Envoy of Climate Change) and Jairam Ramesh (the Environment Minister) were at loggerheads. Shortly after the Summit, Shyam Saran resigned from his post and the Prime Minister invested the Environment Minister with complete authority as far as climate change negotiations are concerned. Even to this date, differences are conspicuous. The bureaucrats admit that though the systems are in place, there are differences that need to be sorted out (Suri, 2011).

## Conclusion

The global climate change regime is divided along various lines despite agreeing on issues that require global attention. It is very clear that India has repeatedly showed its leadership calibre in order to lead the South-South cooperation since 1972. However, its views have not been given as much seriousness as they deserved. One of the biggest roadblocks is the varied levels of development that the countries in the South are in. However, India has tried to accommodate the fears and interests of all the countries. Yet there remains trust deficit that has divided the developing countries leaving India to make a number of concessions right from the beginning. Hence, India now has to change its strategy to revamp and revitalise the international environmental order as the current one still favours the North in a big way. South-South cooperation faces several challenges and the only way these can be resolved is to cooperate outside the purview of the negotiations fully as the negotiations themselves have become a relic of absurdity and shortsightedness. This would be a pragmatic step that would serve India's interests in the end. Furthermore, this cooperation at the technological and financial levels could be translated into a united stand at the international negotiations so that the upcoming environmental regimes would be equitable and futuristic.

## References

Astrid Caldas, 'Bottom-up and top-down approaches in climate adaptation strategies', *dotWild* (9 March, 2011), see http://experts.defendersblog.org/2011/03/bottom-up-and-top-down-approaches-in-climate-adaptation-strategies, accessed on 26 September, 2011.

Cleo Paskal and Scott Savitt, 'Copenhagen consequences for the U.S., China and India', http://www.upiasia.com/Politics/2010/01/04/copenhagen_ cons equences_ for_the_us_china_and_india/1537/, accessed on 16 August, 2011.

Climate Avenue, 'Clean Energy or Renewable Energy in China', 17 March 2011, http://www.climateavenue.com/China.en.renewable.htm, accessed on 17 September 2011.

David Batty, 'Copenhagen reaction: delegates speak', *guardian.co.uk* (19 December 2009), see http://www.guardian.co.uk/environment/2009/dec/19/copenhagen-reaction-delegates-speak, accessed on 16 August 2011.

'Fight Global Warming by Going Vegetarian', PETA, for more information, see http://www.peta.org/issues/animals-used-for-food/global-warming.aspx, accessed on 25 September 2011.

Fiorella Gardella, 'The Meat Industry: Is it Worth it?', Seminar in Global Sustainability (University of California, Irvine: March, 1999), see http://www.dbc.uci.edu/sustain/global/sensem/MeatIndustry.html, accessed on 25 September 2011.

Hyung-Kwon Jeon and Seong-Suk Yoon, 'From International Linkages to Internal Divisions in China: The Political Response to Climate Change Negotiations', *Asian Survey* (University of California: November-December, 2006), v. 46, n. 6, see http://www.jstor.org/stable/4497211, accessed on 4 August, 2011.

Interaction with Cleo Paskal, Associate Fellow, Chatam House, at Manipal University, on 20 September 2011.

Interaction with Navdeep Suri, Joint Secretary, Public Diplomacy Division, Ministry of External Affairs, Government of India, at Manipal University, on 7 October, 2011.

Interaction with Niu Qingbao, Consul General of the People's Republic of China, at Manipal University, on 14 October 2011.

Interaction with Prodipto Ghosh, Member, Prime Minister's Council on Climate Change and Distinguished Fellow, Earth Science and Climate Change at The Energy Research Institute (TERI), on 3 January, 2012.

Interaction with Shyam Saran, Former Prime Minister's Envoy of Climate Change and Former Foreign Secretary, at Research and Information System for Developing Countries (RIS), on 15 June, 2011.

'Jairam defends nuancing India's position at Cancun', *The Hindu* (Mangalore: 25 December 2010), http://www.hindu.com/2010/12/25/stories/2010122567091800.htm, accessed on 6 February 2011.

James Jacob, 'Copenhagen and the 'BASIC' countries: some basic differences between the Indian and Chinese emissions', *Current Science* (Indian Academy of Sciences, Bengaluru: August 2010), v. 99, n. 4, see http://www.ias.ac.in/currsci/25aug2010/439.pdf, accessed on 14 August, 2011.

Louise Gray, 'Durban climate change: the agreement explained', *The Telegraph* (11 December 2011), see http://www.telegraph.co.uk/earth/environment/climatechange/8949099/Durban-climate-change-the-agreement-explained.html, accessed on 15 December 2011.

Matthew Paterson, *Global Warming and Global Politics* (Routledge, London: 1996).

Mukund Govind Rajan, *Global Environmental Politics: India and the North-South Politics of Global Environmental Issues* (Oxford University Press, Oxford: 1997).

Navroz K. Dubash, 'Towards a Progressive Indian and Global Climate Politics', *CPR Climate Initiative* (Centre for Policy Research, New Delhi: September, 2009).

Navroz K. Dubash and Lavanya Rajamani, 'Beyond Copenhagen: Next Steps', *Climate Policy* (Earthscan, London: 2010), issue 5.

Neil Carter, *The Politics of Environment (Second edition)* (Cambridge University Press, New York: 2007).

Niranjan Ramakrishnan, 'Corruption and Its Discontents', *Countercurrents.org* (26 August, 2011), see http://www.countercurrents.org/rama260811.htm, accessed on 18 October, 2011.

Nitin Sethi, 'India lost a lot, gained little at Cancun climate summit', *The Times of India* (New Delhi: 8 June 2011), see http://articles.timesofindia.indiatimes.com/2011-06-08/global-warming/29633648_1_ cancun-climate-kyoto-protocol-rich-countries, accessed on 9 June 2011.

Nitin Sethi, 'UN official lobbied for private biofuel firms during India visit', *The Times of India* (2 July 2011), see http://articles.timesof india. indiatimes.com/2011-07-02/developmental-issues/29729976_ 1_bio-fuels-jatropha-iata, accessed on 20 August 2011.

Press Trust of India, 'EU, India clash at Copenhagen over 'climate justice', *Hindustan Times* (12 December 2009), see http://www.hindustantimes.com/EU-India-clash-at-Copenhagen-over-climate-justice/Article1-485727.aspx, accessed on 5 February 2011.

United Nations Framework Convention on Climate Change, see for more information, http://unfccc.int/kyoto_protocol/items/2830.php, accessed on 13 March 2011.

'Why India must push for ethanol blended petrol', *Rediff Business* (27 July 2010), http://business.rediff.com/slide-show/2010/jul/27/slide-show-1-why-india-must-push-for-ethanol-blended-petrol.htm, accessed on 27 July 2011.

# 9

# Impact of Climate Changes in Africa: A Review of Human Security and Conflicting Threat

*Saubhagyalaxmi Misra*

## Introduction

The concept of security predates state. It is nothing short of incessant search for security against the adversary that has led human beings toward the emergence of state.

Security is survival. Human behaviour is directed towards a sense of security, as the base, on which the apex, self actualisation stands. The survival instinct of human beings, for which nations, individuals are craving for, is nothing but the first stage of human search for security. In recent years, the concept of security has undergone a fundamental transformation from a narrow Machiavellian definition and identification with perceived military threats from other states, to a much broader spectrum of threats emanating from economic, political, social and now even environmental factors. Security is symbolically used to protect an issue from a threat.

Security is a basic and key concept in the social sciences that is used in intellectual traditions and schools, conceptual frameworks, and approaches. The term 'security' is associated with many different meanings that refer to frameworks and dimensions, apply to individuals, issue areas, societal conventions, and changing historical conditions and circumstances. Thus, security as an individual or socio-political value stripped of any independent meaning is always related to a context, a specific individual or societal value system, and its realisation.

The word security is derived from the Ancient Greek 'Se-Cura' and literally translates to 'without fear'. 'Security' is therefore the state of being secure, or the actions employed to achieve that state, i.e., to be secure is to be without fear of harm.

Today's world is confronted with the concept of human security. Policies and institutions must respond to these insecurities in stronger and more integrated ways. The state continues to have the primary responsibility for security. However, as security challenges become more complex and various new actors attempt to play a role, the stimulus has ripened to broaden the definition of security. The focus must broaden from the state to the security of people - to human security. Human security means protecting vital freedoms. It means protecting people from critical and pervasive threats and situations, building on their strengths and aspirations, which can give people the building blocks of survival, dignity and livelihood.

Freedom is connected with security - freedom from want, freedom from fear and freedom to take action on one's own behalf. To do this, it offers two general strategies: protection and empowerment. Protection shields people from dangers. It requires effort to develop norms, processes and institutions that systematically address insecurities. Empowerment enables people to develop their potential and become full participants in decision-making. Protection and empowerment are mutually reinforcing, and both are required in most situations. When a society is dreadful, people need a shield and by the norms and policies of the state, they want protection. Within a safe and secured environment, they think of organising themselves and gradually empower themselves.

Human security complements state security, furthers human development and enhances human rights. It complements state security by being people-centred and addresses insecurities that have not been considered as state security threats. By looking at 'downside risks', it broadens the human development focus beyond 'growth with equity'. Respecting human rights are at the core of protecting human security.

**Toronto School of Thought (The Idea of Homer-Dixon)**

The term 'Environmental Security' refers to a range of concerns arising out of:

- Concerns about the adverse impact of human activities on the environment, seen as an asset to be inherited by next generations.
- Concerns about the direct and indirect effects of various forms of environmental change (especially scarcity and degradation), which may be natural or human-generated on national and regional security.
- Concerns about the loss of security individuals and groups (from small communities to humankind) experience due to environmental

change such as water scarcity, air pollution, global warming, and so on.

- Environmental security is the existence of two different agendas: a scientific agenda and a political agenda. Although they overlap, the scientific agenda is embedded in the mainly natural sciences and nongovernmental activity. The political agenda is essentially governmental and intergovernmental. It consists of public decision-making process and public policies that address how to deal with environmental concerns. Environmental security concerns the maintenance of the local and planetary biosphere as the essential support system on which all other human enterprises depend. Thus in the environmental sector, two different kinds of referent objects represent two wings within the environmental movement: the environment itself and the nexus of civilisation and environment.

**Global Climate Change**

Climate change has consistently been securitised by environmentalists. Assessing the severity of the threat posed by a build-up of greenhouse gases, especially carbon dioxide, that might cause the earth's temperature to rise and cause traumatic climate change requires much more complex response mechanisms than halting or reducing ozone depletion. Climate change of this magnitude, environmental advocates have argued, may lead to depletion of fresh water resources, sea level rise, erosion, loss in wetlands, agricultural productivity, biodiversity, air quality and human health. The UN Framework Convention on Climate Change and subsequent Kyoto Protocol in December 1997 committed the industrialised nations to specify, legally binding reductions in emissions of six 'greenhouse gases'. While climate change may pose only an intermediate threat to the health of Americans, it constitutes an existential danger for many developing countries. Global warming also threatens to increase incidence of malaria and other insect-borne diseases and may cause more summer heat-caused deaths.

The environmental insecurity comes from four major areas such as:

- Crisis is global and widespread
- Present rate of exploitation of resources threatens the sustainability of development
- Danger of toxicity increases due to current style of development
- Persistent and expanding trend is in morbidity.

The climate change issue is no more restricted itself either to speech acts or an action plan, but the store house of the politics of developed and developing countries. Before anyone enters into the politics of global climate change, it is imperative to understand the science behind such changes.

The growing global concern in the scientific community is that human activities are leading to increasing emissions of greenhouse gases, which in turn is leading to heat being trapped in the earth's atmosphere and subsequent increase in temperature. Emission of greenhouse gases is one of the major factors leading to climate change; other than human activities including stratospheric ozone depletion and emissions of other particles like sulphur dioxide and aerosols are also leading to climate change. Burning of fossil fuels and other industrial activities led to an increase in carbon dioxide levels in the atmosphere. Increase in the Earth's average temperature will adversely affect agriculture, forestry, natural habitats, water resources, and sea level.

**Forms of Climate Change**

The effects of climate change are strongly evident: melting of glaciers and permafrost; unexpected release of dammed water; decrease in sea ice extent; strong warming in the Arctic; thermal expansion of ocean waters; and rainfall and extreme temperature changes. It is expected to see loss of Greenland's ice along with the West Antarctic ice, leading to sea level increases of 16 to 23 feet. It is also expected to flood the harbours along the East Coast of USA, lowlands of Florida and the islands. This is particularly alarming since one-third of Americans live in coastal areas.

Effects of climate change are visible. Another hazard is that, with melting of the permafrost lid comes up with a large amount of trapped methane, which is a much more harmful gas than carbon dioxide for climate change.

Climate change will pose a bigger concern with the passage of time, with effects getting more serious. An increase in greenhouse gas emissions will lead to an increase in the average temperature of the Earth, change in precipitation, increase in sea levels as well as change in storm patterns. The magnitude of these challenges will depend on the factors such as increase or decrease in greenhouse gases and response of climatic features like temperature, precipitation and sea level to greenhouse gas emissions change in climate based on natural phenomena.

**Impacts of Climate Change**

'The attention to climate change will be the first priority. It is a major challenge because the hardest things to change are habits of thinking and it is precisely these that stand in a way of a healthy future for a human race, not only with regard to climate change, but also many other global problems' said Maurice Strong. It is also a priority because some changes require a long lead-time, so that it is important to start early in order to avoid disaster. Climate change is one such problem and it is urgent to begin addressing it.

Climate change and agriculture are inter-related and over the next century may have significant effects on the crop production and food availability. It is speculated that by 2050 there would not be any glacier in the world. The melting of ice will result in frequent floods and significant rise in sea level.

It is expected that the effect of global warming will threaten 50 per cent of land in Nile Delta by 2020. In South Asia loss of many staple products such as rice, millet and maize could attain a proportion of 10 per cent. Therefore, IPCC estimated that GDP in the developing and the less-developing countries would decline by 1.4 – 3 per cent due to climate change. Thus, the direct impact of climate change on agriculture and food supply includes shortage in grain production resulting in less availability of food items especially to the economically poor people, changes in agriculture imputes such as fertilizer and pesticides, soil erosion, soil drainage and low fertility level. The IPCC has predicted that GHGs will cause temperature to increase from 1.5-5.8 C and the sea water level by 15-19 cm by 2100. There would be floods in the coastlines. Hence, the agriculture will be affected. Ultimately, people will strive for their livelihood or may move to urban areas.

The changing temperature and rainfall patterns and increasing carbon levels will undoubtedly have important effects on global agriculture and on food security. Therefore, the developing countries should come forward and emerge as a leader in solar energy and show the world a way out of global warming.

The effects of global warming have the potential to be manageable for most nations because changes have been gradual so far, and are projected to be similarly gradual in the future. Abrupt slowing of the ocean's thermohaline conveyor, which could lead to harsher winter, quality depreciation in soil moisture, and more intense winds in certain regions that can provide a significant fraction of the world's food production.

The climate change scenario could potentially de-stabilise the geo-political environment, leading to skirmishes, battles, and even war due to resource constraints such as; food shortages due to decreases in net global agricultural production, decreased availability and quality of fresh water in key regions. The changes in the precipitation patterns will cause more frequent floods and droughts.

As the UNFCCC defines it, 'climate change refers to a change of climate that is attributed directly or indirectly to human activity that alters the composition of the global atmosphere and that is in addition to natural climate variability observed over comparable time periods.'

The "consequences" of climate change affect quality of life and have indirect political effects. Due to resource constraints, it could lead to war and tension in politics. Other resource constraints are the following:

- With the reduction in agricultural production due to unfavourable climatic conditions caused by climate change, it could lead to food shortages. Climate change would induce a shift in precipitation causing lack of availability of fresh water in regions, and floods and droughts.
- With excessive storms and rising water levels, access to energy supplies would become difficult, making nations adopt a defensive and offensive strategy to protect and acquire energy resources.
- With resources scarce as climate change becomes more visible, it will be managed through economic, political and diplomatic means such as trade embargos, taxes, and treaties after which states will be pressured for action with an increase in desperation for resources.

As General Anthony Zinni, former commander of American forces in the Middle East said in a 2008 National Intelligence report, 'We will pay the price later in military terms, and that will involve human lives.' This is because climate change will lead to a new rule of chaos and human insecurity. It will bring with it famine, drought, health epidemics, pandemics, natural disasters, resource scarcity, and human displacement.

In Al Gore's words, 'We risk fanning the flames of failed stoicism and offering glaring opportunities to the worst actors in our international system. In an interconnected world, that endangers all of us.' Al Gore pointed out, the conflict in Darfur could be viewed as an indirect consequence of climate change. With the change in climate, a lot of farmland was lost leaving herders to compete over dwindling resources. By then, the desert had grown by 60 miles and rainfall had diminished by 30 per cent. This event led to the tension and risk of the territories.

As per the fourth IPCC assessment report, there are distinct consequences of climate change on various constituents of life: ecosystems, food, society, health, and water.

**Ecosystems**

Many ecosystems are expected to be affected by the consequences of climate change, pollution, fragmentation of natural systems, and overexploitation of resources: flooding, drought wildfire, insects, and ocean acidification. Experts expect net carbon uptake by terrestrial ecosystems to peak before mid-century and weaken, thus demonstrating a huge climate change. If global temperatures rise by 1.5 to 2.5 degrees centigrade, 20 to 30 per cent of flora and fauna will face the danger of extinction. Such a rise in temperature, along with change in concentration of carbon dioxide levels, will lead to changes in ecological interactions, shifts in migration patterns, and negative consequences for biodiversity and other factors, like food and water, that are necessary for natural ecosystems

**Food**

The IPCC has been able to report with medium confidence the effect of few-degree rise in temperature on crop productivity, but they believe that it, '...is projected to increase slightly at mid to high latitude for local mean temperature increases of up to one to three degree Centigrade, depending on the crop, and then decrease beyond that in some regions' (medium confidence).

**Coasts**

These regions are expected to face the worst of climate change; mostly displacement, flooding, and coastal erosion due to climate change, and an increase in sea level. In addition, human activities, like land reclamation, will add to the stress on these regions. The IPCC states with very high confidence, 'By the year 2080s, many millions more people than today are projected to experience floods every year due to sea level rise. The number affected will be largest in the densely populated and low-lying mega-deltas of Asia and Africa, while small islands are especially vulnerable.'

**Settlements**

As mentioned in the section above, settlements in coastal and river flood plains, societies with activities closely linked to climate sensitive doings, and areas prone to extreme weather will be the ones worst affected by change in climate. Poor communities that lack resources for displacement are likely to face the worst of climate change.

**Health**

Some of the after effects of climate change would be: Increase in malnutrition, increase in deaths, diseases and injury due to extreme weather events, diarrhoeal diseases, increased occurrence of cardio-respiratory diseases due to high concentration of ground level ozone because of the depletion of the ozone layer, altered displacement of infectious diseases

There are certain benefits of climate change as well: a rise in temperature will reduce the number of deaths from exposure to the cold, and changes in transmission malaria and other infectious diseases. It is expected that the health benefits of a rise in temperature will exceed the negative effects.

**Water**

One of the most visible effects of climate change will be the melting of snowcaps and glaciers. These structures are crucial for fresh water availability and the reductions of these layers will put a lot of strain on water resources. According to IPCC, 'The negative impacts of climate change on freshwater systems outweigh its benefits (high confidence). Areas in which runoff is projected to decline face a reduction in the value of the services provided by water resources (very high confidence). The beneficial impacts of increased annual runoff in some areas are likely to be tempered by negative effects of increased precipitation variability and seasonal runoffs shifts on water supply, water quality and flood risk.'

Additionally, a rise in sea level (salt-water resources) would affect the salinity of ground water to further strain on water resources. An increase in temperature of water bodies will affect physical and ecological properties of freshwater lakes and rivers. One of the other water-related threats from climate change is the risk of flooding, which poses challenges to settlements, physical infrastructure, water quality, health, and safety. It is likely, according to an IPCC assessment, that 20 per cent of the world's population will live in areas in risk of flooding in the 2080s.

**Countries and Climate Changes Challenges**

It is essential to understand that no country solely dominates emissions and all countries in the world have to work together to reduce emissions. Many countries, based on social, economic and political interests, are taking steps towards combating climate change. African nations are the worst victims of climate changes as the human security is affected vehemently.

**African Nations**

Africa is severely affected by adverse climatic conditions; they have had to face scorching heat, droughts and floods. With a low per capita income and poverty, it is hard for these nations to make sizeable contributions towards combating climate change. The IPCC has concluded that Africa is vulnerable with the changes in climate. More specifically, there are six areas which would be hard-hit in:

- Water resources: Global warming will lead to reduction in soil moisture and droughts, in turn, leading to much lower agricultural produce, thereby, imposing huge economic constraints on the nation and leading to a lot of human suffering.
- Food: With the reduction in water, there will be a lower production of food leading to higher food costs and human suffering.
- Natural resources: Climate change will lead to losses in plant and animal life, will also affect their natural habitat and migration patterns.
- Human living conditions: Lack of the above factors will lead to a much lower lifestyle; floods and droughts induced by climate change will lead to illnesses, epidemics and aggravate suffering. With the lack of available health services and economic resources to devote, these illnesses will have to go poorly treated.

- Coastal flooding: one of the biggest consequences of climate change is rising water levels and flooding. Flooding along the coastal regions would deeply displace communities and lead to economic losses.
- Although African nations face far-reaching consequences with climate changes and have ratified the Kyoto Protocol, they have not taken any steps to reduce emissions in the first commitment period of 2008-12.

**Thinking through Linkages between Climate Change and Security**

Africa is one of the most vulnerable continents to climate variability and change because of multiple stresses and low adaptive capacity. Some adaptation to current climate variability is taking place, however, this may be insufficient for future changes in climate (IPCC 2007: 10).

There are at least seven ways that climate change is likely to affect security in its narrow and wider meanings. Some effects are already evident and will become very clear in the human and climatic short run (2007-20). They will increase and others will manifest themselves in the medium term (2021-2050); whilst in the long run (2051-2100), they will all be active and interacting strongly with other major trends. The end of the petroleum economy for many producing and consuming nations, possible financial and economic crisis, a larger population of humans, and a much more urbanised humanity – far in excess of the 50 per cent now living in small to very large cities. All these processes will be accompanied by redistribution of population nationally and internationally. Such redistributions typically have significant gender dimensions; for example, extreme event impacts can lead to male out migration in search of work, culminating in an increase in women-headed households – a group often considered particularly vulnerable.

Poverty and resource management has to be controlled if Government can take pre-cautionary measures actively. Human insecurity can be minimied if environmental stresses can be reduced. The role of climate change adaptation policies are to be active for future conflict prevention. Africa, with its history of ethnic, resource and interstate conflict, is seen by many as particularly vulnerable to this new type of security threat despite being the continent least responsible for global greenhouse gas emissions. Africa is going to face problems such as water scarcity, reduction in agricultural yield and damages to coastal areas. As a result the carrying capacity in Africa will be affected which will lead to migration. The countries in Africa, which are in a vulnerable stage, will have eco-political problems. At the state and regional level, adaptation programmes must be taken into consideration, which can be feasible.

Climate change involves the interactions of many systems such as the atmosphere, hydrosphere, cryosphere, and biosphere that are immensely complex in their own right. Thus, a recurrent theme in IPCC reports is the significance of thresholds and non-linearity. When human systems are added to the mix, complexity escalates. (Economic politics, The Environment, Dams and Climate Change, 2009: 3)

In February 2004, John Reid MP, then British Secretary of State for Defence and now Home Secretary, argued that climate change might spark conflict between nations. He forecast that violence and political conflict would become more likely in the next 20 to 30 years as climate change turned land into desert, melted ice fields and poisoned water supplies. He listed climate change alongside the major threats in future decades, including terrorism, demographic changes, and global energy demand. 'As we look beyond the next decade, we see uncertainty growing; uncertainty about the geopolitical and human consequences of climate change... Impacts such as flooding, melting permafrost and desertification could lead to loss of agricultural land, poisoning of water supplies and destruction of economic infrastructure... More than 300 million people in Africa currently lack access to safe water; climate change will worsen this dire situation' (Reid, J, 2004: 2).

Climate change impacts support the call for more dams. The loss of glaciers and snowpack mean that natural water flows are going to become more variable. More and larger dams could help to smooth that out, as well as compensate for how our current hydroelectric infrastructure will face challenges because of decreased summer water flow. Dams have virtues as a consistent source of energy for electrical generation. Their variable output also means they can be used to balance out production from sources like wind farms and solar facilities. With pumped hydroelectric storage, dams can also save energy at times when production exceeds demand, and do so in a way that is reasonably efficient (Economic politics, The environment, Dams and climate change, 2009:3) http://www.sindark.com/2009/03/26/dams-and-climate-change/

## Climate Change and Food Insecurity

It is aptly said that food and livelihood may lead to another war. It is estimated that between 1980 and 2001 there were 95 attempted coups in "Africa-33" of them successful. Food and livelihood pressure due to climate change will lead to populist and/or military coups in a number of countries. The access to food, in many African countries and regions is affected because of climate variability and change. The area suitable for agriculture, the length of growing seasons and yield potential, particularly along the margins of semi-arid and arid areas, are expected to decrease. This would further adversely affect food security and exacerbate malnutrition in the continent.

Women are responsible for around 70 per cent of household food production in sub-Saharan Africa, often on the basis of informal resource rights. Climate-induced changes in crop and livestock production could threaten those rights, as well as affect the gendered division of labour with negative effects on women's and men's incomes, livelihoods and household security (Thomas F. Homer Dixon, 1994).

The climate-food connection will not only affect the poor. In some industrial and industrialising nations, as climate change creates *new patterns of food production* — new exporting and new importing zones – access to supplies and the energy to import them may become strategic concerns and lead to international conflict.

**Health Impacts on Climate Change**

Health impacts of climate change including epidemics and insect outbreaks will have a similar effect, also compounding food and livelihood crises. This is beginning, but it will accelerate in the medium term. As we have seen with SARS and avian influenza, disease does not respect national boundaries in a globalied world. Changing climate may bring many epidemiological surprises as vector habitats change, sometimes quite rapidly.

Projected climate change-related exposures are likely to affect the health status of millions of people, particularly those with low adaptive capacity as following:

- increases in malnutrition and consequent disorders, with implications for child growth and development;
- increased deaths, disease and injury due to heat waves, floods, storms, fires and droughts;
- the increased burden of diarrhoea disease;
- the increased frequency of cardio-respiratory diseases due to higher concentrations of ground level ozone related to climate change; and,
- altered spatial distribution of some infectious disease vectors.
- Food-Fuel Conflict and Human Security.
- Large-scale investments in biofuels in the medium and long term as a substitute for green house gas-producing petrol-based energy sources may have the perverse effect of taking considerable land out of food production and diverting food grains, thus raising food prices and eroding biodiversity. The maize (corn) price in Mexico has already increased because of demand for grain by US ethanol plants, and protests have resulted. If this occurs, then food and livelihood pressures will increase and the desperation of many rural and urban people will increase national and regional insecurity. Rapid expansion of sugar production in Brazil and Africa palm in

Colombia, as energy feed stocks; have been a major source of displacement of small farmers from their lands. Also, recent research has shown the energy and carbon efficiency of biofuel production to be variable and often lower than with the burning of fossil fuels.

**Risks of Rising Temperatures**

If temperatures continue to rise, it is expected an additional 393,000 battle deaths by the year 2030 in sub-Saharan Africa, argues Marshal B. Burke. Such alarmist claims arouse interest among international organisations and national agencies with security issues, ranging from national governments and their intelligence communities to international bodies, such as the United Nations Security Council.

More than two-thirds of the countries in sub-Saharan Africa ('Africa' hereinafter) have experienced civil conflict since 1960, resulting in millions of deaths and monumental human suffering. Understanding the causes and consequences of this conflict has been a major focus of social science research, with recent empirical work highlighting the role of economic fluctuations in shaping conflict risk. Combined with accumulating evidence on the potentially disruptive effects of climate change on human enterprise, such as through possible declines in global food production and significant sea level rise, such findings have encouraged claims that climate change will worsen instability in already volatile regions.

Those most getting affected by climate changes are those in the poorest areas of the world although they are least responsible for it. Societies with more climate-sensitive economies, largely in the developing world, will be most affected as climate change acts as a 'threat multiplier' and adds to existing burdens. Decreasing availability of resources due to regional effects of climate change — like drought and desertification —

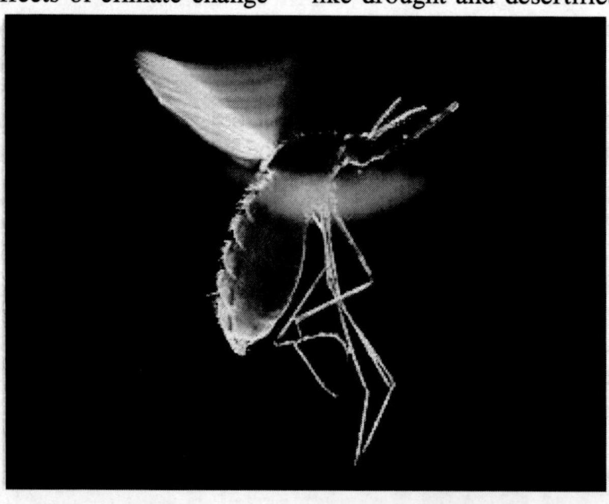

leads to intensified competition for these resources. In addition, this, when compounded by pressures like rapid population growth, tribalism and sectarianism (as in Darfur and Somalia), could result in armed violence. It could also result in the spread of diseases due to rising temperatures and change of ecosystems.

Rising temperatures will transmit diseases such as malaria to regions where it is currently absent, for example in the highlands of East Africa. In addition, short-term impacts of climate-related hazards are worrisome: waterborne diseases and new breeding sites for insects that carry disease, such as mosquitoes, can claim lives in the aftermath of a natural disaster. An example is Honduras, where in 1998, in the aftermath of Hurricane Mitch, an additional 30,000 cases of malaria were reported. Thankfully, the application of insecticide-treated bed nets has proven an effective measure to reduce the risk.

Presenting human security perspectives on climate change, this volume raises issues of equity, ethics and environmental justice, as well as our capacity to respond to what is increasingly considered the greatest societal challenge for humankind. Written by international experts, it argues that climate change must be viewed as an issue of human security, and not an environmental problem that can be managed in isolation from larger questions concerning development trajectories, and ethical obligations towards the poor and to future generations. The concept of human security offers a new approach to the challenges of climate change, and the responses that could lead to a more equitable and sustainable future. Climate Change, Ethics and Human Security will be of interest to researchers, policy makers, and practitioners concerned with the human dimensions of climate change, as well as to upper-level students in the social sciences and humanities interested in climate change.

**Food Security in a Changing Climate**

Climate change directly affects food security and nutrition. It undermines current efforts to protect the lives and livelihoods and end the suffering of the over 1 billion food insecure people and will increase the risk of hunger and malnutrition by an unprecedented scale within the next decades. By 2050, the risk of hunger is projected to increase by 10 – 20 per cent, and child malnutrition is anticipated to be 20 per cent higher compared to a no-climate change scenario.

Climate change will have further devastating consequences over the next 20 years as yields of staples in some regions decline dramatically. Wheat yields may fall by more than 20 per cent in Brazil, for example, while maize production in central and West Africa may fall by more than 10 per cent, it says. In Guatemala, climate change has affected smallholders dramatically in the last two years. The country is particularly vulnerable to climate change and extreme events, thanks to its geographical position in an earthquake and hurricane zone.

The experience of 260 families living on the settlement of Guadalupe in the Suchitepéquez region of the Pacific coast is typical. They lost much of their staple crop last year and the year before. There is more rain in winter now, and the rains that used to come in April now only come towards the end of May. Rains have recently been accompanied by increasingly violent storms, followed by prolonged drought. Deforestation and diversion of rivers by plantation owners producing food or bio fuel for export have exacerbated the problems.

Smallholders have had to switch harvesting seasons from winter to summer so the young plants have a better chance of survival, but that means irrigating and using seeds more resistant to drought. Farmers need technical assistance to adapt, but this is precisely the sort of help that previous policies have seen cut from many developing "countries" government spending under pressure from the international financial institutions.

**Climate Changes and Vulnerability**

Through much research, Africa has been widely acknowledged as one of the regions most vulnerable to climate change. To address those shortcomings, the CCAPS programme has sought to identify which areas in Africa are most vulnerable to climate change, and why these are happening.

There has been a new policy brief from CCAPS researchers; Josh Busby, Kaiba White and Todd Smith focusing on what drives climate security vulnerability in Africa, and examine potential policy responses. The study reports that policymakers should focus first on the regions and countries where the effects of climate change could lead to a large number of deaths and these include areas characterised by not only high physical vulnerability to climate hazards, but also dense populations, low household and community resilience, and weak governance.

However, parts of the continent, such as the Democratic Republic of the Congo (DRC), Sudan, and Somalia, may be less physically exposed to climate-related hazards, but more vulnerable due to other factors, such as low household and weak governance and community resilience. Within these regions, support for basic governance capacity and conflict resolution is as important as disaster prevention and targeted assistance for risk reduction such as early warning systems, better building codes, and drought resistant agriculture.

Other parts of the continent, such as South Africa and the coastal region of Northern Africa, have high physical exposure to climate hazards, but are likely less vulnerable than other regions due to more robust social and political resources. Support for these countries should come in the form of information sharing to help African leaders mobilise domestic resources most effectively.

'Climate Change is a serious threat to the agricultural field in the African Continent,' says Joemat-Pettersson to Bua News. The Food Security threat is one of the greatest challenges facing the African Continent due to climate change. As Africa has the responsibility to feed the world, they are faced with the enormous climate change constraints such as drought, floods and dreadful diseases that block them from even feeding themselves. The objective of the African Ministerial Conference on Climate-smart Agriculture is to share their perspective amongst the leadership, to explore challenges and to grasp new opportunities for agriculture in Africa.

Global warming and the rise of sea levels could threaten fisheries and shrimp production in the African continent. DRC ambassador to South Africa, Bene M'Poko, believes that developed countries are the main cause of climate change and should commit to an absolute approach to curb the release of greenhouse gases. He believes that these developed countries are also responsible for deforestation in the African continent in a number of ways, and believes that as Africans, they deserve to be compensated by those countries. He also states that Cancun gave them hope, but the developed countries failed to show a stronger commitment to reduce greenhouse gases to avoid dangerous climate change. He believes that this time, Africa should become more active, become the leaders of discussion, but also follow protocol.

**Human Security and Darfur**

There has been resource competition, relative deprivation and dwindling subsistence opportunities, etc., mostly conflicts between tribes and other groups in Darfur for as long as history records in the face of rapid population growth. Perhaps it was history of resource conflict that led the UN Secretary-General to identify climate change as one of the root causes of the Darfur conflict.

The effects of climate change on natural systems have been well documented, and humans will inevitably be affected. The increase in the length of drought cycles in Darfur combined with the southward expansion of the Sahara has amplified the effect of human behaviour on the environment. Small-scale but widespread farming and cattle grazing by the groups in Darfur has accelerated desertification.

Environmental degradation in Darfur is not a new process, nor is it even one that has accelerated in any meaningful way in the past ten years. What is it then about environmental change in Darfur that has led to conflict now? Today's Darfur conflict has an analogue in the 1987 FurArab war, which was one of resource competition; as drought and corresponding famine sent the largely Arab pastoralists of North Darfur southward in search of greener pastures, competition with Fur and other farmers was inevitable. Climate change not only forces migration but can also trigger conflict. However, the environmental degradation and corresponding migration in Darfur were not sufficient conditions for conflict. Conflict is rather the result of environmental pressures together with the breakdown of social structures designed to mitigate traditional resource conflict.

In the early 1970s Sudan's President Nimeiri took steps to consolidate power in Khartoum. One such measure was to abolish the tribal based Native Administration in Darfur that had served as a venue for aggrieved parties to air complaints and build compromises that mitigated in equality.

The location of where environmentally induced migration and conflict might occur is slightly more easily answered than whether it will occur and to what extent. Demographic stress (and possible conflict) can be categorised on a scale from:

Very high risk
- Large 'youth bulge'
- Rapid urban growth
- Low cropland or fresh water availability

1. High risk
   - Large youth bulge
   - Either rapid urban growth
   - Or low cropland/freshwater availability
2. Low risk
   - Low cropland or fresh water availability
3. No risk
   - None of these factors

This process identifies Africa and, to a lesser extent, the Middle East as primary areas of potential conflict and tension arising from rapid population growth and environmental stress. There are other 'hot spots' in Asia and northern South America where conflicts arising from environmental change are much less likely than conflicts generated by religious, ethnic and other rivalries.

It is clear that environmental stress in the 'high risk' nations is likely to be a factor in increasing population location and potential conflict. Twenty-five countries – of which the majority are in Africa – have been identified as falling in the highest risk category of civil conflict in the next two decades, and the likelihood continues to increase over time. All these countries have low cropland availability per person, half of them have fresh water availability problems and all are ranked amongst the poorest nations in the world.

Environmental stress is likely to be linked to conflict indirectly but significantly impacts will come directly from declining resources and conflicts over those resources from the tensions created by populations who are displaced or who move seeking improved life chances in other regions. However, most of the conflicts will embed environmentally induced conflict under the guise of religious, ethnic or civil conflict. It has been and will continue to be difficult to identify any simple or straightforward links between the environment, environmentally induced migration and conflict.

**Support to Adaptation**

There is no doubt that desertification and drought altered the migration patterns of pastoralist tribes into new areas. There is also no doubt that the altered migration patterns, combined with migration that is more permanent by people in North Darfur looking for viable land for subsistence, led to conflict. The Janjawid militia were almost certainly enticed with promises of viable land belonging to those they would force to flee.

As the climate continues to change and local environments degrade to such an extent that people feel they must move elsewhere in order to survive, the first step is to address the underlying need for migration. Migration because of climate change is less a result of the underlying change than a reflection of the ability of people and communities to adapt. The deployment of programs designed to maximise environmental resources may mean, for example, the difference between having to migrate in the face of decreased seasonal rainfall, and being able to adapt to short-term cycles made more severe by climate change. International aid providers should work with states to identify those groups at greatest risk of environmentally forced migration–not just those where conflict is an obvious risk - and develop long-term aid and development programmes to allow people to live in a way that is at least consistent with traditional standards of living without having to migrate.

Just as climate change occurs overlong cycles, much of the migration resulting from environmental degradation will occur slowly but consistently over long periods. It is these situations that are the most neglected, and arguably the most dangerous types of environmental migration. As in Darfur, changing climate coupled with changing migration patterns threatens to bring groups of people into conflict, potentially creating a cycle of violence and displacement that can easily spread, intensify and exacerbate local environmental conditions. It is easier to integrate hundreds of peopled who are replaced as a result of environmental degradation than it is to resettle, return or integrate hundreds of thousands as a result of violent conflict created by inadequate responses to the initial migration. It is for this reason that special care should also be taken to build up those local social institutions that allow for dialogue and dispute reconciliation where migration patterns make such disputes more likely.

The causes of the Darfur crisis can be traced to not only environmental degradation and the overwhelming of local resources but also to social degradation and the inability or unavailability of local dispute reconciliation mechanisms to handle environmentally driven migration. As climate change continues to alter local environments, the international community must be prepared not only to provide the means for communities to develop in order to adapt but also to provide new host communities with the social and political resources to integrate those who have no choice but to find greener pastures elsewhere.

**SOMALIA**

**Climate Change and Security Research**

Climate change is one of the greatest challenges of the 21st century. There is scientific evidence of the human impact on climate change but

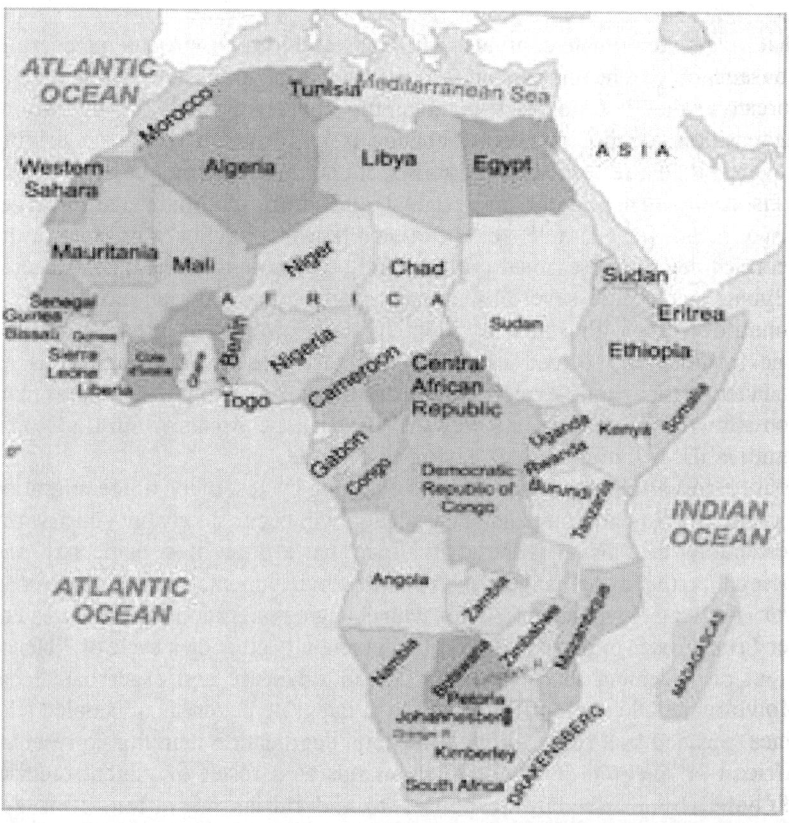

research on feedbacks between climate change and human interactions is in its initial stages.

**Conclusion**

In recent years, the study of climate change and security has become the focus of both scholarly work and policy-making. A number of policy reports and individual statements proposed alarmist claims about the impact of climate change on security. In 2007 and under the leadership of the United Kingdom, the United Nations Security Council held its first session on climate change and security.

Others are more cautious to make direct links between climate change and security such as the Intergovernmental Panel on Climate Change (IPCC) in its Forth Assessment Report or the UK-Treasury commissioned Stern Report on the economics of climate change. However, can the empirical evidence support these claims? My research on climate change and security aims at examining potential security implications that are most likely to result from the impact of climate change.

Yet, there is another dimension to research on climate and security that relates to the area of energy security. Attempting to mitigate the consequences of climate change is likely to cause new conflicts, when forests are cleared for palm oil plantations, food prices rise because of increasing demand for bio fuels and more droughts, or when non-democratic governments promote the development of nuclear energy as a means to substitute for fossil fuels.

Human security in Africa is at a fatal state. Its natural resources and economy are not stable, putting its people in a very unfortunate plight. Climate change is badly affecting this region of the world. Today, the country is facing scarcity of water resources, food, and natural resources. The lack of these essential factors worsens its human living conditions and causes coastal flooding. With its history in ethnic, resource and interstate conflicts, Africa may be seen as vulnerable to the threats of climate change and the insecurity that it brings to the country affected, even though the country is least responsible for global greenhouse gas emissions.

Because of the human insecurity in Africa, migration is one of the best choices the citizens have. The natural problems the nation is currently facing may turn into a bigger difficulty-an eco-political difficulty, or maybe war, within not only the state, but also participation of the other states is very possible as well.

Climate change is responsible not only for causing damages to Mother Nature, but also for conflict in some critically damaged countries, since people are competing for resources just to sustain their survival. Africa has been identified to belong in the category with the highest risk of civil conflict. It has low cropland availability per person and freshwater scarcity. Hence, undoubtedly Africa's human security is very low. Death, disease and danger have destroyed this region.

## References

Thomas Homer-Dixon, 'On the Threshold: Environmental Changes and Acute Conflict', *International Security,* Vol. 16, 1991, pp. 76-116.

Thomas F. Homer-Dixon, '*Environmental Scarcities and Violent Conflict: Evidence from Cases*', Peace and Conflict Studies Programme, University of Toronto, International Security, Vol. 19, No. I (Summer 1994), pp. 5-40.

Constantin-Horia Barbu, Camelia Sand, Letitia Oprean, '*Lucian Blaga*', University of Sibiu, Sibiu Romania Economic politics, The environment, Dams and climate change, 2009: 3 Reid, J, 2004:2.

Economic politics, The environment, Dams and climate change, 2009:3.

http://www.sindark.com/2009/03/26/dams-and-climate-change/

Thomas F. Homer Dixon, 1994.

# 10

# Environmental Goods and Services: Challenges and Opportunities for Bangladesh in the Context of Climate Change

*Sumitra Mitra* and *Saptarshi Basu Roy Choudhury*

## 1. Introduction

Climate change is one of the most serious threats to the world at present. The main reason behind this climate change or global warming is the excessive emission of greenhouse gases (GHGs). [Author: check sequence of tag numbers] Along with some natural reasons like volcanoes and ocean currents, some of the human activities e.g. burning of fossil fuels, industrial emission and deforestation have also radically increased GHG emissions more than the ability of the environment to absorb these gases. As a result, climate change is taking place and affecting people globally in an adverse manner. It is endangering biodiversity, human health, food and fresh water supplies and coastal and low-lying areas. The problem of climate change requires to be addressed in international agreement due to its global public-bad nature. Unilateral action or action by a group of countries would be inadequate. Therefore, from Kyoto Protocol to the recently concluded Copenhagen Summit, negotiations are taking place at the international level to bring different countries under the same umbrella, in order to make them abide by their own share of responsibility to reduce the impact of climate change that is taking place. There is however, a huge debate on the share of responsibilities because the developed countries are the historical emitters whereas the developing countries are the worst victims.

The importance of climate change has been reflected in the international trade of goods and services also. Environmental Goods and Services (EGSs) as a subset of goods and services were singled out for attention in the negotiating mandate adopted at the Fourth Ministerial Conference of the World Trade Organisation (WTO) at Doha in November 2001. Paragraph 31 of Doha Ministerial Declaration suggests the elimination or reduction of trade restrictions and distortions by the developing countries for a smoother access of environmental goods and services. This will environmentally benefit the developing countries through a reduction in their GHG emissions and in achieving sustainable development goals. EGS liberalisation is argued to have positive effect on both trade and environment. The World Bank study (2008) on International Trade and Climate Change argues that, developing countries have to deal with poverty and other social challenges. As a result, climate change may rank as a low priority to them, whereas their economies are more dependent on climate-sensitive sectors like environment and forestry and thus more vulnerable to impacts of climate change. They lack the infrastructure or resources to respond to the resulting change in climate. This gap can be met by EGS liberalisation.

In the era of climate change, the justification for developing an EGS industry to cater to domestic environmental needs is evident. However, this justification does not translate easily to its trade. This is especially true for trade between countries with different environmental problems and those that are at different stages of development. As most of the EGS industry is located in developed countries, it would be reasonable to assume that exports of EGS would flow from developed to developing countries.

Thus, developing countries are somehow sceptical about their gains. The list of EGSs mainly includes technologically advanced, sophisticated products and exports of developed countries. Hence, many developing country policy makers argue that these negotiations would primarily benefit the economies of developed countries, which are looking for new markets and might generate important tariff revenue losses for some developing countries. At the same time, negative impacts of liberalisation on vulnerable industries in developing countries have also been cited as a major criticism (Howse and Bork, 2006). Developing countries could benefit from EGS liberalisation if affordable green technologies lead to environmental improvements. However, the latter is not guaranteed. Moreover, green technology does not always match the needs of importing developing countries, which operate in a different environment and at a different scale (Araya, 2003). Thus, trade liberalisation in EGS cannot deliver the expected benefit unless and until there is proper infrastructure as well as supportive regulation (Jha, 2008).

However, Hamwey (2005) talks about dynamic gains of developing countries due to trade liberalisation of EGSs. 'According to him, although at present developed countries possess a significant trade surplus and developing countries a large trade deficit for the EGS, but interestingly developing countries' exports are growing at extremely fast rates. The present trade balances alone thus fail to take into account the dynamic gains of the developing countries. Jha (2008) also admits that dynamic comparative advantage appears to be shifting in favour of the developing countries for a number of goods. With a shifting dynamic comparative advantage, at least in the medium to long term, developing countries are likely to benefit from tariff liberalisation. Nevertheless, as developed countries already have low tariffs, developing countries may find it more beneficial to focus on non-tariff barriers.

Jha (2008) points out that while environmental problems arise in almost all developing countries, trade in EGS is restricted to only a handful of countries. Thus, not all environmental hotspots are serviced by trade in environmental goods (EGs). The main reason for lack of trade is the absence of a viable market. The much enthusiasm witnessed in the environmental goods sector is not true for the environmental service sector. The comparative advantage of developing countries in this area also needs to be carefully examined.

In the light of this background, our paper attempts to explore the opportunities and challenges for Bangladesh in the context of EGS liberalisation. Bangladesh is one of the least developed countries and at the same time highly affected by the climate change. In the next section we explore the ongoing debates over issues on how to define EGS, whether to use negative or positive approach for their trading, what all goods and services whose market should be enhanced, how to liberalise trading etc. All these will reveal the issue of developed and developing countries' interests. Section 3 gives a brief overview of the environmental goods and services sector of Bangladesh. The opportunities and challenges of EGS sector of Bangladesh are analysed in Section 4 while Section 5 concludes.

## 2. Debate

The inevitable debate arises from the very fact that Paragraph 31 of Doha Ministerial Declaration provides no universally accepted definition for EGSs. Environmental goods have been conceptualised in two different ways. The first one is the conventional notion that focuses on treating a specific environmental problem through the end use of that particular good. This classification includes goods such as wastewater treatment or air-pollution-control equipment that have an environmental end use; i.e., they directly address an environmental problem. The second

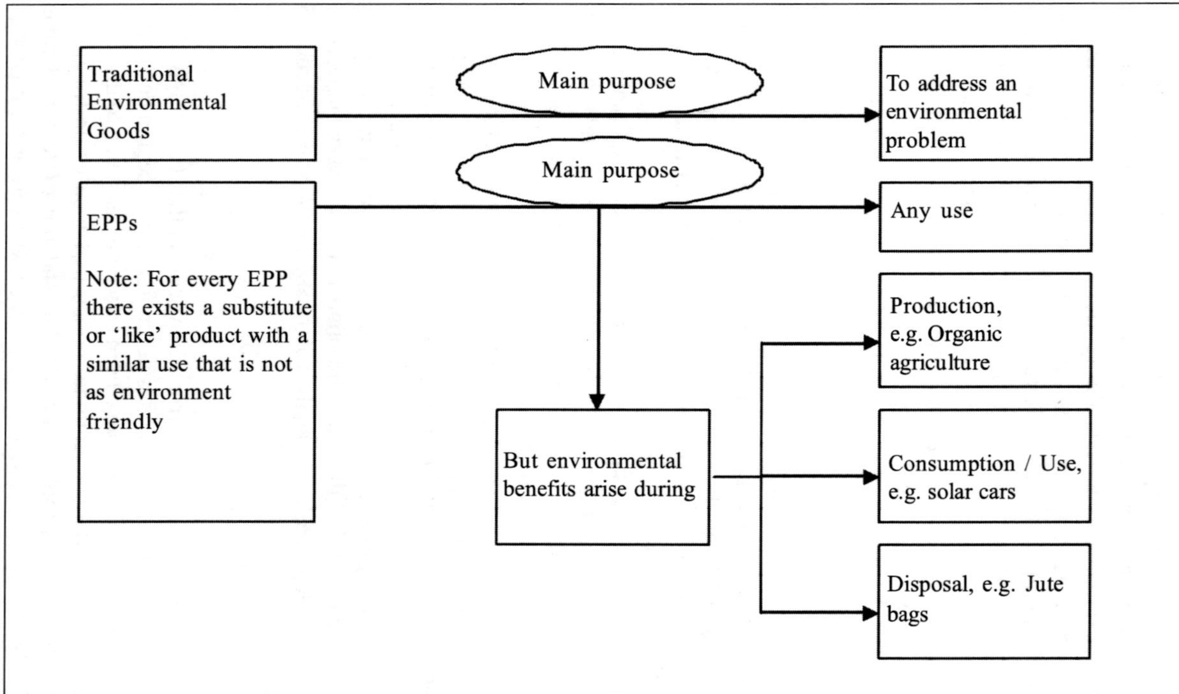


**Diagram 1: Traditional Environmental Goods vs. Environmentally Preferable Products (EPPS)**

*Source*: Sugathan et al. (2007).

conceptualisation is broader and includes environmentally preferable products (EPPs). The United Nations Conference on Trade and Development (UNCTAD) defines EPPs as products that cause significantly less 'environmental harm' at some stage of their life cycle than alternative products that serve the same purpose, or products whose production and sale contribute significantly to the preservation of the environment (UNCTAD 1995). In this case, the primary purpose of the product or service is usually not to remedy an environmental problem. The environmental benefits may arise during and because of the production method, during the course of its use, or during the disposal stage of the product (Sugathan et. al., 2007). The two conceptualisations of environmental goods are shown precisely in the following diagram.

Coming to the definition of environmental service, it consists of those activities, which measure, prevent, limit, and correct environmental damage to air, water, soil, and problems relating to waste, noise, and ecosystems (Organisation for Economic Cooperation and Development [OECD]). Alternatively, the United States (US) environment industry defines environmental services as all revenue generating service activities, which are related to compliance with environmental regulations, environmental assessment, analysis, and protection, pollution control, waste management, remediation, provision and delivery of environmental resources like water, recovered materials, and energy, and activities for improving energy and resource efficiency, increasing productivity and enabling sustainable economic growth. Thus, broadly speaking, environmental services can be defined as those service activities, which reduce environmental risk, minimise pollution and enable efficient resource use (Sawhney and Chanda, 2003).

In international trading, generally both negative and positive approaches are taken into account to give special treatment to the environmental goods. A negative approach is to identify specific goods and services that countries require to ban for trading in order to improve the environment. For example, in the Montreal Protocol (signed in 1987 to control the depletion of the Ozone Layer) parties to the treaty were required to ban trade with non-parties in ozone-depleting substances, such as chlorofluorocarbons and products containing them (e.g. refrigerators). However, when it comes to carbon abatement and climate change mitigation, it is less straightforward to identify products that should be banned. Nearly every product or technology causes environmental harm or affects the climate to some degree. Therefore, in this case a positive approach, which seeks to identify certain environmental goods and services for enhanced market access, holds some promise (Zhang, 2009).

However, in the positive approach the basic question arises that which EGSs need to be encouraged. Identifying them depends on their definition. Given their conceptual complexities and a lack of consensus on their definition, WTO members have persistently disagreed over how to identify which EGSs should be subject to trade liberalisation. Upto now three approaches have been proposed in the WTO negotiations—list approach, project-based approach and request-offer approach (Zhang, 2009).

The OECD advocates a list-based approach, whereby goods and services on an agreed list will gain enhanced market access through the elimination or reduction of tariffs and non-tariff barriers. Such lists have been produced by the OECD and by the Asia-Pacific Economic Cooperation (APEC) groups. The two lists are more or less overlapping. The main difference between the two lists is that only the OECD list contains minerals and chemicals for water/waste treatment, while the APEC list includes a relatively more extensive set of goods needed for environmental monitoring and assessment. The OECD list also contains a large number of environmentally preferable products (Steenblik, 2005). Taking the OECD or APEC lists of EGs as reference points, a group of countries proposed a list of 153 products in April 2007. Just prior to the United Nations Climate Change Conference in Bali in December 2007, the EU and the US submitted a joint proposal at the WTO calling for trade liberalisation of 43 climatefriendly goods that were identified by the World Bank (2007) from a list of the Friends' 153 products, with the aim of securing a zero tariff for these climate-friendly goods by 2013 (Zhang, 2009).

Many developing countries had consistently expressed concerns about using a list approach, noting that majority of listed products is primarily of export interest to industrialised countries. Another point is related to the issue of dual use of these products. A number of the products submitted by interested members had dual uses; that is, they were used for both environmental and non-environmental purposes. A good example is a pipe, which can be used as an input to a renewable energy plant or wastewater treatment plant but can also be used to transport oil. The question arises that, should a pipe therefore be liberalised as an environmental good or not.

In response to the dual use, India has advocated a project-based approach. According to this approach, each WTO member would designate a national authority to select environmental projects based upon criteria developed by the Special Session of the Committee on Trade and Environment in whose domestic implementation would be subject to WTO dispute settlement. The goods and services required for a selected environmental project would temporarily enjoy preferred market access for the duration of the project. India has argued that the project approach would ensure that the approved EGSs are used strictly for environmental purposes.

Brazil, on the other hand, has suggested a request-offer approach. In this approach, countries would request specific liberalisation commitments from each other on products of interest to both of them and then extend tariff cuts.

The next section gives a brief overview of EGS sectors of Bangladesh. This will help us to understand the opportunities and challenges of the country in the context of EGS liberalisation.

## 3. Overview of Bangladesh's Environmental Goods and Services Industry

### *Environmental Goods Industry*

Until last decade, Bangladesh remained a net importer of environmental goods. It imported as many as 125 items on the OECD list of 164 environmental goods of about US$184 million. The share of this was about 2.2 per cent of total imports to Bangladesh in 2000. Table 1 shows the classification of environmental goods imported to Bangladesh.

**Table 1: Bangladesh's Imports of OECD Classified Environmental Goods (2000)**

| *Product* | *Import (US$ '000)* |
|---|---|
| **(A) Pollution Management** | |
| 1. Air pollution control | 29,754 |
| 2. Wastewater management | 67,637 |
| 3. Solid waste management | 27,140 |
| 4. Remediation and cleanup | 2,581 |
| 5. Noise and vibration abatement | 25,311 |
| 6. Environmental monitoring, analysis and assessment | 10,492 |
| **(B) Cleaner Technologies and Products** | |
| 1. Cleaner/resource efficient technologies and processes | N/A |
| 2. Cleaner/resource efficient products | 5,915 |
| **(C) Resources Management Group** | |
| 1. Indoor air pollution control | N/A |
| 2. Water supply | 262 |
| 3. Recycled materials | N/A |
| 4. Renewable energy plant | 2,095 |
| 5. Heat/energy savings and management | 13,434 |
| 6. Sustainable agriculture and fisheries | N/A |
| 7. Sustainable forestry | N/A |
| 8. Natural risk management | N/A |
| 9. Eco-tourism | N/A |
| 10. Other | N/A |
| **Grand Total** | **184,621** |

*Source*: Khatun (2004).

***Note:*** Data on exports of environmental goods is not available.

**Table 2: Total and Environmental Imports of Bangladesh (2000)**

| | |
|---|---|
| Number of OECD Environmental Products imported | 125 |
| Import Volume of Environmental Products (US$ '000) | 184,621 |
| Total Import Volume of Bangladesh (US$ '000) | 8,360,000 |
| Imports of Environmental Products as a percentage of Total Imports | 2.21 per cent |

*Source:* Khatun (2004).

As evident from the figures, import in the pollution management category is the highest, which is 88.2 per cent of total imports of environmental goods into the country. This group consists of goods such as air pollution control, wastewater management, solid waste management, remediation and cleanup, noise and vibration abatement and environmental monitoring, analysis and assessment. The shares of import in the other two categories are quite low. The resource management category is 8.6 per cent and the cleaner technology and products group is 3.2 per cent of total imports of environmental goods.

***Environmental Services Industry***

The service sector is a significant sector in respect to GDP as well as employment generation in Bangladesh. Being a less developed country, these two are important indicators. The service sector accounts for nearly 60 per cent of the GDP of Bangladesh and generates more than a quarter of its total employment.

**Diagram 2: Sectoral Share (per cent) in GDP of Bangladesh**

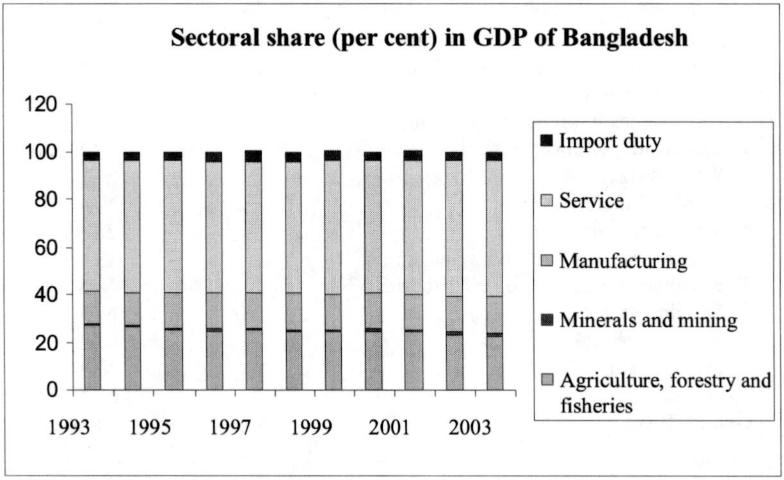


*Source:* Compiled by the authors.
*Data:* Raihan and Mahmood (2007).

**Diagram 3: Share of Sectoral Employment in Total Employment (2000)**

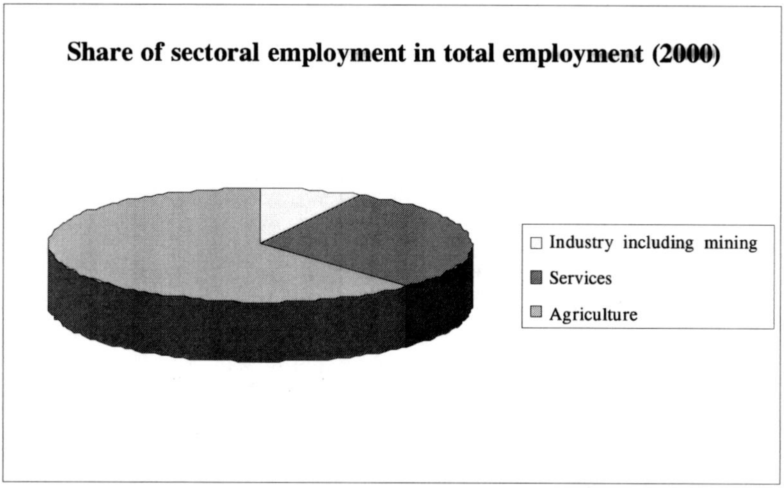


*Source:* Compiled by the authors.
*Data:* Raihan and Mahmood (2007).

Although there is significant contribution of the overall service sector to the country's GDP, trade in service sector is limited (as shown in the following table). This implies that there is substantial, unexploited

**Diagram 4: Remittances Earned by Bangladeshi Temporary Labour Working Abroad**

*Source:* Compiled by the authors.
*Data:* Raihan and Mahmood (2007).

**Table 3: Service Trade of Bangladesh**

| | *2000* | | *2001* | | *2002* | | *2003* | |
|---|---|---|---|---|---|---|---|---|
| Sub-Sectors | Receipts | Payments | Receipts | Payments | Receipts | Payments | Receipts | Payments |
| **Computer and Related Services** | **2799.8** | **347.2** | **2213.2** | **336.5** | **2792.9** | **300.3** | **4199.1** | **164.2** |
| Rental/Leasing Services without Operators | 0.0 | 3328.9 | 17.7 | 5331.6 | 0.0 | 1409.6 | 0.0 | 1179.3 |
| **Telecommunication Services** | **41531.6** | **7071.2** | **13791.0** | **7572.6** | **22870.9** | **4835.9** | **9237.2** | **7083.1** |
| Audiovisual Services | 8.1 | 0.0 | 4.5 | 0.0 | 416.0 | 478.0 | 3629.6 | 156.7 |
| Construction and Related Engineering Services | 132.1 | 219.3 | 0.0 | 0.0 | 252.9 | 6650.0 | 2362.0 | 8485.6 |
| Distribution (merchanting and other services) | 14842.6 | 3541.8 | 5910.4 | 2730.0 | 5091.2 | 3934.8 | 3740.9 | 2520.6 |
| Education Services | 2508.9 | 25355.2 | 3114.3 | 29031.6 | 2582.2 | 29009.4 | 3010.7 | 28840.0 |
| All insurance and insurance-related Services | 3415.9 | 4448.9 | 2171.6 | 3901.7 | 3072.9 | 8143.5 | 3383.6 | 6450.6 |
| **Life, accident and health insurance Services** | **614.7** | **153.7** | **665.9** | **150.4** | **764.1** | **139.4** | **750.2** | **142.6** |
| **Non-life insurance Services** | **2660.4** | **4294.4** | **1162.7** | **2883.3** | **2141.1** | **3290.4** | **2041.8** | **8.1** |
| Reinsurance and retro-cession | 140.8 | 0.8 | 343.0 | 868.0 | 167.7 | 4713.7 | 591.6 | 6299.8 |

*(Contd. )*

*(Contd. Table)*

| | *2000* | | *2001* | | *2002* | | *2003* | |
|---|---|---|---|---|---|---|---|---|
| Sub-Sectors | Receipts | Payments | Receipts | Payments | Receipts | Payments | Receipts | Payments |
| Reinsurance and retro-cession | 140.8 | 0.8 | 343.0 | 868.0 | 167.7 | 4713.7 | 591.6 | 6299.8 |
| Banking Services | 15797.5 | 29057.0 | 9074.7 | 18977.2 | 9859.6 | 12794.8 | 18791.0 | 9081.1 |
| Health Related and Social Services | 0.0 | 835.6 | 0.0 | 698.4 | 0.0 | 452.3 | 0.0 | 523.0 |
| Tourism and Travel-Related Services | 48828.7 | 269293.8 | 45780.8 | 236030.1 | 51174.2 | 58126.2 | 52386.4 | 84717.2 |
| Recreational, Cultural and Sporting Services (other than audiovisual services) | 312.2 | 0.0 | 516.9 | 11.3 | 62.4 | 55.6 | 35.7 | 205.8 |
| News agency Services | 132.9 | 1116.2 | 200.1 | 909.1 | 77.6 | 684.3 | 115.5 | 603.6 |
| Research and Develop-ment Services | 74.7 | 3599.3 | 22.7 | 5957.8 | 97.5 | 3686.9 | 252.9 | 3150.1 |
| **Maritime Transport Services** | **80618.0** | **21747.9** | **65198.3** | **23454.2** | **68842.0** | **23272.6** | **67569.3** | **26851.4** |
| Air Transport Services | 25107.7 | 192919.4 | 19439.3 | 228309.6 | 15294.5 | 202130.7 | 30096.6 | 249843.5 |
| Rail Transport Services | 9.2 | 0.0 | 319.9 | 0.0 | 0.0 | 0.0 | 0.0 | 0.0 |

*Source*: Raihan and Mahmood (2007).

potential in the entire service sector as well as in the environmental service sector.

The above table shows that Bangladesh has export interest in four services sub-sectors, which generally enjoy a positive balance of payments: computer and related services; telecommunication services; insurance services; and maritime transport services.

The Temporary Movement of Natural Persons (TMNP) or the temporary labour migration abroad is the greatest export interest to Bangladesh, which already exports service providers to many countries under bilateral arrangements. TMNP is especially important for Bangladesh since remittance inflows from its citizens working abroad have contributed significantly to the stability of the country's balance of payments in recent years.

Although entire service sector is of much importance for the country, environmental service sector has not been singled out and emphasised much in case of Bangladesh.

## 4. Opportunities and Challenges

Since there is no widely accepted and uniform definition of Environmental Goods and Services (EGSs), the concern for Bangladesh is therefore twofold. On one hand, Bangladesh might face revenue loss and deficit from trade liberalisation of EGSs. If there is tariff reduction, the country might lose tariff revenue. Bangladesh at present is already in a situation of trade deficit in both the environmental goods as well as service sector. If this deficit increases as a result of EGS liberalisation then the country might reach a vulnerable situation. On the other hand, the country has to draw up a list of exportable EGSs of its own to gain the benefits from growing export potential. The basic economic and trade theory suggests that the list has to be built up according to the comparative advantage of Bangladesh. Regarding the formation of its own list of EGSs, Bangladesh has to consider many things. Firstly, Bangladesh needs to identify its own comparative advantage and prepare a list from where it can have export earnings. Secondly, it has to identify different barriers, which can act as hindrance to its own export earnings. Thirdly, the country has to build up its absorption capacity of the clean technology, which might flow into the country to a large extent due to EGS liberalisation.

Below we enlist the opportunities and challenges of the country in the context of EGS liberalisation.

### *Opportunities*

Trade policies in Bangladesh have always supported the import of technologies that are environment friendly and less pollution intensive. Liberalisation of EGSs can provide much needed capital and technological expertise in environmental management, since the domestic sector has poor investment in clean R & D. If tariffs were reduced on the import of such cleaner technologies, it would definitely improve environmental quality.

In the climate change era, the labour abundant developing countries like Bangladesh have production potential of even manufacturing goods because labour intensive and less capital and energy intensive technologies are more environment friendly. This will make more resources available to protect the environment and enable efficient resource use. If production of manufactured goods will shift to labour abundant Bangladesh, then it will also have a spill over effect on employment as well as income generation. This as a result that might increase the willingness among the citizens to pay for environmental improvement.

Apart from the production shifting of traditional manufacturing products, the newly defined Environmentally Preferable Products (EPPs) is regarded as a good opportunity for Bangladesh to enhance its export performance. A prominent example is raw Jute and jute products, in which Bangladesh has a comparative advantage. At present Bangladesh is the largest jute producer in the world. Bangladesh Jute Mills Corporation (BJMC), a public corporation in Bangladesh, is the largest state owned manufacturing and exporting organisation in the world in the jute sector.

The potential direct and indirect gains from liberalisation of the environmental services sector might also come up as a good opportunity for Bangladesh. The expansion in the environmental services sector can provide employment opportunities to the unskilled as well as skilled labour in the country, as some of the environmental segments are labour intensive like waste management, sanitation, consulting, training etc. For a labour abundant country like Bangladesh, thus growth of the environmental service sector would also mean increased employment generation.

However, these potential benefits face some serious challenges.

***Challenges***

It is true that liberalisation of EGSs will provide required capital and technological expertise in environmental management, but for that, the importing country has to improve its absorption capacity of the advanced technology. Given the poverty and poor literacy level of Bangladesh, this is a major challenge. At the same time, there are doubts over the definition of a cleaner technology. Since technology is changing rapidly over the time, technologies considered as cleaner today may not be considered cleaner tomorrow. The sustainability of the cleaner technology has to be judged very carefully.

There is ambiguity over the definition of EPPs. There are high chances that EPPs from Bangladesh are refused by the developed countries on the basis of some other international standards. Production process of Bangladesh is expected to involve technologies, which are not environment friendly or even use of unskilled as well as child labour. Bangladesh also needs to identify non-tariff barriers such as registration requirements, approval of health authorities in these products. It might be difficult for Bangladesh to respond to such requirements since those involve additional costs. All these might reduce competitiveness of Bangladesh in EPPs.

Although the service sector is important for Bangladesh, no data has been found regarding environmental service sector. However, it is expected that being a labour abundant country, Bangladesh will be benefited from labour intensive service sectors within the country as well as abroad. However, the following diagram shows that the Bangladeshi labour who are absorbed abroad, are mainly unskilled and semi-skilled. If the environmental service sector uses sophisticated clean technology, which demands skilled labour, then these unskilled labours might not be required.

**Table 5: Temporary Service Providers of Bangladesh Going Abroad**

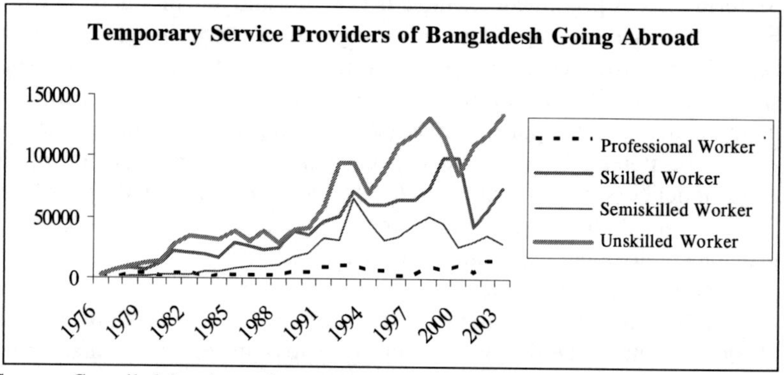


*Source:* Compiled by the authors.
*Data:* Raihan and Mahmood (2007).

At the same time, there are other challenges like poverty, corruption, inadequate transport and poor port related facilities, inadequate technological advancement etc. These are some major reasons why Bangladesh's economy is struggling to keep pace with the global economy at present. In addition, there are various natural calamities Bangladesh experiences every year that hamper the economy to a great extent. Every year Bangladesh loses many resources due to the excessive flood, rain and cyclone.

### *Way Forward*

At present jute production as well as trade is centred on the Indian subcontinent. The major producing countries of jute are: Bangladesh, India, China, Thailand, and Myanmar. Bangladesh is the largest exporter of raw jute, and India is the largest producer as well as largest consumer of jute products in the world. Therefore, the local price of raw jute in Bangladesh is the international price. Again, the local price of jute products in India is the international price.

Jute production in Bangladesh and India has an interesting history. Before partition in 1947 undivided India was the world's main supplier of jute and jute products. As a result of partition, the main jute growing areas remained in Bangladesh (the then East Pakistan) and most of the jute mills remained in India (mainly in the state of West Bengal). Therefore, partition was a big blow to the jute industry in India. Later on, when plastic and nylon appeared as cheaper substitutes of jute, the industry got its second blow. Because of these, most of the jute mills in India at present suffer from under-utilisation of capacity and economic loss.

In the era of climate change, the real cost of non-degradable plastic and nylon products are much higher compared to the jute products because of their cost of environmental degradation. As a result, the market of jute is gaining back its importance as an EPP. Therefore, there is a way forward, if India and Bangladesh decide to join hands to exploit the benefits arising from the growing importance of jute. The challenges of standard, registration process etc. of Bangladesh regarding the jute products can be overcome by India because at present India is one of the highly growing countries. The under-utilisation of capacity of Indian jute mills can be overcome by the high yield of jute production in Bangladesh. Further research is needed in this particular area.

## 5. Conclusion

Climate change mitigation gives rise to developed-developing countries debate, the former being the historical emitters and the latter, the worst victims. In the context of trade liberalisation in EGSs also, the same debate is prominent. The list of EGSs as suggested by OECD and APEC countries are mainly technologically advanced sophisticated products and exports of developed countries. At present most of the developing countries are net importers of EGSs. Bangladesh being a developing country is no exception. However, economists argue that developing countries will have dynamic benefits from this trade liberalisation of EGSs in the coming years. Studying the opportunities and challenges of Bangladesh, we found that being a labour abundant country it has a potential gain of production shift of manufactured goods from developed countries because labour intensive technologies are environment-friendly. This explanation holds for labour intensive environmental services also. Particularly regarding jute and jute products Bangladesh has a good opportunity because jute is one of the EPPs as defined in WTO negotiations. Bangladesh being the largest producer and exporter of jute thus holds a huge potential

However, there are several shortcomings for the country also. The smoother access of technology does not merely guarantee its absorption. Given Bangladesh's high poverty and poor literacy levels, the technological absorption is a big cause of worry. Chances are high that the jute export of the country might get refused by developed countries for not following all international trade norms and standards. Bangladesh struggles to keep pace with the global trade standards due to the use of unskilled labour, child labour and technologies, which are not environment-friendly. If India and Bangladesh join hands in the production of jute and jute-related goods then these problems can be overcome. Before the partition of 1947, undivided India's jute industry had a glorious history. However, the infamous divide and rule policy of the British affected both the countries' jute sector. In addition to this, cheap substitutes like plastic and nylon posed hindrance to the development of this segment all the more. Currently jute is gaining back its importance due to its environment-friendly properties. This is high time for India and Bangladesh to give a serious thought to cooperation and repeat the golden history of their jute industry all over again.

## References

Araya, M. 2003. WTO Negotiations on Environmental Goods and Services: Maximising Opportunities? Global Environment and Trade Study. Yale Center for Environmental law and Policy. Accessed February 2010 at www.ppl.nl/bibliographies/wto/files/1550.pdf.

Hamwey, R. M. 2005. Environmental Goods: Where Do the Dynamic Trade Opportunities for Developing Countries Lie? Centre for Economic and Ecological Studies (Cen2eco) Working Paper prepared to support discussions at the Hong Kong Trade and Development Symposium and the Sixth WTO Ministerial Conference in Hong Kong in December 2005. Geneva, Switzerland.

Howse, R. and P. van Bork, 2006. Options for Liberalising Trade in Environmental Goods in the Doha Round. Trade and Environment Series Issue Paper No. 2. International Centre for Trade and Sustainable Development (ICTSD). Geneva, Switzerland.

Jha, V. 2008. Environmental Priorities and Trade Policy for Environmental Goods: A Reality Check. Environmental Goods and Services Series Issue Paper No. 7. International Centre for Trade and Sustainable Development (ICTSD). Geneva, Switzerland.

Khatun, F. A. 2004. Environmental Debates in the WTO: Defining Bangladesh's Interests. Paper 35. Centre for Policy Dialogue (CPD) Occasional Paper Series. Dhaka, Bangladesh.

Raihan A. and M. Mahmood 2007. Opportunities and Risks of Liberalising Trade in Services. Country Study on Bangladesh. Trade in Services Series Issue

Paper No. 3. International Centre for Trade and Sustainable Development (ICTSD). Geneva, Switzerland.

Sawhney, A. and R. Chanda 2003. Trade in Environmental Services: Opportunities and Constraints. Working Paper No. 102. Indian Council for Research on International Economic Relations (ICRIER), New Delhi, India.

Steenblik, R. 2005. Environmental Goods: A Comparison of the APEC and OECD Lists. OECD Trade and Environment Working Paper No. 2005-04.

Sugathan, M., Claro, E., Lucas, N. Marconini, M. and E. L. Fuentes 2007. Trade in Environmental Goods and Services and Sustainable Development: Domestic Considerations and Strategies for WTO Negotiations. Environmental Goods and Services Series. Policy Discussion Paper. International Centre for Trade and Sustainable Development (ICTSD). Geneva, Switzerland.

UNCTAD, 1995. Environmentally Preferable Products (EPPs) as a Trade Opportunity for Developing Countries.UNCTAD/COM/70. Geneva: UNCTAD.

World Bank Study, 2008. International Trade and Climate Change: Economic, Legal, and Institutional Perspectives.

Zhang, Z. 2009. Liberalising Climate-friendly Goods and Technologies in the WTO: Product Coverage, Modalities, Challenges and the Way Forward. Munich Personal RePEc Archive Paper No. 18343.

# 11

# Climate Change Debate in Nepal: Policies and Challenges

*Anjan Kumar Sahu*

## Introduction

Lately, the discourse of security has been heavily slanted in favour of 'climate security' rather than environmental security. The discourse has redefined the nature and understanding of intra-state as well as interstate security concerns. Some countries continue to give optimum premium on linkage between national security and climate change, and some gives subsidiary importance to the linkage. However, the discourse of national security and climate change linkage has been primarily more vigorous in developed countries rather than in developing countries; and the empirical findings of the association, though disputed, has been found in developing countries. Scholars from developed countries argue that national security of developed countries like the US will face new challenges like threat of environmental degradations emanating from developing countries (Homer Dixon Autumn 1991, 1994; Kaplan 1994; Butts 1994, cited in Elliott 2004: 210). However, the security discourse in developing or least developed countries (LDCs) depends on both national and international debate on the nature of the issue. At the international level, climate change has been dubbed as a security issue (United Nations Security Council 2007); however, at the national level, the security aspect of climate change varies among the nations. Nevertheless, climate change precariousness in South Asia, which comprises both developing and LDCs, is an interesting area of study for scholars to look into a region which is blessed with replete of water and diversity of natural eco-system and subjected to enduring inter-state and intra-state wars. The warring nature has implications for the present and future climate change security situations in the region.

South Asia is one of the most militarised and securitised regions in the world and the issue of climate change may exacerbate the existing problems. There has been an animated debate on putting climate change within a national security framework of the South Asian countries because out of 32 conflicts in the region 18 are based on natural resource conflict (Rustad and Binningsbo 2012: 9). In India, the linkage has been well-established (Sahu 2011), however, and in other countries like Nepal the intersection has been in the background or invisible. The focus of the paper is, therefore, to flesh out the security aspect of climate change in Nepal, one of the LDCs in South Asia.

At the outset, it explains the debate on causative factor causes environmental degradation in Nepal or to analysing environmental degradation or climate change can play an independent role for other anomalies. Secondly, it analyses the recent trends and policies of the Government of Nepal (GoN) to contend with the perils. The next section explicates a debate to intertwine human security with national security as far as Nepal's climate change policy is concerned. This debate is an attempt to help both policy-makers and political leaders to formulate their policies according to the present climate change politics at the national as well as international levels.

## Three Debates: Socio-Economic, Political and Environment

### *Socio-economic Situation*

According to Human Development Report 2011 (2011: 129), Nepal is one of the low human development countries and ranks 157 out of 187 countries in the Human Development Index (HDI). There has been a wide chasm both in income inequality and in urban and rural poverty. The existing poverty compels the people to depend on natural resources for their livelihood and it puts excessive burden on eco-system. The socio-economic deprivation drives poor people to exploit natural resources wantonly without heeding the sustainable and judicious use of the resources. This leads to wanton consumption of resources, deforestation, soil erosion and degradation of environment. Many scholars argue that conservation of forest is the best strategy to protect the forest and water resources from the destructive practice of local people. At the same time, the socio-economic deprivation propels poor people to prevent the government initiated environmental projects, which might have beneficial for the poor or local people. For example, the rise of unemployment and discrimination led to the people to protest against the Melamchi Megha Drinking Water Project as the project was unable to fulfil the demand for employment of local people (Upreti 2007: 20). The interlinking of poverty with environment is not a new factor. The relationship between the two has been established by a number of scholars like Dabelko et. al. (2000, cited in Najam 2004), Najam (2004), Bhurtel and Ali (see the references) and others. Dabelko et. al. (2000) argue that 'poverty is a key factor in causing tension, unrest and, eventually, conflict' (cited in Najam 2004: 161). According to Najam (2004: 161).

All across South Asia, poverty emerges as the key variable, both in terms of defining environmental degradation and outlining human insecurity. Importantly, poverty is both the casual motivator of environmental stress as well as the most important manifestation of human insecurity.

The socio-economic dimension of Nepalese society has implications for environmental degradation as the nature of society and economy has been vitiated by the entrenched control of powerful political elites with the support of military. The enduring and conflicting nature of society has given political and military rulers to cement their plenary power and use natural resources without heeding the needs and aspiration of the destitute or forest-dwellers. The control of resources has resulted into poverty and underdevelopment. Political ecologists like Peluso and Watts (2001) and Rustad and Binningsbo (2012: 4) have ardently argued for an equal distribution of natural resources among the people and they blame political or powerful elites who control and regulate the natural resources for their vested interests. In Nepal, the Maoist and poor people are continuously demanding for accessing the natural resources including land and water and protesting against the 'entrenched discrimination' (Rotberg and Swain 2007: 8), both socially and economically. According to Rustad and Binningsbo (2012: 4).

Many rebel groups, such as the insurgents in Nepal, justify their fighting and attract supporters by highlighting unfair distribution of natural resource revenues and ownership as a cause of violence and promise, for example, land reform if they get into power.

| *Areas* | *Population below the poverty line* | | | *Poverty Gap* | | *Intensity of Poverty* | |
|---|---|---|---|---|---|---|---|
| | *2003/04* | *2008/09* | *2009/10* | *2003/04* | *2008/09* | *2003/04* | *2008/09* |
| Urban | 9.55 | 7.63 | --- | 2.20 | 1.70 | 0.70 | 0.56 |
| Rural | 34.62 | 28.54 | --- | 8.50 | 6.89 | 3.10 | 2.40 |
| **Nepal** | **30.85** | **25.39** | **25.4** | **7.60** | **6.10** | **2.70** | **2.12** |

*Source*: Central Bureau of Statistics, cited in Ministry of Finance 2011: 91-95.

### *Political Instability*

Political nature and system of a country shapes environmental policies. A stable political system can prevent or minimise both internal and external traditional and/or non-traditional threats like environmental insecurity. Democracy, transparency and participation of the people are the sine-qua-non for implementing environmental policies. According to Buzan et. al. (1998: 8), 'Political security concerns the organisational stability of states, systems of government and the ideologies that give them legitimacy.' However, the Government of Nepal has failed to provide political security and political instability and insecurity is the common phenomenon. The recent history of Nepal politics shows that there has been dearth of democratic political engagement of political parties and the people. The state-controlled political system led to the failure of governance structure and the rise of Maoist group in 1996. The Maoist group challenged the legitimacy of the political system of Nepal and the armed struggle between the Government and the disgruntled Maoist continued till the political arrangement came into being in 2006 resulted into Comprehensive Peace Agreement (CPA). The unsettled political insecurity in Nepal fomented environmental degradation. The armed struggle in the forms of training, landmines, use of forest resources for the purpose of production and purchase of armaments led to an unprecedented use and abuse of natural resources.

Both the Maoist and the ruling party have been vying for the control of state machinery in the name of people, security and development. For example, the failure of the institutional machinery and the officious and centralised system of administration led to the unsuccessful and ineffective role of Community Forestry User Group (CFUG). Both the Maoist and the government tried to regulate the CFUG to counter each other and, as a result, the affirmative objectives of CFUG became a matter of contention and derision (Upreti (2006: 151).

It has also been found that the Community Forests of the country are becoming the shelters of both conflicting parties. The insurgents train and take refugee inside the Community Forests. When they are present inside the forest, access to the forest is restricted and those who do enter run the risk of being labelled a 'spy'. If the state military notices the presence of the insurgence inside the forest, they also circulate the prohibitory order to the CFUGs for not entering the forest... In addition, it has been found that the state military has established the military camps inside the Community Forests.

**Community Forestry**

| | |
|---|---|
| Number of Community Forestry User Groups | 15256 |
| Area Handed Over Community Forests | 1350655 Hectares |
| Number of Beneficiary Household | 1782550 |

*Source*: Ministry of Finance 2011: 172.

Absence of effective institutions or process of institutionalisation and entrenchment of bureaucratic practices of government also encourage political leaders and bureaucrats to favour large hydro projects as it fetches hefty amount. The linkage between large hydro-projects and corruption is a common feature in Nepal and the 402 MW Arun III and the 300 MW Upper Karnali Hydro-Electric Project are clear cases of such instances (Upreti 2007: 25).

The water resources policy of Nepal is full of confusion and inconsistency...It is operating on an ad-hoc basis, depending upon the interests of the governing political party, or even invisibly controlled by the nexus of individual influential political leaders and external forces. Water resource institutions in Nepal are undermined by political tradeoffs corruption, over-staffing, incompetence and inadequacy in providing services to people (Urpeti 2007: 30).

### *Environmental Degradation*

Socio-economic and political factors are not the sufficient causes of environmental degradation. Environmental degradation can cause social, economic and political instability and this dimension has been more clearly corroborated with the rise of the intensity of climate change related threats. Presently, both developed and developing countries have been looking at the climate change induced socio-economic and political instability. In Nepal 92 per cent of the economically active, people directly dependent on natural resources for their livelihood (HDR 2011: 54). The natural resources of the country are linked particularly with forest and water. The 'State of the Environment 2001' report published by the United Nations Environmental Programme indicates that within the last 15 to 16 years the country has lost 24 per cent of the forest coverage (Bhurtel and Ali, see the References). The present and potential climate change variability may cause migration, low productivity, variation in water flow, political instability, energy insecurity and extreme events like flood, droughts etc. As far as energy is concerned, the energy consumption in Financial Year (FY) 2009-10 rose by 5.5 per cent as compared to FY 2008-09 totalling 9,911 tons of Oil Equivalent (TOE) (Ministry of Finance 2011: 164). There is a report that the shortfall of electrical energy by 200 MW in rainy season and 600 MW in the winter season forced the nation for up to 18 hours load shedding in the winter of 2009 (Ministry of Environment 2010: 23). Thus, the government has the target of producing up to 2,035 MW hydropower electricity by 2017 and up to 4,000 MW by 2027 (WECS 2011: 19). Likewise, the climate change impact may have ramifications for the already existing food insecurity situation in the country. The total food deficit for the mountain, mid-hill and the *Terai* region is at 127, 483, and 125 thousand tons respectively (Ministry of Environment 2010: 43). To meet the national food demands Nepal imports over 485 thousand tons of foods every year (Ministry of Environment 2010: 43). In his speech at the High-Level segment of the United Nations Climate Change Conference (COP 17 and CMP 7) at Durban, South Africa, in 2011, the Minister for Environment of Nepal, Mr. Hemraj Tater enunciated that:

A recent study has ranked Nepal as the fourth climate vulnerable country. Over 40 per cent of the total population, about 12 million people, are identified as vulnerable to climate change and are at risk. During the last decade alone, over 7 thousand people died due to climate-induced disasters such as floods and landslides with an estimated loss of US $ 5.3 billion. The intensity, frequency and scale of impacts are alarming...Despite our limited capabilities; we have recognised adaptation as a national 'survival strategy' to protect our people, livelihood and ecosystems.

Lately, Climate change adaptation is the top priority of the government as the adaptation issue has been considered as a 'survival strategy'. The potential threat has forced the Government of Nepal to formulate a climate change policy, which is different from socio-economic and political issues, though latter issues are invariably interlinked. Opschoor has rightly observed that 'ecological stress and the consequence thereof may exacerbate tension within and between countries' (cited in Gleditsch 1998: 382). The following tables show the climate change induced mortality in Nepal and this gives an impression that the threat is an empirical reality and its independent analysis must not be muddled up with other issues. At the end, it can be said that the foregoing factors have their own significance and closely connected, however, recently, the debate

**Mortality Due to Extreme Temperature Events**

| *STROKE* | *2004* | *2005* | *2007* | *2008* |
|---|---|---|---|---|
| HEAT | 20 | 80 | 60 | 75 |
| COLD | 9 | 10 | 18 | 30 |

*Source*: DOHS (2004-2009), cited in Ministry of Environment 2010: 69.

**Climate Induced Disaster and Mortality in Nepal**

| *Events* | *1998* | *1999* | *2000* | *2001* | *2002* | *2007* |
|---|---|---|---|---|---|---|
| Epidemics | 840 | 1207 | 141 | 154 | 151 | 200 |
| Avalanches | 0 | 5 | 0 | 0 | 0 | 0 |
| Fires | 54 | 39 | 38 | 26 | 11 | 12 |
| Flood/landslides | 273 | 193 | 173 | 196 | 441 | 216 |
| Earthquakes | 0 | 0 | 0 | 1 | 0 | 0 |
| Thunderbolts | 17 | 16 | 23 | 36 | 3 | 46 |
| Wind and hailstorm | 6 | 6 | 3 | 2 | 3 | 0 |
| Total | 1190 | 1466 | 378 | 415 | 609 | 474 |

*Source*: Ministry of Environment 2010: 70.

has slanted in favour of climate change induced socio-economic and political challenges.

**Recent Trends and Policies**

In Nepal, the climate change debate became a national agenda in 2009 and that led to the formation of a twenty-five member Climate Change Council under the chairmanship of Prime Minister in July 2009. Subsequently, the government established the Climate Change Management Division in the Ministry of Environment and the ministry formulated the National Adaptation Programme of Action (NAPA) in 2010.

The GoN approved the Climate Change Policy, 2011, on 17 January 2011 as climate change challenges have been perceived as an impediment to socio-economic development of the country. It is 'an urgent necessity to address the issue' (Government of Nepal 2011: 1) and it is 'urgently required' (Government of Nepal 2011: 4) to formulate and implement climate change related polices to minimise and mitigate the 'existing and likely impacts in different ecological regions' (Government of Nepal 2011: (1) The Prime Minister of Nepal enunciated that:

> Despite our daunting economic and geographical challenges as a least developed and land locked country and limited technical and institutional capacities, we have identified national framework of actions for sustainable development with a multi-stakeholder approach. We have created a policy framework, plan of action, institutional set-up, monitoring and evaluation arrangements and high-level co-ordination mechanism at the central level. As part of our national initiatives, we have launched Sustainable Community Development and Community Forestry Programmes, rural energy programme as well as inclusive governance. Likewise, we are giving priority to hydropower development and sustainable tourism in Nepal as part of our green growth strategy.

The government's main goal is to give climate change issue as an independent significance as the impacts have implications for socio-economic and political stability. The main objectives of the policy are:

1. To establish a Climate Change Centre as an effective technical institution to address issues of climate change and also strengthen existing institutions.
2. To implement climate adaptation-related programmes and maximise the benefits by enhancing positive impacts and mitigating the adverse impacts.
3. To reduce GHG emissions by promoting the use of clean energy, such as hydroelectricity, renewable and alternative energies, and by increasing energy efficiency and encouraging the use of green technology.

4. To enhance the climate adaptation and resilience capacity of local communities for optimum utilisation of natural resources and their efficient management.
5. To adopt a low-carbon development path by pursuing climate-resilient socio-economic development.
6. To develop capacity for identifying and quantifying present and future impacts of climate change, adapting to climate risks and adverse impacts of climate change.
7. To improve the living standard of people by maximum utilisation of the opportunities created from the climate change-related conventions, protocols and agreements (Government of Nepal 2011).

Nevertheless, policy relating to water is the main priority of the government because water relates to economy, energy, agriculture, and livelihood security of the people. There are about 6,000 rivers in Nepal with a total drainage area of 194,471 sq. km. Out of this, 74 per cent lies within the country. Thirty-three of these rivers have a drainage area that exceeds 1000 sq. km (WECS 2011: 2). Nepal has the capacity to produce 83,000 MW hydropower and this potential energy production has given impetus to the government and bureaucrats to go for large, medium and small hydro power projects. Out of this estimated potential, 114 projects with a combined capacity of 45,610 MW have been identified as economically feasible. The recent estimate at 40 per cent dependable flows-for the run-of-river (ROR) hydropower potential in Nepal stands at 53836 MW. Some hydropower projects are already built up and some are in the pipeline (WECS 2011: 15). There are also major hydropower project co-operation between India and Nepal. Currently, four major storage projects

**Estimated Runoff from the Rivers of Nepal**

| *Sl. No.* | *River* | *Length (km)* | *Drainage Area* | | *Estimated Runoff ($m^3$/sec)* | |
|---|---|---|---|---|---|---|
| | | | *Total* | *Nepal* | *From all Basin* | *From Nepal* |
| 1. | Mahakali | 223 | 15,260 | 5,410 | 698 | 247 |
| 2. | Karnali | 507 | 44,000 | 41,890 | 1441 | 1371 |
| 3. | Babai | 190 | 3,400 | 3400 | 103 | 103 |
| 4. | West Rapti | 257 | 6,500 | 6,500 | 224 | 224 |
| 5. | Narayani | 332 | 34,960 | 28090 | 1753 | 1409 |
| 6. | Bagmati | 163 | 3,700 | 3,700 | 178 | 178 |
| 7. | Sapta Koshi | 513 | 60,400 | 31940 | 1658 | 878 |
| 8. | Kankai | 108 | 1330 | 1330 | 68 | 68 |
| 9. | Other River | - | 24921 | 24921 | 1001 | 1001 |
| | Total | | 194,471 | 147,181 | 7125 | 5479 |

*Source:* WECS 2003, cited in WECS 2011:8.

are proposed as Indo-Nepal cooperative initiatives: the Chisapani-Karnali (10,800 MW), the Pancheswor (680 MW), the Budhi Gandaki (600 MW) and the Sapta Koshi High Dam (3600 MW) (WECS 2011: 16). There are reports that the total required investments for Nepal's water sector over 25 years are estimated to be in the order of Rs. 1,000 billion, excluding investments needed to develop sufficient hydropower capacity for export and large multipurpose projects such as Pacheswar (WECS 2002: xxv).

**Existing Hydropower Plants**

| *River Basin* | *Name* | *Capacity (MW)* |
|---|---|---|
| Bagmati | Kulekhani I* | 60.0 |
| | Kulekhani II* | 32.0 |
| | Panauti | 2.4 |
| Rapti | Sunkoshi | 10.5 |
| | Indrawati | 7.5 |
| | Khimti | 60.0 |
| | BhoteKoshi | 36.0 |
| Kankai | Puwa | 6.2 |
| | Trisuli | 24.0 |
| | Devighat | 14.1 |
| | Gandak | 15.0 |
| | Marsyangdi | 69.0 |
| Narayani (Gandak) | | |
| | Middle Marsyangdi | 70.0 |
| | Andhikhola | 5.1 |
| | Chilime | 20.0 |
| | Kali Gandaki A | 144.0 |
| | Modi | 14.8 |

*Source:* NEA 2009, cited in WECS 2011: 16.

**Large Hydropower Projects with Feasibility Study**

| *Project Name* | *Capacity (MW)* | *Cost million USD* | *Year of Study* |
|---|---|---|---|
| Karnali (Chisapani) | 10,800 | 7,666 | Updated in 2001 |
| Pancheswor | 6,480 | 2,980 | 1995 |
| West Seti | 750 | 1,098 | 1997 |
| Arun III | 402 | 859 | 1991 |
| Upper Tamakoshi | 309 | 464 | May 2005 |
| Upper Karnali | 300 | 454 | 1998 |
| Dudhkoshi | 300 | 690 | 1998 |
| Andhi Khola | 176 | 463 | 1997 |
| Tamur-Mewa | 101 | 191 | 1998 |

*Source:* WECS 2003, cited in WECS 2011: 17.

**Human Security and National Security: An Intersection**

The enduring political bargaining between the Maoist group and the Government, failure of the government for providing basic facilities, rise of socio-economic inequality and the present climate change threats are the prime factors for human insecurity in Nepal. To keep the Nepalese from the freedom from hunger and other basic facilities and engendering freedom from fear, both internal as well as external, shall be the main priority of the government. Though it is observed that government is trying hard for providing basic facilities by initiating number of poverty alleviation programmes and other socio-economic development issues while formulating climate change policies, the government has failed to make the climate change as a political or security issue.

However, the issue of hydro-projects may get a political push in the light of climate change crisis because large hydro-projects in South Asia are primarily politico-economically driven. The GoN has ambitious project for building large hydropower projects and production of hydropower, but rise of episodic and violent social and environmental movements are against such large projects. Large projects cause human insecurity in the forms of displacement, livelihood insecurity, sedimentation, and emission of Green House Gases (WCD 2000). The report of the Word Commission on Dam (2000: 176) mentions that 'in some cases, such as Norway, Quebec, Brazil or Nepal, a high level political choice made in favour of hydropower has driven subsequent choice of technology (large dams) and project development'. Dixit and Gyawali (2004: 207) also echoed the same view that both the left and right political parties have betrayed the people of Nepal by building large hydropower projects and this unitary solution has been used as a mechanism in the name of economic and social benefits of the people.

According to Najam (2003: 68), in order to look into the environmental degradation in South Asia, the nature of human security needs to be understood within the framework of sustainable development. However, the use of the term sustainable development itself a contested and a debatable issue and the paper consciously avoid the use of the term. As far as Nepal is concerned, people should feel the freedom from the climate change induced socio-economic deprivation and internal and external insecurities. Political leaders in Nepal concede that climate change is a 'threat', but the threat is framed for the' survival of people', not for the survival of the state. According to the Minister of Environment of Nepal, Mr. Sharma, snows and glaciers are retreating, sea levels are rising, human lives and civilisations are at great threat... We must act with utmost urgency and due diligence to save lives of millions and their precious civilisations. Indeed, addressing the challenges of climate change is a moral obligation and existential imperative.

However, the feeling of insecurity from climate change induced external threats has not been a major policy of policy-makers and political leaders. They are unable to make it a political and public issue and unable to foresee the potential conflict over the water resources of the region. The relentless rise of China and India will certainly put pressure on Nepal for rethinking and reconfiguring of climate change policy. According to Najam (2003: 68), 'the challenge of environmental security in South Asia is both principally a challenge at the domestic level as well as a challenge common to the region.' The 'moral obligation' to contend with climate change threats exacts metamorphosis of moral obligation to politico-security obligation. However, though the linkage between national security and climate change has not been a clear manifestation in the climate change policy of the government, the inclusion of Ministry of Foreign Affairs and Ministry of Home Affair in the Climate Change Council has shown a latent and muted move on the security aspect of climate change. The Ministry of Home Affairs is also the nodal agency to co-ordinate the activities of different departments relating disaster management. The inclusion of the Ministry of Foreign Affairs and Home Affairs, the former deals with external relations with other countries and the latter has the onus of both internal and external security, indicates an inclination for linking security with climate change threats. However, the climate change security policy of the government primarily sees the challenge stemming primarily from internal security, not external security.

The mutual supportive role of security and climate change has been the crux of the present climate change debate. Nepal's territorial contiguity with India and China, and the potential competition and conflict over water between India and China shall be an opportunity for Nepal to rethink its climate change policy in the light of regional water security because nations assert their rights over water and build 'empire of water' for the sake of 'national pride' (Lahiri-Dutt 2008: xx-xxi). China can use water as a strategic weapon to secure its water supply and divert the water flow towards the arid region of the country that has implications for the entire region of the South Asia. The geo-political significance of water becomes an object of military activities in the region, as it is invariably an important component of a nation's power (Lahiri-Dutt 2008: xxx). Scholars from Nepal like Rajan (2009: 33) avers that securitisation of socio-economic, political, ethnic and environment is the sine-qua-non for dealing with the problems in Nepal and this requires an urgent response of the state. However, this argument of securitisation has four problems. Firstly, the securitisation of all the issues at a time dilutes the importance and independence of climate change. Secondly, securitisation of all the issues at a time dilutes the significance of either socio-economic or political or environment and muddles up the issues. Thirdly, the immediate state response may give political and military rulers excuse to exercise their plenary and coercive power, which can cause militarisation of society, polity, economy and environment; and finally, the excessive role of state may prompt to act on the cause of national security based on military security and defeats the very purpose of human security. For example, earlier Nepal's rulers viewed the forest as a strategic resource both for the offensive and defensive purposes and forest 'was used to secure their power and position through 'buying' allegiance and supporting weapons of manufacture' (Donovan 2011: 241-42). Rustad and Binningsbo (2012: 4) also underlined the same argument that natural resources provide an important source for rebel and the government to further their offensive strategies both in civil war or inter-state conflicts like Casamance in Senegal, timber export in Burma and illegal levying logging taxes in the Philippines. The securitisation of climate change may further militarise the forest or water in the name of adaptation and mitigation of climate change threats. According to Donovan (2011: 256):

> 'The important role of forests in maintaining the hydrological regime, soil stability, biological diversity, and wildlife habitat, though appreciated by local communities, appears to have been largely ignored in favour of income generation. Nepal's position, both geographic and political, vis-à-vis the British administration India, facilitated such a strategy.'

Climate change or environmental issues are in the national agenda, but the securitisation of the issue needs to be integrated with democratic process. For example, when military and executives use excessive power and bring climate change legislation without consulting the affected people or visiting the affected sites, people of the country put question mark over the intention and functions of political and military leaders. The democratisation and deliberation of climate change issues and policies can give ample latitude to political leaders and military leaders to carry out adaptive and mitigating activities (Sahu 2012). Securitisation and democratisation shall be the priority of political leaders while dealing the climate change challenges that can only fulfil the objectives of both human and national securities. Securitisation causes state security and democratisation breeds human security and the intermingling of the two serve the purpose of both human and national security. However, few scholars argue that non-traditional security threats can be staved off using non-traditional means (Naham 2004: 5; Deudney 1990). However, this understanding of non-traditional threats like climate change prevents and disregards the importance of military technology and efficacious role of military institutions. For example, in 1997, the Nepalese government set up temporary army camps at different locations along the Rolwaling Valley downstream of the lake to warn people through Nepal's main radio station that a glacial lake outburst flood event had occurred (Matambo and Shrestha, see the references). Upreti (2007: 35) argues that the water policy of Nepal has to be 'integrated with the foreign policy into a holistic vision' and needs new institutional arrangements. The new institutional arrangements mean redefining the role and functions of institutions dealing with military and foreign policy. The role of army or military cannot be avoided while dealing the unforeseen and unprecedented threats like climate change. Military and non-military means are useful for tackling such threats.

Securitisation = State Security (Military)
Democratisation = Human Security (Non-military)
Securitisation and Democratisation = State Security + Human Security
Military) (Non-military)

**Conclusion**

A country specific analysis on climate change issue is not an easy task to understand the complexity and the politics of the issue. However, the issue has been framed as politico-security issue at the international level and the complexity of international climate security will be more shambles and muddle up unless we study the internal complexity of a particular country and a region.

South Asia is one of the versatile regions as far as its economy, polity, security and environment are concerned and Nepal has its own distinction. The national and transnational water linkage has been an additional development in the recent climate change discourse both in South Asia and in Nepal. Taking into account the challenges, the GoN has undertaken number of climate change policies and has emphasised the independent role of climate change as a field of study and its repercussions on political economy. However, these policies are more centralised and narrowed in its use and scope. Policies are politically motivated and restricted to minimise the human insecurity at the national level, though the government has failed to provide minimum human security.

However, one thing that misses in the climate change policy of the government is absence of security aspect of the issue. This may be the lack of interest or deliberate ignorance of the political leaders or people. The attempt to give importance to human security utterly misses the national security of the country. The national security aspect of climate change is an empirical reality and the people and political leaders have to accept the reality and to act on it. The security aspect of climate change needs intersection of people and political leaders in the decision-making mechanisms, and both human and national security through democratic process.

## APPENDIX
## Composition of Climate Change Council

| | |
|---|---|
| Right Honourable Prime Minister | Chair |
| Honourable Deputy Prime Minister | Vice-Chair |
| Honourable Minister for Environment | Member |
| Honourable Minister for Forests and Soil Conservation | Member |
| Honourable Minister for Finance | Member |
| Honourable Minister for Foreign Affairs | Member |
| Honourable Minister for Agriculture and Cooperatives | Member |
| Honourable Minister for Energy | Member |
| Honourable Minister for Industry | Member |
| Honourable Minister for Health and Population | Member |
| Honourable Minister for Home Affairs | Member |
| Honourable Minister for Local Development | Member |
| Honourable Minister for Law and Justice | Member |
| Honourable Vice-Chair, National Planning Commission | Member |
| Honourable Member (Environment), National Planning Commission | Member |
| Chief Secretary, Office of Prime Minister and Council of Ministers | Member |
| Experts (8) nominated by the Government of Nepal | Member |
| Secretary, Ministry of Environment | Member Secretary |

## References

Bhattarai, R. 2009. 'The Concept of Human Security and Changing Security Dynamics in Nepal', in R. Bhattarai and R. Cave (ed.), *Changing Security Dynamics in Nepal: A Collection of Essays*, Nepal: Nepal Institute for Policy Studies.

Bhurtel, J. And S. H. Ali, 'The Green Roots of Red Rebellion: Environmental Degradation and the Rise of the Maoist Movement in Nepal', [Online: web] Accessed on 24 June 2012, URL: http://www.uvm.edu/~shali/Maoist.pdf

Buzan, Barry *et. al.* 1998. *Security: A New Framework for Analysis*, Boulder-London: Lynne Rienner.

Deudney, D. 1990. 'The Case Against Linking Environmental Degradation and National Security', *Millennium: Journal of International Studies*, 19(3): 461–76.

Dixit, A. and D. Gyawali. 2004. 'A cultural theory perspective on environment and security issues in Nepal', in R. Thakur and E. Newman (ed.), *Broadening Asia's Security Discourse and agenda: Political, Social, and Environmental Perspective*, Tokyo: United Nations University Press.

Donovan, D. G. 2011. 'Forest at the Edge of Empire: The Case of Nepal', in Deepak Kumar *et. al.* (ed.). *The British Empire and the Natural World: Environmental Encounters in South Asia*, New Delhi: Oxford University Press.

Elliott, L. 2004. *The Global Politics of the Environment*, New York: Palgrave.

Gleditsch, N. P. 1998. 'Armed Conflict and The Environment: A Critique of the Literature', *Journal of Peace Research*, Vol. 35, No. 3.

Government of Nepal. 2011. Climate Change Policy, 2011, [Online: web] Accessed on 20 June 2012, URL: http://www.moenv.gov.np/newwebsite/

Homer-Dixon, Thomas F. Fall 1991. 'On the Threshold: Environmental Changes as Causes of Acute Conflict', *International Security*, 16 (2): 76-116.

Homer-Dixon, Thomas F. Summer 1994. Environmental Scarcities and Violent Conflict: Evidence from Cases', *International Security* 19 (1): 5-40.

Human Development Report. 2011. *Sustainability and Equity: A Better Future for All*, New York: UNDP.

Kaplan, R. D. February 1994. 'The Coming Anarchy', *The Atlantic Monthly*, [Online: web] Accessed on 4 January 2008, URL: http://www.TheAtlantic.com

Lahiri-Dutt, K. 2008. 'Introduction: Placing Water First', in K. Lahiri-Dutt and Robert J. Wasson (ed.), *Water First: Issues and Challenges for Nations and Communities in South Asia*, New Delhi: Sage.

Matambo, S. and A. B. Shrestha. 'World Resource Report Case Study. Nepal: Responding Proactively to Glacial Hazards', Washington DC: World Resource Report. [Online: web] Accessed on 15 July 2012, URL: http://www.worldresourcesreport.org

Ministry of Environment. 2010. *National Adaptation Programme of Action (NAPA)*, Kathmandu: Government of Nepal.

Ministry of Finance. 2011. *Economic Survey 2010-11, Vol. I*, Government of Nepal, [Online: web] Accessed on 20 Jane 2012, URL: http://www.mof.gov.np/contentFiles-Content-dW0=

Najam, A. 2003. 'The Human Dimensions of Environmental Insecurity: Some Insights from South Asia', ECSP Report, Issue 9. [Online: web] Accessed on 4 July 2008, URL: http://www.wilsoncenter.org/sites/default/files/najam.pdf

Najam, A. 2004. 'Sustainable and human security: Lessons from South Asia', in R. Thakur and E. Newman (ed.), *Broadening Asia's Security Discourse and Agenda: Political, Social, and Environmental Perspective*, Tokyo: United Nations University Press.

Peluso, N. L. and M. Watts. 2001. 'Violent Environments', in N. L. Peluso and M. Watts (ed.), *Violent Environments*, Ithaca and London: Cornell University Press.

Rotberg, F. and A. Swain. 2007. 'Introduction', in F. Rotberg and A. Swain (ed.), *Natural Resources Security in South Asia: Nepal's Water*, Singapore: Institute for Security and Development Policy.

Rustad, S. A. and H. M. Binningsbo. 2012. 'A price worth fighting for? Natural resources and conflict recurrence', *Journal of Peace Research*, [Online: web] Accessed on 15 July 2012, URL: http://jpr.sagepub.com/content/early/2012/07/09/0022343312444942.abstract

Sahu, A.K. 2011. 'Securitisation of Climate Change: A Case Study of India's National Action Plan on Climate Change (NAPCC)', *The IUP Journal of History and Culture*, Vol. V, No. 4.

Sahu, A.K. 2012. 'Climate Change and National Security: An Intersection', *The IUP Journal of International Relations*, Vol. VI, No. 3.

United Nations Security Council. 2007. Security Council Holds First-Ever Debate on Impact of Climate Change on Peace, Security, Hearing Over 50 Speakers. *Department of Public Information: News and Media Division*, April 17. [Online: web] Accessed on 24July 2008, URL:http://www.un.org/News/Press/docs/2007/sc9000.doc.htm/

Upreti, D. R. 2006 .Conflicts in natural resources management- Examples from community forestry'. [Online: web] Accessed on 4 July 2008, URL: http://www.boku.ac.at/oega

Upreti, D. R. 2007. 'Changing Political Context, New Power Relations and Hydro-Conflict in Nepal', in F. Rotberg and A. Swain (ed.), *Natural Resources Security in South Asia: Nepal's Water*, Singapore: Institute for Security and Development Policy.

Water and Energy Commission Secretariat (WECS) (2011), *Water Resources of Nepal in the Context of Climate Change*, Kathmandu: Government of Nepal.

World Commission on Dams (WCD) (2000), *Dams and Development: A New Framework for Decision Making*, London and Sterling: Earthscan Publication.

# 12

# Climate Change vis-à-vis Human Security in Maldives: Prospects and Perspectives

***Subhashree Patnaik***

*Maldives is imperiled by climate change. On most world maps, the Maldives Islands are tiny specks in the middle of the Indian Ocean. But in a few years, those dots may disappear.*

*Dr Mohamed Waheed, Vice President, Maldives*

**Introduction**

If glittering tourists with their colourful fantasies throng into Maldives mushrooming magnificent resorts boosting its spectacular economic growth, or its energetic fishing industry places the country better than its counterpart like Bangladesh, Nepal, India, Pakistan, etc. in prospects of human development index (UN Human Development Index, 2011). if after 30 years of draconian autocratic rule, the country experiences stability by promising democracy, the country experiences no major civil or political crises since then, generating new hopes of personal as well as political human security, a deep sense of fear and suspicion. However, of a prospective monster, the monster of environmental crisis and its implicating human insecurity grasps the whole country compelling the president Mohammed Nasheer searching lands in India and Sri Lanka. Yes, the monster is approaching; perhaps after causing devastating human insecurity in this beautiful country will swipe away its signature from the global map. At this point a study on prospects and perspectives of climate change vis-à-vis human security in Maldives becomes significant. Even issues of climate change have received a significant part in the contemporary governmental policy of Maldives only because climate change has been turned into a major threat even for the very existence of the country. A brief peep into the country's current determinants of human security is, therefore necessary to excavate the prospective implication of climate change on the process of human security.

The main objective of this article is to reveal how global climate change has affected the human security in Maldives. To demonstrate it, the paper is manifested in five ramifications, i.e., nuances of the relationship between climate change and human security, the paradise panorama, the paradise almost lost, the paradise to be regained and a conclusion. At the outset, the paper glides us to understand the nuances of the relationship between climate change and human security. The second part attempts to spell out the geography, population, economy and present political scenario in Maldives, etc. Then it proceeds to show how climate change is impacting the present situations in Maldives such as economy, geography, health, water, biodiversity. Furthermore, it will highlight the perspectives of climate change in Maldives under the heading paradise to be regained. In this significant part, it will enlist the steps taken by the Maldivian government to combat climate change, and last but not the least its future perspectives in the concluding remarks.

**The Paradigm Shift**

The scenes are dimming of the time, when the world was torn apart through the swords and gunpowder; the territories were constructed, expanded and deleted. When human beings of flesh and blood were treated as mere goods of presentation and were made to somebody according to the whims of the kings or queens, when capable were made incapables and then they were thrown to the concentration camps. Gone are the days of imperialism, when people of a particular region are bombarded because somewhere, at some part of the world, a whimsical ruler wanted to impress his girlfriend, or because some other rulers decided to join the adventure of this man at a kitty party. Gone are the days when people shook with fear at the Russian ships or American aeroplanes. However, Russian ships have been dismantled and American planes have been scattered with the demise of Soviet Union, the end of cold war, and emergence of power hubs like China, Japan, India, etc. The world population has started desedimenting the layers of fears and frustrations from tanks and missiles. Globalisation, liberalisation, information technology, etc. have radically ridiculed the territorial significance. At the same time the gradual and steady strengthening of international organisations like UNO have frustrated national importance. The role, importance and therefore arrogance of a nation state associated with human security have been proved ridiculous. Rather, the popular perception of the nation as the caretaker has been turned into as an exploiter.

History flashes the fact that since its organisation, the modern state has been associated with the national security by deploying military and deterring the external aggression. Down through the century, national security has been construed by means of military extravagansa. History has also been the witness to the fact that in the name of national security, myriad state patronised atrocities have been strategically rationalised. Political choices and alternatives have been sacked in the name of anarchy and therefore, threats to human security.

Even the security studies literature has traditionally restricted its remit to the discussion of threats to a state's security emanating from human, intentional, activities – such as those threats posed by the military activities of other states or terrorist groups. The explanation for this can be found in the way in which security has usually been characterised as the effort to protect a population and territory against organised force while advancing state interests through competitive behaviour'(Dabelko G and D. Dabelko,1995:3-13). According to Dalby 'cold war versions of security have usually been understood in spatial terms as moves of exclusion and protection is a spatial exercise in distancing and boundary making' (Dalby Simon,1992:503-522).

During the last couple of decades, things, however, radically have gone otherwise. It all dates back to 1970s that witnessed a colossal non-military modalities coming from activities like poverty, population growth, environmental degradation, etc. The above paradigm of security, otherwise known as 'realist' paradigm, that associates national security with only guns and tanks has been proved futile to address these new challenges properly. On the other side, a more holistic approach that transcends the smokes of gunpowder has taken the centre stage. The new paradigm of security aims at destroying the cocoon of traditional view of security and tries to understand its impetus through more social, economic and environmental concerns (Gaan, N. 2004). This view stands with the fundamental assumption that the dimensions of security can be construed as significantly through the growing poverty, economic crackdowns, precarious scarcity of natural resources and social conflicts thereafter, environmental degradation and resulting intra and inter-state conflicts, etc. as it is understood through the military modalities. Three spectacular processes are widely responsible for this paradigm shift. Firstly, the disintegration of Soviet Union was a crucial event in the evolution of the security concept, particularly for the way in which this reduced the threat of full-scale nuclear confrontation between the superpowers and paved space for the discussion of other potential threats to national security. Second the growing awareness of the interdependence of economic relations throughout the world, as manifested in a series of global, economic and financial crisis, has led to an awareness that destabilising changes in economic variables in one part of the world can lead to significant levels of instability and insecurity in distant regions. Third, increasing evidence has emerged that certain environmental problems could endanger the existence of whole communities, as well as exacerbating existing social problems such as poverty, mortality, morbidity, overpopulation, and so on. This consideration has gained particular salience in the light of successive reports on the likely future impacts of environmental degradation and climate change on coastal zones; small islands states and the developing world in broader terms (IPCC, 2001; UNDP, 2007; WCED, 1987).

The contemporary security regime focuses more on 'humans' rather than 'state'. National boundaries are considered less important than the lives of human beings. So today's concept of security is 'people-centered'. It is concerned with how people live and breathe in a society, how freely they exercise their many choices, how much access they have to market and social opportunities and whether they live in conflict or in peace (UNDP, 1994).

According to the United Nations Secretary General (former) Kofi Annan, 'while security policy had traditionally focused on the defense of the territory from external attack it has now come to embrace the protection of community of individual from internal violence. The need for a more human-centered approach to security is reinforced by the continuing dangers that weapons of mass destruction, most notably nuclear weapons, pose to humanity: their very name reveals their scope and their intended objective, if they were ever used' (Anan, Kofi, 2000L: 43—44). That the securities of life and livelihood, pleasure and aspirations, freedom and dignity have become more important than the security of national boundary can be felt from this view point. National security in terms of secrecy and military muscles has turned its hue in protecting human security. Though the portrait of human security had been shaping in discussions of security during most of the post-Cold War era, the United Nations Development Report, 1994 stood as a milestone in this arena, which introduced human security and its allocative spirit. In the *United Nations Development Report,* human security is distinguished from traditional security in five key ways. First, rather than being a concern with weapons, human security is tied to *notions of human life and dignity*. Second unlike traditional security, which was bounded by the borders of individual states, human security is presented as a universal concern *unconstrained by territorial borders*. Most threats, which endanger human security, are common to most people, although it is acknowledged that the intensity of the threat may differ. Third, it is argued that all of the components of human security are *interdependent*. Different threats to human security are related, mutually enforcing, and likely to have global repercussions. Fourth, the provision of human security is thought to be much easier to achieve through *early prevention* rather than through intervention at later stages. Finally, unlike realist security, which is state-centred human security focuses on individuals (UNDP, 1995; Current History 94: 229).

Broadly human security as envisioned in the United Nations Development Report 1994 include seven key elements like economic security, food security, health security, environmental security, personal security, community security and political security (i.e., human rights) (ibid), which are under six major threats such as unchecked population growth, disparities in economic opportunities (both within and between states), migration pressures, environmental degradation, drug trafficking, and international terrorism (*ibid*). Among them, threats of environmental degradation can seriously jeopardize the process of human security. Sensing the peril of environmental threat on human security, a dramatic summary of the problem followed in 2007. Evidence is fast accumulating that, within our children's lifetimes, severe droughts, storms and heat waves caused by climate change to international security just as dangerous - and more intractable - than the arms race between the United States and the Soviet Union during the Cold war or the proliferation of nuclear weapons among rogue states today (Dixon Homer, 2007).

That the perils of environmental degradation on human security are unprecedented and immense is of no doubt. So also, the attitude and outlook towards environmental security is worth praising. In 2004, for the first time in its 103-year history, the coveted Nobel Prize was awarded for work to preserve the environment. In their citation of Wangari Maathai, the Nobel committee said, 'Peace on Earth depends on our ability to secure our living environment.' The Nobel Peace Prize 2007 was awarded jointly to Intergovernmental Panel on Climate Change (IPCC) and Albert Arnold Al Gore Jr. 'for their efforts to build up and disseminate greater knowledge about man-made climate change and to lay the foundations for the measures that are needed to counteract such change' (IPCC, 2007). These awards reflect the growing consciousness that a healthy, bio-diverse environment is an essential aspect of political stability and international security. The contemporary global intelligentsia is now sincerely serious on environmental security as they were never before.

The concept of environmental security, therefore, though in nascent stage has taken significant position in contemporary security and strategic debates. It has emerged to shower awareness of the threats by ever-increasing environmental problems. In one hand thinkers like Winnedfeld and Morries and Moller define environmental security within the context of the traditional security strategy by establishing three main types of linkages such as (i) the impact of environmental problems on the likelihood of inter-state conflicts and wars, (ii) the impact of environmental problems on human survival in its broadest sense ; and (iii) the impact of wars on environmental degradation (Hans Gunter Brauch, 2003: 563-72). These thinkers endeavour to free the environmental issues from the clutches of technocrats and place them among the overall concept of national security, as they could pose more threats to that security than the traditional military sources. On the other hand, thinkers like Nazam (Najam Adil, 2002) and Thakur aims at depoliticising the concept and approaches it from a technical perspective. Whatever difference these definitions might take, the fact remains unaltered about the linkage of human security with environmental security (Selim Mohammed EI Sayd, 2004).

Generally, whereas the first tradition dominates the 'Northern' discourse on environmental security, most 'Southern' analysts and policy makers subscribe to the second tradition. This is understandable given the fear in the South that environmental issues could be used to justify intervention in the domestic affairs of southern countries (*ibid*).

Environmental security, according to Cyril Obi, focuses on controlling a range of threats or contradictions that arise due to the interaction between human beings and nature. This can either be in the form of the extraction of natural resources or their transformation into food, goods and services for human security and commercial purposes. Environmental degradation arising from the extraction and transformation of these resources constitutes a threat, however, because it may lead to the depletion of such resources in terms of either quantity or quality. Resource extraction and transformation are dependent on access, control, ownership and power, and are therefore linked to political economy and security. Overall, environmental change constitutes a major threat to human security (Obi C, 2000).

Environmental changes usually referred to changes in earth's environment, land use and land cover, bio-diversity, atmospheric composition and climate (Page E, 2000) pose serious threat to human security aggravating poverty, social conflict, refugee problem, etc. Changes in environmental variables (such as those affecting access to food and fresh water) trigger social-political effects (such as increasing competition for scare resources) which in turn trigger insecurity enhancing violent conflicts amongst those affected (Dixon, Homer, 1999: 6—8).

In his 2006 book, *The Upside of Down*, he notes that environmental dent deteriorates deprived societies from the bottom up and top down. Apart from other strains – like friction between ethnic groups, external economic stuns, or uncontrollable mega cities – it can scrimmage a society's social fabric; corrode its community, urban, and government institutions; and foster civil violence, including riots, insurgency, guerilla war, and even ethnic cleansing. We see these results everywhere in poor countries (*ibid*, 2006: 14).

From the holistic point of view, the environment is perceived as a living organism that supplies the essential resources for sustenance of human life. Currently, the rapid and unprecedented rate of global climate change, a direct result of human interference on the environment, unsustainable development intimidates these vital resources by limiting both their quantities and qualities. Taking the path of Homer-Dixon's analysis, a series of high profile scientists, commentators and security agencies have warned of national security threats posed by environmental problems such as climate change.

David King, former Chief Scientific Advisor to the UK Government, argued in 2004, "climate change is the most severe problem that we are facing today - more serious than the threat of terrorism" (King, 2004: 176: King, D. A., 2004: 176—77). Ban Ki Moon attributed past, and potentially future, conflicts in Africa in part to climatic changes in the region (2007: 1) (Ban K.M. (2007). Most recently, two major US Government reports came to similar conclusions:

Climate change and energy are two key issues that will play a significant role in shaping the future security environment... While climate change alone does not cause conflict, it may act as an accelerant of instability or conflict, placing a burden to respond on civilian institutions and militaries around the world (US Government, 2010a:84-5) US Government Quadrennial Defense Review Report, 2010).

More broadly, an influential European Commission report describes '...climate change as a threat multiplier which exacerbates pre-existing trends, tensions and instabilities underlying intrastate and interstate violent conflict such as competition for energy resources; migration; water and food shortages; and environmental disasters' (European Commission, 2008: European Commission, 2008).

Environmental change invites direct and immediate effects on wellbeing and livelihoods. For example, water scarcity may not cause war but still engender insecurity by contributing to dehydration-related death, reducing food production, and undermining livelihood opportunities. These changes may also extend into the future to impact the lives of generations to come. Water resources, again, provide an illustrative example of these different types of effects and their complex interactions. Over two billion people live in water-stressed river basins and that figure is likely to rise to 3.5 billion, or one-half of the world's population, by 2025. Water scarcities have multi-scale effects, for example when river basins are trans-boundary, multi-subject effects (from families without access to safe drinking water to corporations who must pay higher costs for water use), and multiple impacts (such as by undermining sectoral production in agriculture and industry, as well as contributing to desertification in vast ecological areas) (Khagram S. 2003: 294).

There are numerous interconnections among people (social systems) plants and animals (natural systems), and livelihood opportunities (economic systems) that frequently are linked to political empowerment. Forest and tree provide a number of environmental services, including the regulation of the flow of water between soils and the atmosphere, the prevention of soil erosion and the provision of habitat to 'the largest collection of biodiversity of any ecosystem on the planet'. In 1998 alone, "forest clearing was blamed...for worsening flooding in China that killed 3,000 and caused $20 billion in damage", a significant price to pay in human and financial terms (World watch Institute, 2002).

The environmental impacts on human survival, well-being and dignity in particular and human security in general are enormous. The links between people, nature and economies are inescapable when looking at environmental security and environmental risks as they relate to human security. Aspirations for security and development therefore must be directed towards efforts to protect individuals from environmental threats. They must also be based on practical steps to seize upon the opportunities presented by the environment, in recognition of its inherent value, and its deep connections to human beings, societies and economies.

After the Second World War, the emergent nuclear powered countries and multipolarisation of the world power have turned the process of human insecurity caused by military warfare irrelevant. At the same time, rapid industrialisation and consumerism have brought about severe environmental insecurity in particular and human insecurity in general across the globe due to drastic climate change. The process of environmental insecurity is temporal and spatial. As a macro-driver of many kinds of environmental changes such as coastal erosion, declining precipitation and social moisture, increased storm intensity, and species migration, climate change poses risks to human security (McCarthy J. et. al., 2001). People who are more dependent on climate sensitive production, distribution and consumption system are more liable to environmental insecurity or in turn human security. For example, in case of our present study about Maldives, the major part of its developmental process depends on tourism and fisheries, which might be jeopardised by a minimal degree of climate change. Of course, environmental change scarcely affects human security in isolation of broader social factors. As described by Barnett and Adger, these factors, include, among other things, poverty, the degree of support (or conversely discrimination) people receive from the state, their access to economic opportunities, the effectiveness of decision making processes that govern people's lives, and the extent of social cohesion within and surrounding vulnerable groups (Barnett Jon et. al., 2007: 6410). In case of Maldives, however, all these social factors seriously succumb to environmental factors.

**The Paradise Panorama**

The republic of Maldives is a tiny, beautiful dotted country, consists of a chain of coral atolls, 80-120 km wide and 860 km long, which lies on the submarine ridge in the mighty Indian Ocean. There are 26 natural atolls extended over a total land area of about 300 square kilometres. The country comprising a total 1192 islands is known for the lowest country in the world and the most beautiful paradise for honeymooning tourists due to its natural sceneries, attractive sea beaches and stupendous seaside resorts. According to world tourist Award 2006, Maldives is the world's leading diving destination and Indian Ocean tourist destination (World Travel Award, 2006).

This glittering tourism industry and the addition of fishery exports make Maldives the wealthiest country in Asia having per capita income 3000 US dollar. Tourism has proved itself the largest industry in the Maldives comprising of 28 per cent. GDP and more than 60 per cent of Maldives' foreign exchange receipts. It pumped the current GDP per capita to expand 265 per cent in the 1980s and a further 115 per cent in the 1990s. Over 90 per cent of government tax revenue flows in from import duties and tourism related taxes (Maldives, 2011). Activities related to fishery also facilitate the economic growth of the country tremendously. The country's infrastructural development is mainly dependent on the tourism industry and its complementary tertiary sector such as transport, distribution, real estate, construction, etc. Its charming multi-storeyed resorts, attractive shopping malls and charismatic ferries indeed clown it to be the paradise in the eyes of mainly European tourists. Moreover, in the last decade, real GDP growth averaged around spectacular 6 per cent per year except for 2005, when GDP declined due to heavy damage on the tourism following the Indian Ocean tsunami and in 2009 when GDP shrank by three per cent as tourist arrivals declined and due to the wake of financial crisis capital flows plunged very badly (Maldives Economic Profile, 2011).

Even Maldives has proved itself the only Muslim country where democracy grew naturally from within without any external interference and by minimal bloodshed. Recent two decades of Maldivian history has not experienced major religious rivalry or civil war. With the current democratic government headed by a president who is otherwise known as stalwart of human rights activism, the country looks forward to bench a milestone towards the establishment of political and civil security in Maldives.

However, all these existing scenes in Maldives are under a serious threat; the threat is not from the political aggression from the neighbouring countries nor from the result of a civil war, nor due to the economic slowdown in the region. Had the threat been intruded from external or civil war, or a short time economic collapse, the country could have combated them with its own effort. The threat, however, has been received from the result of climate change, which can swipe away the very existence of this country from the global map, and the world is going to experience the worst phenomena of paradise lost.

**Impacts of Climate Change: Paradise Almost Lost**

Down the temporal linkages, the country has stood vulnerable to climate change. Being a low-lying country, the country has experienced bitterly the impacts of climate change. The most spectacular example is the 2004 Indian Ocean tsunami which flooded most of islands heavily causing the death of 82 people and severe damage to social infrastructure like schools, transports and obviously sea-side resorts (Hermann M. Fritz et. al., 2006). The plight of poor people living under severe difficulties since the tsunami tells the sad story of this paradise lost. The displacement politics, huge real estate competition, severe damage on tourism, shortage of fresh drinking water, poor urban congestion, disappearance of mango trees, frequent flooding altogether herald that this paradise is almost lost or going to be almost lost.

Maldives, being a low-lying country is very much vulnerable to multifarious impacts of climate change. It is noteworthy that though Maldives contributed 0.001 per cent to global emissions of GHGs (Zahir Maeed M, 2011), natural disasters associated with sea level rise and changes in the temperatures and precipitation patterns are likely to threaten it seriously. Moreover, climate changes will also impact the socio-economic development of the country adversely giving rise to multiple social stresses, since most of the economic activities are heavily dependent on the coastal ecosystem.

Though sea rise has been proved to be the major threat to Maldivian mayhem, the country is also prone to the negative impacts of rising air and sea surface temperatures and changes in precipitation patterns. As a whole climate change threats the following socio-physical factors of human security in Maldives:

- Sea-level rise, land loss and beach erosion
- Infrastructure damage
- Damage to coral reefs
- Impacts on the economy
- Food security
- Water resources
- Human health
- Bio-diversity

## Sea Level Rise, Land Loss and Beach Erosion

The average land area in Maldives is less than 1metre above mean sea level. Being so low-lying, therefore, these islands of Maldives are very much prone to inundation and beach erosion. 50 per cent of all inhabited islands and 45 per cent of tourist resorts are facing varying degrees of beach erosion (Haq Saleemul et. al. 2005; South Asia Report, 2011). The most plausible example is the Male International Airport on the Hulhul Island. The swing of climate change and expected sea rise would add fuel to the problem of beach erosion. Studies on climate change expect that even 1metre rise in sea level would cause the loss of the entire land area of Maldives. The Intergovernmental Panel on Climate Change (IPCC) in its Third Assessment Report estimates a projected sea level rise of 0.09 m to 0.88 m for 1990 to 2100 (Maldives: State of the Environment, 2002). If continues the world will loose Maldives about 2100 year.

To add, entire Maldivian islands are built of coral reef and coral sand, which are primarily built and sustained by continuous ecological and physical processes in the coral reef ecosystem, which in turn is liable to process of wave direction and sea surface temperature. The natural forces like wind waves, current tides and also extreme events such as storms and tsunamis will badly hit the country and give rise to further land loss and beach erosion.

## Infrastructure Damage

The projected rise in sea level will bring in grave damage to the human settlement, industry and vital infrastructures like houses, schools, roads and hospitals since all these infrastructures in Maldives lie very close to seacoast. For instance, one of the most vulnerable structures in Maldives is the Male international Airport, which is the only international Airport in Maldives and the gateway to the Maldivian tourists and the world.

## Damage to Coral Reefs

The famous Maldivian coral reef with its entire splendour augments the country's tourism and fishery industries. This benevolent shield of the country, however, is sensitive to sea surface temperature that will adversely impact fishery industry and tourism. To name the case, unusual high sea surface temperature in 1998 caused mass bleaching on coral reefs in the central regions of the Maldives (Maldives, 2001: First National Commission of the Republic of Maldives to the UN Framework Convention on Climate Change).

## Impacts on the Economy

Maldives' major economic activities are dependent on tourism and fisheries. While the white sandy sea beaches which are the main attraction for tourists are under serious attack due to sea beach erosion, the beautiful and protective coral reef would be severely bleached which will add to plight for tourism industry. Similarly, studies show that tuna fishing is also sensitive to monsoon activity and other weather activities like EI Niño and La Niña (Adam M.S et. al., 1996). Since major of the country's economic activities are due to these two industries, climate change would adversely affect the economic security of the country.

**Food Security**

Due to the poor soil quality in the Maldives, most of the staple food items except tuna fishes and coconut are imported mainly from India, Thailand even as far as from Germany. Changes in weather conditions mainly due to storm events, sea transport would be disturbed, thereby threatening food security in this island nation. For example in 1990 due to high winds and storms, the transport system in the country was disrupted and many islands in the country faced shortage of food for 10 days in June (Maldives, 2001). Even sea level rise accompanied by salt-water intrusion on groundwater will pose a threat to the little agriculture practiced by the rural farmers in Maldives.

**Water Resources**

The population of the Maldives mainly depends on ground water and rainwater as a source of fresh drinking water. However, the climate and hydrology, sea level movement and human activities, inundation of land and associated saltwater intrusion due to the predicted sea level rise, pollution by solid waste and other pollutants would reduce the available fresh water lenses and thus reduce the available fresh ground water of these islands. It is estimated that nine per cent of the population had no access to safe drinking water (MPND and UNDP, 1998). Though, the country is expected to have frequent precipitation patterns, the change of temporal patterns and amount of rainfall to the country may cause more frequent flooding when the island receive heavy rainfall to the freshwater lakes found in some islands.

**Human Health**

Notable relations to changes in climate have been seen for dengue and dengue hemorrhagic fever in the country. Poor sanitation and increase of rainfall would result in out breaking water born diseases like diarrhoea. Though Maldives has eradicated Malaria from its soil, the adverse climate change may cause outbreak of this fatal disease in the country. Adverse weather condition disrupts the medical access of most of the Maldivian.

**Bio-diversity**

The Maldives' coral reefs are globally significant being the 7th largest in the world covering an area of 8,920 km and inhabiting over 1,900 species of fish, 187 coral species, 400 species of molluscs and 350 species of crustacean (Vanishing Maldives, 2011). The islands house over 583 species of plants and 170 species of birds (Asian Development Bank, 2007). Rising ocean levels threatens the entire existence of Maldives and its plethora of biodiversity. The human population can be easily relocated to other neighbouring countries, but the world is certainly going to lose this inhuman treasure of bio-diversity since cataloguing or restoring these species seems to be highly implausible.

**External Factors**

The plight of Maldives due to climate change is further aggravated by the several external factors. One of the serious factors is the oil leaks or spills from large tankers carrying colossal quantities of crude oil crossing the Indian Ocean and its adjacent semi enclosed seas. Oil spills from normal tanker operations and occasional large-scale tanker catastrophes have had deleterious effects on phytoplankton and zooplankton, both necessary parts of the food chain of commercial fisheries (Encyclopaedia Britannica, 2011). Furthermore, due to the oil spill, Maldives sophisticated and delicate marine ecosystem gets disturbed which affects Maldives whole existence phenomenally. Most sea birds in Maldives are at the door of extinction since 'most birds affected by an oil spill die unless there is human intervention' (Dunnet G et. al. 1982). Marine mammals face the same fate. Floating oil coat prevents sunlight to penetrate into the water limiting the photosynthesis of marine plants and phytoplankton, which directly affects lives of Maldivian who are highly dependent on marine ecosystem for their sustenance. Oil spills affects Maldives marine ecosystem, its flora and fauna, its birds and mammals, its environmental climate and of course its life sustaining programmes in general.

**Perspectives: The Paradise to be Regained**

The above discussion very amply indicates that though climate change is a threat to environmental security to a country, in case of Maldives it shakes the whole concept of human security in totality. According to UNDP report (1994) and as propounded by Dr Mahbub ul Haq, the notion of human security engulfs seven kinds of security concepts into its fold, such as economic security, environmental security, political security, food security, health security, personal security, and community security. However, in case of Maldives, climate change is not only a threat for environment but also a threat to socio-economic, political and personal security. Therefore, issues relating to climate change in Maldives have been purely comprehended as issues relating to human security. That is why; both the government of Maldives and the international consciousness have to be very serious about possible perspectives in minimising the risk of climate change.

Of course, the story of Maldives' socio-economic development throughout 1990s and the early part of 21st century has been phenomenal. Its rapid economic growth exemplified by the average GDP growth of 7.6 per cent per year in the period of 1976-2004, more than 10 per cent per year for most of the 1980s, surpassing the performance of all other South Asian economies (World Bank, 2006). It has achieved the most cherished goal of universal primary education; child and maternal mortality rates have been declined fast and eradicated diseases such as Malaria.

Despite rapid growth, the Maldivian population still realises high poverty in particular regions such as the Northern Atolls not only due to aggregate shocks like the recent impacts of the tsunami and potential future environmental risks of global warming leading to a rise in sea level but also due to individual shocks like health risks and joblessness. Entire Maldivians have suffered and are still living in continuous fear of future suffering due to climate change.

Maldives takes proper cognizance about the severity of the problem of climate change. Maldives endeavours to take every possible effort to expedite the drafting of the framework of convention in the commonwealth heads of government meeting. In its own effort, a number of conferences, declarations like Male declaration on Global warming and sea level rise where an action group to coordinate a joint approach on the issues of climate change, global warming and sea level rise have been convened.

The most exemplary effort was taken by the President Narsheed when he unveiled the first ever plans to make his country carbon neutral by 2020. The plans enlist a number of measures including a new renewable energy, electricity generation and transmission infrastructure with large wind turbines, half a square kilometre of rooftop solar panels, and a biomass plant burning coconut husks. The National Communications (NC), National Adaptation Plan of Action (NAPA) have endeavoured to address the issue of Climate change in Maldives. To make a comparison, the total emission of carbon dioxide is at 4.1 tones per Maldivian vis-à-vis 23.5 tones per person per year in the United States (Rodriguez Diva, 2010). The above data shows that the contribution of Maldives to global emission of GHGs is insignificant. Maldives drives intensely to reduce GHG further by using high efficiency generators, increasing awareness on the use of huge energy efficiency appliances, increasing the use of renewable energy sources, etc.

Maldives also aims at promoting awareness about the climate change through using schools as demonstrating unit. Despite financial and technological hurdles, 63 local residents under the Human Resource Development Project in 2001 have been trained in monitoring and assessing the changes in their island environment (Das P.K, 2010). Serious consideration of availability and management of data has been taken. For the past 15 years, Maldives have achieved tremendous pace in protecting environmental security. Environmental studies have been incorporated in the primary and middle school curriculum. The president's Environmental Award Programme is an annual award that recognises outstanding contribution made by individuals and groups involved in environmental protection. The World Environment Day, the World Clean-up Day Programme, and the Clean Maldives Campaign have been practiced with high seriousness. The innovative 'The Safer Island Development Programme' (SIDP) is the exemplary one in combating the natural disaster and providing long-term infrastructural sustainability to the country.

In its international initiatives towards environmental protection, the Maldives is a party to the United Nations Framework Convention on Climate Change (UNFCC) and of course, is the first country to sign the Kyoto Protocol (Maldives, 2001). UNDP has always been a partner with the Government of Maldives on environmental matters. Maldives also is underway in adopting climate change activities by implementing the recommendations of Davos declaration, as part of the UN World tourism Organisations (UNWTO) strategy (Maldives prepares for Climate Change, 2008).

**Conclusion**

Maldives is a victim of environmental crisis leading to human insecurity for which it is hardly responsible. The total emigration of GHG from Maldives is merely 0.001 per cent of world average (Zahir Maeed M. 2011). However, it is the worst affected country in the world by climate change. Since the problems caused by environmental crisis do not squeeze themselves into one region; since these problems have multinational penchant or effect, it would be rather foolish to consider Maldives problems as regional. Large-scale displacement in Maldives may cause congestion in neighbouring countries like India, Sri Lanka and Australia. It may create grave problems regarding human rights of environmental refugees. It may also cause socio-political confusion and frustration in these neighbouring countries, which in turn may pose threat to human security in those countries. Threat to food security in Maldives also adversely affects the food security in the exporting countries like India, Thailand, and as far as Germany. After all, cases regarding Maldives are also true in cases of other countries in the world. Issues regarding climate change and human security in Maldives need collective focus and action plan on the part of global authorities.

## References

Adam, M.S. and R.C. Anderson. 1996. 'Yellowfin tuna Thunnus albacares in the Maldives', Expert Consultation on Indian Ocean Tunas, 6th Session, Colombo, Sri Lanka.

Annan Kofi. 2000).*United Nations Millennium Report*, Chapter 3.

Asian development Bank. September 2007. *Maldives: Environment Assessment*, http://www.adb.org/Documents/Assessments/Country-Environmental/MLD/Environment-Assessment.pdf

Ban, K.M. 2007. 'A Climate Culprit in Darfur', *Washington Post*, Saturday 16 June: http://www.washingtonpost.com/wp-dyn/content/article/2007/06/15/AR2007061501857_pf.html

Barnett Jon and W. Neil Adger. 2007. 'Climate change, human security and violent conflict', *Political Geography,* Vol. 26.

Dabelko, G. and D. Dabelko. 1995. 'Environmental security: issues of conflict and redefinition, *Environmental Change and Security Project Report*, Washington, DC: the Woodrow Wilson Center, spring.

Dalby Simon, 'Ecopolitical discourse: 'environmental security' and political geography', Progress in Human Geography, October 1992, Vol. 16, No. 4

Das P.K. 2010. *Climate Change and Education, Genesis (Pvt.) Ltd: Maldives.*

Dixon Homer, T. 1999. 'Environment, Security and Violence', Princeton: Princeton University Press.

Dixon Homer, T. 2006. 'The Upside of Down', Washington, DC: Island Press.

Dixon Homer, T. 2007. 'Terror in the Weather Forecast', New York Times, 24 April: 1-2 (http://www.nytimes.com/2007/04/24/opinion/24homer-dixon.html).

Dunnet, G., Crisp, D., Conan, G., Bourne, W. 1982. 'Oil Pollution and Seabird Populations', (and Discussion). Philosophical Transactions of the Royal Society of London.

Encyclopaedia Britannica, 'Environmental impact of human activity', http://www.britannica.com/EBchecked/topic/285876/Indian-Ocean/22785/Environmental-impact-of-human-activity.

European Commission. 2008. *Climate Change and International Security,* http://www.consilium.europa.eu/ueDocs/cms_Data/docs/pressdata/EN/reports/99387.pdf.

Gaan Narottam. 2004. *Environmental Security - Concept and Dimensions*, Gyan Publishing House: New Delhi.

Giambrone Andrew. September 22, 2010. 'Sinking nation seeks help', http://www.yaledailynews.com.

Hans Gunter Brauch, et al. 2003. 'Security and Environment in the Mediterranean', Springer: Berlin

Hermann M. Fritz, Costas E. Synolakis, and Brian G. McAdoo. 2006. 'Maldives Field Survey after the December 2004 Indian Ocean Tsunami', http://mahabghodss.net/NewBooks/www/web/digital/nashrieh/Journal%20of%20earth%20qua/Vol.%2022-2006/Vol.22% 20Issue% 20S3%20June% 202006/Maldives%20Field%20Survey.pdf

Huq Saleemul and Hannah Reid. June 2005. 'Climate Change and Development: South Asia Report', http://pubs.iied.org/pdfs/G00055.pdf

IPCC. 2001. *Climate Change 2001: Impacts, Adaptation, and Vulnerability*, Cambridge: Cambridge University Press.

Khagram Sanjeev et. al. July 2003. 'From the Environment and Human Security to Sustainable Security and Development', *Journal of Human Development* Vol. 4, No. 2,

King, D.A. 2004. 'Climate Change Science: Adapt, Mitigate, or Ignore?' *Science* 303.

Maldives, http://www.exchangerate.com/country-information/republic-of-maldives.html

Maldives. 2001. *First National Communication of the Republic of Maldives to the United Nations Framework Convention on Climate Change*, Ministry of Home Affairs, Housing and Environment *Maldives: State of the Environment, 2002,* http://www.rrcap.unep.org/pub/soe/maldives_climate.pdf

Maldives Economy Profile. 2011. http://www.indexmundi.com/maldives/economy_profile.html

*Maldives Prepares for Climate Change*. 5 June 2008. http://climate-l.iisd.org/news/maldives-prepares-for-climate-change/

Maldives wins the 'World's Leading Dive Destination' award and 'Indian Ocean's Leading Destination' award at the World Travel Awards 2006 http://www.mymaldives.com/blog/page/17/

McCarthy, J., Canziani, O., Leary, N., Dokken, D., and White, K. 2001. 'Climate Change 2001: Impacts, Adaptation and Vulnerability', Cambridge: Cambridge University Press.

MPND and UNDP. 1998. 'Vulnerability and Poverty Assessment 1998'. Malé: Ministry of Planning and National Development and United Nations Development Programme.

Najam, Adil, ed. 2002. 'Environmental and Human Security in South Asia', University Press of America: United States.

Obi, C. 2000. 'Globalised Images of Environmental Security in Africa', *Review of African Political Economy*, Vol. 27, No. 83.

Page, E. 2000. 'Theorizing the link between environmental change and security', *Reciel*, 9(1):33–43.

Parry, M.L., Canziani, J.P. and Palutikof, J.P. 2007. 'Technical Summary', in M.Parry, O. Canziani, J.P. Palutikof, P. van der Linden and C. Hanson (eds.) *Climate Change 2007: Impacts, Adaptation and Vulnerability*, Cambridge: Cambridge University Press.

Rodriguez Diva. 25 November 2010. 'Maldives heading toward 2020 carbon neutrality',http://www.climateactionprogramme.org/news/maldives_heading_toward_2020_carbon_neutrality/

Selim Mohammad El-Sayed. 'Environmental Security in the Arab World', paper prepared for presentation at the Meeting of the *International Studies Association*, 17-20 March 2004, Montreal, Canada.

The Nobel Peace Prize. 2007. Intergovernmental Panel on Climate Change, Al Gore, http://nobelprize.org/nobel_prizes/peace/laureates/2007/

UNDP. 1994. *Human Development Report.*

United Nations Development Program (UNDP). 1995. 'Redefining Security: The Human Dimension', *Current History* 94 (592).

United Nations Development Programme (UNDP). 2007. *Human Development Report 2007/2008: Fighting Climate Change: Human solidarity in a divided world,* Basingstoke: Palgrave.

*Vanishing Maldives; The Land Of Biodiversity*, http://monsoonsky.blogspot.com/

WCED. 1987. *Our Common Future*, Oxford: Oxford University Press.

World Bank. 2006. 'Social Protection in the Maldives: Options for Reforming Pensions and Safety Nets', A Report.

World watch Institute. 2002. State *of the World 2002*, W. W. Norton and Company, New York.

Zahir Maeed M, 'Threats and cure for a climate struck destination - An outline' http://ecocare.mv/threats-and-cure-for-a-climate-struck-destination-%E2%80%93-an-outline.

# 13

# Sea Level Rise in Bay of Bengal: Challenges for India

*Ramakrushna Pradhan*

One of the most profound effects of climate change has been the sea level rise due to melting of glaciers and increase in thermal temperature of the sea surface since the mid 19th century. Coastal zones with complex geographical entities encompassing diverse human activities and ecological processes in interconnected maritime and terrestrial spaces are among most vulnerable of all environmental concerns significantly threatening these interdependent systems. The rise in sea levels resulting in disastrous consequences like, intensification of tropical cyclones, larger storm surges, increasing sea-surface temperatures and growing acidification of surface water increasingly endanger the food security, water security, health and livelihoods of coastal population. In this context, this paper attempts to study the Sea level rise in Bay of Bengal and its multiple challenges for India. Sea level rise is of great concern to India in view of the damages that occur along the east coast of the country from the cyclones that form in the Bay of Bengal, exacerbation of flooding, beach erosion and salt water intrusion to rivers and ground water aquifers endangering the life and habitats of one-quarter of the population of India living along the coasts. Any increase in frequency and severity of these extreme climate events or change in coastlines as projected is likely to have serious effects and can cause greater harm to the populace. Hence, in the mean time the urgent need is to respond to the coming disasters through various policy processes like, mitigation, adaptation and protection.

The concept of sea level rise is quite straightforward. Some 97 per cent of all the water on Earth is now in the oceans (Douglas, Kearney and Leatherman; 2001: xii). Most of the rest is found in glaciers, much of it in Antarctica and Greenland. Some 20,000 years ago at the peak of the last ice age, much more water was in ice and the sea level was more than 100 meters lower than it is today. The glaciers began to melt; the oceans began to fill, and the shorelines were pushed back as the sea level rose. The process continues, and the results are obvious. It is now believed that the ocean volume has been increasing since the middle of the 19th century at a rate equivalent to raising sea level almost 2 mm/yr and continues to rise at a rapid rate now than it was a century ago (*ibid*: xiii). Why is the volume of the oceans increasing?

**Sea Level Rise**

One of the most certain reasons of sea level rise has been the consequences of human-induced climate change. It has been projected that global warming could raise the sea as much as one meter in the next century by heating ocean water, which would then expand, and by causing mountain glaciers and parts of ice sheets to melt or slide into the oceans (Church and Gregory 2001: 639–93). Thus, the sea could reach heights unprecedented in the history of civilisation. The immediate concern therefore is to know what climate change does mean and how it got related to sea level rise.

**The Relationship between Climate and Sea Level**

The earth's climate is dynamic and always changing through a natural cycle. What the world is more worried about is that the changes that are occurring today have been speeded up than expected. These changes are being studied as climate change. Climate change is a change in the statistical distribution of weather over periods of time that range from decades to millions of years. It can be a change in the average weather or a change in the distribution of weather events around an average (for example, greater or fewer extreme weather events). Climate change may be limited to a specific region, or may occur across the whole Earth. Carbon dioxide, methane, nitrous oxide together known as 'Greenhouse Gases' have resulted into global warming. However, ever since the Industrial Revolution began about 150 years ago, man-made activities have added significant quantities of GHGs to the atmosphere. The atmospheric concentrations of carbon dioxide, methane, and nitrous oxide have grown by about 31 per cent, 151 per cent and 17 per cent, respectively, between 1750 and 2000 (IPCC 2001). High levels of industrial pollution and a number of man-induced processes have resulted in climate change. The various natural and human causes of the catastrophe are responsible for the drastic shift in average weather, global warming and variations in solar radiation.

One of the most pronounced effects of climate change has been melting of masses of ice around the world. Glaciers and ice sheets are large, slow-moving assemblages of ice that cover about 10 per cent of the world's land area and exist on every continent except Australia. They are the world's largest reservoir of fresh water, holding approximately 75 per cent (National Snow and Ice Data Centre, http://nsidc.org).

Over the past century, most of the world's mountain glaciers and the ice sheets in both Greenland and Antarctica have lost mass. Retreat of this ice occurs when the mass balance (the difference between accumulation of ice in the winter versus ablation or melting in the summer) is negative such that more ice melts each year than is replaced (Ocean and Sea Level Rise, www.climate.org). By affecting the temperature and precipitation of a particular area, both of which are key factors in the ability of a glacier to replenish its volume of ice, climate change affects the mass balance of glaciers and ice sheets. When the temperature exceeds a particular level or warm temperatures last for a long enough period, and/or there is insufficient precipitation, glaciers and ice sheets will lose mass.

While the impacts of sea-level rise are confined to coastal areas, these are the most densely populated land areas on earth, and they support important and productive ecosystems that are sensitive to sea-level changes. Coastal areas and small islands are among the most densely populated parts of the world. They are also the most threatened because of rises in sea level that global warming may cause. The heating of oceans, and melting of glaciers and polar ice sheets, is predicted to raise the average sea level by about half a metre over the next century. Sea-level rise could have a number of physical impacts on coastal areas, including loss of land due to inundation and erosion, increased flooding, and salt-water intrusion. These could adversely affect coastal agriculture, tourism, freshwater resources, fisheries and aquaculture, human settlements, and health. Rising sea levels threaten the survival of many low-lying island nations, such as the Maldives and Marshall Islands.

There could be large decreases in the availability of water in many rivers because of rainfall and snow. The volume in others would increase due to glaciers melting, for example, the rivers originating in the Himalayas. Shifts in water availability could also affect hydropower generation, and industries such as paper, pharmaceutical, and chemical manufacturing, that use large quantities of water. Buildings and other infrastructure would be vulnerable to any increase in the frequency of storms and other extreme events, which could also disrupt transport routes.

## Causes of Sea Level Rise

Climate influences sea level in two ways: (i) Deglaciation, by moving the earth's water between glaciers resting on land and the oceans and (ii) Thermal expansion, by changing the temperature of the ocean water and thus its volume.

## Deglaciation

In the face of global warming deglaciation constitutes the weakest part of the analysis. Although, deglaciation would not directly affect the sea level rise, however, the melting back of this ice can lead to indirect contributions on sea level. For example, the melting back of sea ice leads to a reduction in albedo (surface reflectivity) and allows for greater absorption of solar radiation. More solar radiation being absorbed will accelerate warming, thus increasing the melting back of snow and ice on land. In addition, ongoing breakup of the floating ice shelves will allow a faster flow of ice on land into the oceans, thereby providing an additional contribution to sea level rise.

Melting of glaciers and ice caps significantly contribute to the rise of sea level. IPCC's Fourth Assessment estimated that, during the second half of the 20th century, melting of mountain glaciers and ice caps led to about a 2.5 cm rise in sea level. This is a higher amount than was caused by the loss of ice from the Greenland and Antarctic ice sheets, which added about 1 cm to the sea level. For the 21st century, IPCC's Fourth Assessment projected that melting of glaciers and ice caps will contribute roughly 10--12 cm to sea level rise, with an uncertainty of roughly a third (www.ipcc.ch). This would represent a melting of roughly a quarter of the total amount of ice tied up in mountain glaciers and small ice caps.

Nonetheless, if all the glaciers in Antarctica and Greenland melted, sea level would rise more than seventy meters (over two hundred feet). In the past, enough ocean water has accumulated in glaciers to lower sea level about one hundred and fifty meters (five hundred feet). Although complete melting of land-based glaciers would take thousands of years, partial melting could raise sea level as much as a meter in the next century. Furthermore, glaciers grounded under water could disintegrate more quickly. Two leading glaciologists have estimated that the entire West Antarctic ice sheet (the largest marine-based glacier) could enter the oceans in two hundred years (Hughes et. al., 1979) and five hundred years (Bentley, 1980) raising sea level five to seven meters (about twenty feet). Although a complete disintegration of this marine based glacier will not occur in the near future, parts of it and other ice fields, as well as mountain glaciers, could be vulnerable in the next century.

Although surface waters would warm quickly with rising global air temperatures, the downward transport of heat into deeper ocean layers would warm them much more slowly. Hoffman employed an ocean model that uses diffusion as a surrogate for all heat transport processes to estimate heat transported into the top one thousand meters (3,200 feet) of the ocean.

**Thermal Expansion**

Like air and other fluids, water expands as its temperature increases (i.e., its density goes down as temperature rises). As climate change increases ocean temperatures, initially at the surface and over centuries at depth, the water will expand, contributing to sea level rise due to thermal expansion. Thermal expansion is likely to have contributed to about 2.5 cm of sea level rise during the second half of the 20th century (www.ipcc.ch), with the rate of rise due to this term having increased to about 3 times this rate during the early 21st century. Because this contribution to sea level rise depends mainly on the temperature of the ocean, projecting the increase in ocean temperatures provides an estimate of future growth. Over the 21st century, the IPCC's Fourth Assessment projected that thermal expansion will lead to sea level rise of about 17-28 cm (plus or minus about 50 per cent). That this estimate is less than would occur from a linear extrapolation of the rate during the first decade of the 21st century when all model projections indicate ongoing ocean warming has led to concerns that the IPCC estimate may be too low.

Because water expands when heated, a warmer climate could raise the sea even without any contribution from glaciers. Although a warming of the entire ocean would take several centuries, the upper layers could warm and raise sea level as much as a meter by 2100. This shorter-term effect of a global warming is frequently overlooked.

**Recent Changes in Sea level**

Average sea level over the oceans has never been constant throughout earth history, and it is changing slightly today. The entirety of civilisation has occurred within a single high stand of the sea, and yet global sea level was more than 100 m lower than it is at present only 18,000 years ago (Sea Level Change, National Research Council; 1990: 3). And during the geologic past there have been repeated variations of more than 100 m from present sea level, both during times of intense glaciations and during times of an ice-free Earth.

Relative sea level (RSL; i.e., sea level relative to a fixed point on land) at any particular place in the ocean varies over a wide range of time and space scales. The direct causes of these variations are vertical motions of the land to which the tide gauge or other measuring instrument is attached, and changes in the volume of sea water in which the tide gauge is immersed. Wind-driven waves produce the shortest-period variations in the height of the sea surface. Variations with periods of 12 hours or more are caused by lunar and solar tides. Variations in atmospheric pressure cause inverse variations in sea level—an atmospheric pressure differential of 1 mbar is equivalent to a sea-level differential of 10 mm. A series of depressions in atmospheric pressure can cause a rise in sea level in a shallow ocean basin of 0.3 m or more (Hekstra, 1988: 5). Variations in the runoff of large rivers can result in local 'sea-level' changes of as much as 1 m. In relatively shallow water, large variations in sea level are also caused by offshore and onshore winds that pile up water against the shore or drive it away from the shore. (This process is called wind setup.) In exceptional circumstances, in the North Sea, along the Chinese coast, and in the Bay of Bengal, sea level may rise by 5 m or more in a 'storm surge' under the action of strong winds (Sea Level Change, National Research Council; 1990: 5). Both irregular and seasonal variations in temperature or salinity of the upper ocean layers cause expansion or contraction of the water volume in different regions. These relatively short-term steric changes in sea level may persist for a few days, several months, or even several years, and the magnitude may be as much as 50 to 150 mm (*ibid*).

Sea level change, seemingly so simple and straightforward, is in fact the product of many interrelated processes. Climate, plate tectonics, the cryosphere, and ocean circulation all contribute to changing sea level. Changes occur on all space and time scales, from local to global, and from a few seconds to geologic ages. The relative importance of the forcing functions varies with the time scales of interest. The effects of changing sea level are also broad—on the one hand with direct feedbacks to the causative forcing functions, e.g., albedo change, and on the other hand with effects on other processes such as sedimentation or coastal ecology.

### *Impacts of Sea Level Rise*

The melting back of the glaciers and ice sheets has two major impacts. First, areas that rely on the runoff from the melting of mountain glaciers are very likely to experience severe water shortages as the glaciers disappear. Less runoff will lead to a reduced capability to irrigate crops as freshwater dams and reservoirs more frequently go dry. As the volume of runoff decreases, then the energy, urban, and agricultural infrastructures of surrounding areas are likely to be stressed.

Importantly, the melting of glaciers and ice sheets adds water to the oceans, contributing to sea level rise. And that the projected global warming could raise the sea as much as one meter in the next century by heating ocean water, which would then expand, and by causing mountain glaciers and ice sheets to melt or slide into the oceans. The outcome of such development could be devastating as it would result in the loss of thousands of square miles of land, particularly in low-lying areas, Storm damage will regularly occur. Finally, a rising sea will increase the salinity of marshes, estuaries, and aquifers, disrupting marine life and possibly threatening some drinking water supplies. Fortunately, the most adverse effects can be avoided if timely actions are taken in anticipation of sea level rise.

**Table 1. The main natural system effects of relative sea-level rise, including examples of possible adaptation responses**

| *Natural System Effects* | *Possible Adaptation Responses* |
|---|---|
| 1. inundation, flood and storm damage (a) surge (sea), (b) Back water effect (river) | Dikes/surge barriers, building codes/flood wise buildings, land use planning/hazard delineation |
| | land use planning, managed |
| 2. wetland loss (and change) | realignment/forbid |
| | hard defences, nourishment/sediment |
| 3. erosion (direct and indirect morphological change) | management.<br>coast defences, nourishment, building setbacks |
| 4. saltwater intrusion (a) surface waters, (b) Ground water | saltwater intrusion barriers, change water abstraction |
| | freshwater injection, change water |
| 5. rising water tables/impeded drainage | abstraction<br>upgrade drainage systems, polders, change land<br>use, land use planning/hazard delineation |

*Source*: Nicholis, J. Robert and Tol, S. J. Richards; 2006: 1075.

While there are obviously many challenges to projecting future sea level rise, even a seemingly small increase in sea level can have a dramatic impact on many coastal environments. Over 600 million people live in coastal areas that are less than 10 metres above sea level, and "two-thirds" of the world's cities that have populations over five million are located in these at-risk areas (www.npr.org). With sea level projected to rise at an accelerated rate for at least several centuries, very large numbers of people in vulnerable locations are going to be forced to relocate. If relocation is delayed or populations do not evacuate during times when the areas are inundated by storm surges, very large numbers of environmental refugees are likely to result.

According to the IPCC, even the best-case scenarios indicate that a rising sea level would have a wide range of impacts on coastal environments and infrastructure. Effects are likely to include coastal erosion, wetland and coastal plain flooding, salinisation of aquifers and soils, and a loss of habitats for fish, birds, and other wildlife and plants (www.ipcc-wg1.ucar.edu/wg1/Report)http://ipcc-wg1.ucar.edu/wg1/Report/AR4WG1_Print_Ch10.pdf.

The Environmental Protection Agency estimates that 26,000 square kilometres of land would be lost should sea level rise by 0.66 meters, while the IPCC notes that as much as 33 per cent of coastal land and wetland habitats are likely to be lost in the next hundred years if the level of the ocean continues to rise at its present rate. Even more land would be lost if the increase is significantly greater, and this is quite possible (*ibid*). As a result, very large numbers of wetland and swamp species are likely at serious risk. In addition, species that rely upon the existence of sea ice to survive are likely to be especially impacted as the retreat accelerates, posing the threat of extinction for polar bears, seals, and some breeds of penguins (www.epa.gov/climatechange/effects/eco_animals.html).

Unfortunately, many of the nations that are most vulnerable to sea level rise do not have the resources to prepare for it. Low-lying coastal regions in developing countries such as Bangladesh, Vietnam, India, and China have especially large populations living in at-risk coastal areas such as deltas, where river systems enter the ocean. Both large island nations such as the Philippines and Indonesia and small ones such as Tuvalu and Vanuatu are at severe risk because they do not have enough land at higher elevations to support displaced coastal populations. Another possibility for some island nations are the danger of losing their fresh-water supplies as sea level rise pushes saltwater into their aquifers. For these reasons, those living on several small island nations (including the Maldives in the Indian Ocean and the Marshall Islands in the Pacific) could be forced to evacuate over the 21st century (www.ipcc-wg1.ucar.edu/wg1/Report).

**Effects of Sea Level Rise**

Changes in sea level have many practical consequences, often disruptive but sometimes beneficial. This section describes the physical and environmental effects of sea level rise, and the activities that can be undertaken to prevent or adapt to these effects.

**The Physical Consequences of Sea Level Rise**

The physical consequences of sea level rise can be broadly classified into three categories: shoreline retreat, temporary flooding, and salt intrusion. The most obvious consequence of a rise in sea level would be permanent flooding (inundation) of low-lying areas. Coastal areas with sufficient elevation to avoid inundation would be threatened by a different cause of shoreline retreat: erosion. In fact, the current trend of sea level rise may be causing the serious erosion that is taking place in many coastal resorts. The constant attack of waves causes beaches to take a particular profile, which fluctuates seasonally. Winter storms erode the upper beach and deposit sand offshore, and the calmer spring and summer waves redeposit the sand and restore the beach. However, a rise in sea level alters the relationship of the shore profile to the water level. Because the water nears the beach is deeper than before, more energy is required to move the offshore sand back to the beach. Consequently, some of the material deposited offshore by winter storm waves remains offshore, and a portion of the beach is lost (Bruun, 1962; Schwartz and Fisher, 1979).

Another cause of beach erosion from sea level rise is over wash and the resulting landward migration of coastal barriers. Higher water tables also exacerbate flooding by decreasing the ability of land to drain storm waters. Finally, erosion and deeper water could subject new areas to damaging storm waves.

Sea level rise also causes the salt content of aquifers and estuaries to migrate landward. In coastal aquifers, a layer of freshwater floats on top of the heavier saltwater. The saltwater generally forms a wedge such that the farther inland (the higher the water table), the farther below ground is the boundary between fresh and saltwater. Where sea level rise results in a landward movement of the shoreline, this boundary will move inland as well. Because the level of the water table is itself determined by sea level, a rise in sea level causes the freshwater/saltwater boundary to rise. The landward and upward shift of this boundary implies that certain freshwater wells may become salty. Over pumping of coastal aquifers also has resulted in salt intrusion, however, and in many instances this problem dwarfs the possible impact of sea level rise.

A rise in sea level also would increase the salinity of rivers and estuaries. A decrease in the flow of a river or an increase in the volume of water allows salt to migrate upstream. Because some rivers recharge aquifers, some aquifers might become salty as well. The impacts of storm damage and salt intrusion may be exacerbated or mitigated by the impacts of an enhanced greenhouse effect on climate.

**Environmental Impacts**

Like the physical effects, the environmental impacts of sea level rise fall into the categories of shoreline retreat, salt intrusion, and increased flooding. Salt intrusion is a threat to marine animals as well as vegetation. Many species must swim into fresher water during reproduction. In response to sea level rise, fish might swim farther upstream, but water pollution could prevent such an adaptation from succeeding. Some species, on the other hand, require salty water for their existence.

**Economic Impacts**

The economic impact estimates are measured in terms of the real resource costs to society caused by sea level rise. Real economic losses fall into three categories: (i) the direct losses of economic services from land and capital caused by shoreline retreat and storm damage; (ii) the cost of protection, mitigation, and response measures taken to reduce these losses; and (iii) lost opportunities due to sea level rise (Gibbs, J. Michael; Economic Analysis of Sea Level Rise: Methods and Results, http://earth.epa.gov/climatechange/effects).

**Storm Surges**

A storm surge is a high flood of water caused by wind and low pressure, most commonly associated with hurricanes. The storm surge is the most deadly aspect of a hurricane, responsible for 90 per cent of deaths during hurricanes. Storm surges are different from tidal surges, which are violent surges caused exclusively by the tidal shift in sea level. Storm surge is simply water that is pushed toward the shore by the force of the winds swirling around the storm. This advancing surge combines with the normal tides to create the hurricane storm tide, which can increase the mean water level 15 feet or more. In addition, wind driven waves are superimposed on the storm tide. This rise in water level can cause severe flooding in coastal areas, particularly when the storm tide coincides with the normal high tides.

**Coastal Erosion**

Coastal erosion is the wearing away of land or the removal of beach or dune sediments by wave action, tidal currents, wave currents, or drainage. Waves, generated by storms, wind, or fast moving motor craft, cause coastal erosion, which may take the form of long-term losses of sediment and rocks, or merely the temporary redistribution of coastal sediments; erosion in one location may result in accretion nearby.

**India Specific**

**Climate Change and India**

For the Indian subcontinent, the projected changes by IPCC, based on the General Circulation Model (GCM), project warming of 2-4.7°C, with the most probable level being around 3.3°C by the year 2100 (A1B scenario) (Solomon et. al., 2007: 19-71). Warming is expected to be more marked in the winter half of the year (3.6°C) than in summer (2.7°C), and stronger in the north than in the south. Most scenarios project a decrease in precipitation during the inter dry period and an increase for the rest of the year. At the same time, an increase in heavy rain events is probable, particularly in the north of India. The global sea level rise of 0.1 to 0.9 meters is particularly expected to be high in the Indian Ocean, especially on the west coast.

Indian scientists estimate that the temperatures will increase by 3 to 4°C towards the end of the 21st century. The warming may be about 2.1 to 2.6°C in the 2050s and 3.3 to 3.8°C in the 2080s (DEFRA/GoI, 2005a). The different models/experiments generally indicate the increase of temperature to be of the order of 2-5°C across the country. In case of mean annual temperature, the increase is of the order of 3 to 6°C. The warming is projected to be widespread over the country, and relatively more pronounced over northern parts of India. While the rainfall is projected to increase, there would be variations in the spatial pattern, with some pockets showing increase and others experiencing decline in rainfall. Most models project an increase in rainfall between 10 and 40 per cent from the baseline period (1961–90) to the end of 21st century, with the maximum expected increase in rainfall over north-western and central India. There is very little or no change noted in the monsoon rainfall over a major part of peninsular India (Kumar R. et al., 2006). The climate change would cause global sea to rise between 0.09 to 0.88 m by 2100 and enhance extreme events like excessive rain, flash flood, droughts, cyclones and forest fires.

**Observed effects of climate change in India**

- Rising sea levels: 7,500 km long, densely populated coast line-Flooding mangroves dotting coast lines–Coral bleaching--Extreme weather events-Glacial lake outflow floods affecting downstream village-Melting glaciers affecting water resources-Fall in agriculture productivity-Altered configuration and productivity of forest eco-systems.
- A Chronological report of Natural Calamities (extreme weather events) possibly due to Climate Change (last nine years) (Observed Effects of Climate Change in India, www.greenpeace.org).

In 2010, there occurred sever flood situation in Bihar, Karnataka and Assam. Thunder storm claimed many lives in Mizoram and Assam. Heavy rain brought in untold damages in Arunachala Pradesh. Aila hits West Bengal with swift wind blowing at a speed of 100 kmph. Laila hits Andhra Pradesh causing irreparable damage to life and property in the state. In the years 2009, 2008, 2007 and 2006, floods, storm surges and heavy rained lashed out throughout India and particularly in the coastal India. The states of Tamil Nadu, Andhra Pradesh, Orissa and West Bengal. The year 2005 has been marked as the natural calamity year as the world has witnessed Tsunami in Asia, Hurricanes in North America, Typhoons, earthquakes and locusts around the world. Newspaper Reports over 1,000 dead in severe floods in Bihar, Assam, parts of Haryana and Punjab and parts of Gujarat, while at the same time, Gujarat, Rajasthan, parts of

Haryana, Punjab, Uttar Pradesh and Andhra Pradesh reeled under severe drought. Minor heat wave in parts of Rajasthan was also reported in May 2004, leaving about 50 dead (*ibid*).

India and the sub-continent saw five of the 20 major natural calamities recorded worldwide in terms of victims (Sigma 2003). Uttar Pradesh, Haryana, Punjab, Rajasthan, Gujarat, Bihar, Orissa, Andhra Pradesh: Heat Wave: 1392 dead Assam, Bihar, West Bengal and Orissa (Bay of Bengal): Floods and landslides caused by monsoon rains: 650 dead and 10,00,000 homeless.

Uttar Pradesh, Himachal Pradesh, Rajasthan and Gujarat: Floods and landslides caused by monsoon rains: 373 dead, 9,00,000 homeless. Orissa: Severe floods: 1,400 villages flooded: 360 dead: 12,00,000 homeless, estimated damage – USD 169 million. Parts of North India and Bihar: Severe Floods: 367 dead: 4,00,000 homeless. Cyclonic storms with winds of up to 120 km/h: 4,403 houses destroyed, 13,000 severely damaged: 45 dead; total estimated damage: USD 28 Million (Observed Effects of Climate Change in India, www.greenpeace.org).

Severe Drought Conditions is most of North West, Major parts of North India, North-East India and parts of Andhra Pradesh: Telangana and Rayalseema Region, parts of Tamil Nadu: Crops to the tune of USD 25 Million destroyed, many starvation deaths reported.

Severe floods affected northern and eastern states of the country. The northern and eastern parts of India, particularly Assam, Bihar, Meghalaya, Tripura, and Arunachal Pradesh faced severe floods. More than 10 million people have been affected and almost 300 people were killed due to floods, torrential rains and landslides. More than 20,000 houses have been damaged and the loss of public properties is estimated to INR 1,5 Billion (approximately US$ 30,772,391) (*ibid*). The flood situation in Assam was extremely severe, having hit 19 out of 23 districts. According to official estimates, nearly 3 million people were affected and 5,361 villages inundated. Seventy-six people have died from waterborne diseases, while around 85 deaths were reported due to floods. In addition to this, 2002 saw major of Indian states facing drought conditions and in states like Rajasthan, Gujarat, Madhya Pradesh, it was the 5th consecutive severe drought years. This region also witnessed a bad bout of heat wave, with over 250 people dead (*ibid*).

This year saw one of the country's worst natural calamities with over 15,000 dead in an earthquake measuring 7.7 on the Richter scale rocking parts of Gujarat. That year also saw severe floods in Eastern India, (Orissa), which left over 146 dead. The year 2001 also saw its share of drought in most part of North India and parts of South India (Observed Effects of Climate Change in India, www.greenpeace.org).

**Sea Level Rise in Bay of Bengal**

The projected climate change will affect India particularly severely. Its consequences include a rise in sea level, threatening areas such as the densely populated Ganges delta, changes in the monsoon rains, the melting of the glaciers in the Hindukush-Karakorum-Himalaya region (crucial for the water supply in the dry seasons), and the foreseeable increase in heavy rain events and intensity of tropical cyclones (Parry et. al., 2007). The thousands of glaciers located across the 2,400 km of the Himalayan range are at the epicentre of an emerging crisis.

The Indian subcontinent divides the north Indian Ocean into two semi-enclosed seas, the Arabian Sea and the Bay of Bengal, both of which encroach onto the equatorial regime of the basin. It has 7,500 km long coastline stretching from Gujarat to West Bengal. One quarter of the Indian population live along the country's coasts, and are largely dependent on coastal livelihoods. Climate change effects on sea level can impact coastal areas in two ways – through increase in sea level, and through increased frequency and intensity of coastal surges and storms. Climate change is of concern to India in view of the damages that occur along the east coast of India from the cyclones that form in the Bay of Bengal. Any further increase in the frequency or intensity of tropical disturbances due to climate change in the future could cause increased damages to life and property in the coastal regions.

Sea level changes can be of two types: (i) changes in the mean sea level and (ii) changes in the extreme sea level (Unnikrishnan, Rupa kumar, Fernandes, Michael and Patwardhan; 2006: 326). Mean sea level (MSL) is a measure of the average height of the ocean's surface (such as the halfway point between the mean high tide and the mean low tide); used as a standard in reckoning land elevation. The former is a global phenomenon while the latter is a regional phenomenon. In the past study reveals that the mean sea level along the east coast of India is higher than that along the west coast (Shankar, D and Shetye, R.S.: 2001: 563). This is the only reason along with the monsoon factor that leads to annual cycle of rain, salinity, sea level and cyclonic occurrences in the Bay of Bengal and differentiate it from the Arabian Sea.

Besides causing a sea level rise, there is a concern that global warming may cause changes in the occurrence of extreme events such as cyclones. Extreme events in the coastal regions occur in the form of storm surges, apart from tsunamis. The occurrence of extreme events is more or less a regional phenomenon, caused by changes in regional climate. The occurrence of storm surges is common to the countries surrounding the Bay of Bengal, as the Bay is the hotbed of generation of tropical cyclones. Any further increase in the frequency or intensity of tropical disturbances in future will cause increased damages to life and property in the coastal regions.

**Challenges for India**

Sea-level rise in the Bay of Bengal would impose serious impacts on the natural environment and human society in the coastal zone. Primary impacts of sea-level rise are inundation, exacerbation of flooding, beach erosion, and salt-water intrusion to rivers and groundwater aquifers. These impacts, in turn, cause higher-ordered impacts in a wide range of coastal systems. Since there exist highly productive ecosystems, large portion of the Indian population, and intensive socio-economic activities in the coastal zone, it is crucial to predict the degree and range of the possible impacts of sea-level rise India would have to face in the coming days, months and years.

Sea level rise would cause large-scale emigration from coastal zones to the mainland region. This will result in large numbers of environmental refugees especially from low-lying delta regions. Furthermore, intrusion of sea-water in the ground water and changes in temperature can reduce agricultural and fishing incomes. Populations dependent on coastal fishery and agriculture, which most often include developing countries, are likely to be adversely affected.

The coastal zone, a highly fragile eco system, is an important and critical region for India. It is densely populated and stretches over 800 km with the Arabian Sea in the West and the Bay of Bengal in the East. The area occupied by the coastal districts is around 379,610 km, with an average population density of 455 persons per km, which is about five times the national average of 324 persons per km (GOI, 2004b). Under the present climate conditions, it has been observed that a sea level rise (0.4-2.00 mm) along the Gulf of Kutch and the coast of West Bengal is the highest. Future climate change in the coastal zones is likely to be manifested through worsening of some of the existing coastal zone problems.

**Impacts on the Natural and Human Systems**

One of the most significant impacts of sea-level rise is acceleration of coastal erosion as well as inundation of mangroves, wetlands, and coral reefs. Some of the main climate related concerns in the context of the Indian coastal zones are erosions, flooding, submergence and deterioration of coastal ecosystems, such as mangroves and salinisation. These are mainly caused by, or, exacerbated by, sea level rise and tropical cyclones. The key climate related risks in the coastal zone include tropical cyclones, sea level rises, and changes in the temperature and precipitation. A rise in sea level is likely to have significant implications on the coastal population and agriculture performance of India. A one meter sea level rise is projected to displace approximately 7.1 million people in India and about 5,764 km of land are will be lost, along with 4,200 km of road (GoI, 2004b).

The diverse impact expected as a result of sea level rise includes land loss and population displacement, increased flooding of low lying coastal area, loss of yield and employment resulting from inundation and salinisation. Damage to coastal infrastructure, aquaculture and coastal tourism, due to erosion of the sandy beaches is also likely. The extent of vulnerability however depends not just on the physical exposure to sea level rise and the population affected but also on the extent of economic activity of the areas and capacity to cope with impacts. Natural ecosystems such as mangroves, grasslands and coral reefs are also likely to be affected by climate change. Sea level rise would submerge mangroves as well as increase the salinity of wetlands. The projected sea level rise of 0.1- 0.9 m seems within the ability of the Sunderban mangrove ecosystems, which presently face tidal amplitudes up to 5 m. This may be not true for other mangroves such as the Pichavaram and Muthupet where tidal amplitudes are much lower at 0.64 m (Shukla et. al., 2003). However, increased snowmelt in the Western Himalayas could bring large quantities of fresh water into the Gangetic delta. This would have significant consequences on the composition of Sunderban mangroves species, favouring species that have least tolerance to salinity.

**Tropical Cyclones and Storm Surges**

India has been subject to two cyclonic seasons, during the southwest and northeast monsoons. Cyclones have been observed to be more frequent in the Bay of Bengal than the Arabian Sea due to these wind systems. Consequently, the states of West Bengal, Orissa, Andhra Pradesh, and Tamil Nadu along the Bay of Bengal are the most affected as compared to Maharashtra, Kerala, Karnataka, and Gujarat, which are along the Arabian Sea. Most of the destructive cyclonic storms usually occur during the transition periods: pre-April-May and post- (September-December) monsoon, the latter being the most active period. The coastal districts of Orissa, Andhra Pradesh and Gujarat are most prone to the impacts of cyclone (GoI, 2006). About 1.3 storms form on an average during a year (GoI, 2004b). The east coast of India, which lies in the path of tropical hurricanes from the Gulf of Bengal, is particularly at risk of being damaged by storms and floods (Emanuel, 2005; IPCC, 2007b). The projections by the National Institute of Oceanography (NIO), under the Council of Scientific and Industrial Research (CSIR), Government of India on the impacts of climate change on sea level, to assess the degree to which mean sea level and the occurrence of extreme events may change, showed an increased occurrence of cyclones in the Bay of Bengal, particularly in the post-monsoon period, along with increased maximum wind speeds associated with cyclones and a greater number of high surges under climate change (DEFRA/GoI, 2005b). In addition, the strength of tropical cyclones, which represent a threat to the eastern coast of India and to Bangladesh, could increase. The risk to these areas will be aggravated by the rising sea level.

**Wetland Loss**

Coastal wetland ecosystems, such as salt marshes and mangroves are particularly vulnerable to rising sea level because they are generally within a few feet of sea level (IPCC, 2007). Wetlands provide habitat for many species, play a key role in nutrient uptake, serve as the basis for many communities' economic livelihoods, provide recreational opportunities, and protect local areas from flooding.

As the sea rises, the outer boundary of these wetlands will erode, and new wetlands will form inland as previously dry areas are flooded by the higher water levels. IPCC suggests that by 2080, sea level rise could convert as much as 33 per cent of the world's coastal wetlands to open water (IPCC, 2007).

**Environment**

Home to the world's largest mangrove gene pool and a World Heritage site as declared by UNESCO, the Sunderbans face a threat from global warming and climatic change. It is estimated by a recently released United Nations study that a mere 45 cm rise in the sea level will destroy 75 per cent of forests spread over a 10,000 sq km area in India's eastern state of West Bengal and adjacent Bangladesh (http://www.grida.no/publications/vg/climate/). Bangladesh, one of the world's poorest nations is also the country most vulnerable to sea-level rise. The population is already severely affected by storm surges. Catastrophic events in the past have caused damage up to 100 km inland. It is hard to imagine to what extent these catastrophes would be with accelerated sea-level rise.

There are already 7,000 environmental refugees in the Sunderbans and the numbers can only increase with the sea devouring more islands as a result of global warming and climate change. The School of Oceanographic Studies in Jadavpur University (JU) had conducted a 10-year study in and around the Bay of Bengal and concluded that the sea is rising at 3.14 mm a year in the Sunderbans against a global average of 2 mm, threatening low-lying areas of India and Bangladesh (Dhar, Sujoy; 2007).

"In places like east Sagar Island the sea is raising more due to easterly tilt of subsidence. The rise near the tiger habitat at Pakhiralay is at an average of 5 mm annually, while towards Khulna in Bangladesh it is 10 mm (*ibid*). Mangroves of Sunderbans act as natural buffers against tropical cyclones and also as filtration systems for estuarine and fresh water. They also serve as nurseries for many marine invertebrate species and fish. The shrinking mangrove will leave India and Bangladesh vulnerable to the effects of disturbances in the Bay of Bengal, including tsunamis.

United Nations Intergovernmental Panel on Climate Change (IPCC) has indicated the magnitude of the problem. "Further destruction of the Sunderban mangroves would diminish their critical role as natural buffers against tropical cyclones". The Bay of Bengal is heavily affected by tropical storms, about 10 per cent of the world's tropical cyclones occur in this area and 17 per cent of these sweep the land in Bangladesh (IPCC AR4, www.ipcc.ch). At least 13 sea facing islands are very vulnerable. Bhangaduani and Dalhousie are the two islands respectively losing 15 and 30 per cent mangroves since last 37 years (Dhar, Sujoy, 2007). What is alarming is that both these islands are excellent habitats of the famous Royal Bengal tigers. So there is a reduction in tiger land posing threat to the already threatened big cat population of India.

**Agriculture**

Salinity intrusion due to sea level rise will decrease agricultural production by unavailability of fresh water and soil degradation. Practicing shrimp cultivation in saline water has a drawback, and that is a decrease in rice production due to degraded soil quality. A World Bank (2000) study suggests that increased salinity alone from a 0.3 metre sea level rise will cause a net reduction of 0.5 million metric tons of rice production. Sea level rise affects coastal agriculture, especially rice production in two ways. Salinity intrusion degrades soil quality that decrease or inhibit rice production. When the rice fields are converted into shrimp ponds, total rice production decreases because of decreased rice field areas.

Rice is the staple food of the people of India. Sea level rise will increase flood frequency and flooding duration, affecting rice production. Due to sea level rise, salinity of water and soil will increase, and this will damage rice cultivable land. Because of the shortage of fresh water, rice production will be decreased and wheat production will be affected by salinity increase.

**Health**

Sea level rise may increase the risk of health hazards like diarrhoea, cholera, etc. Cholera is an infectious disease of the small intestine of human beings and is common in the coastal area of India. *Vibrio cholerae* is the causing microbe of cholera that survive longer with salinity level ranging from 2.5 ppt to 30 ppt and need Sodium ion ($Na^+$) for growth (Borroto, 1998). Average salinity of seawater is 35 ppt or 3.5 per cent. For the reason, coastal area is breeding and nursery ground of cholera disease. Water salinity and its distribution in the coastal area are increasing with the increase of sea level rise (IPCC, 2001a).

**Social Implications**

Under the retreat option, resettlement could create major problems. Resettled people are not always well received. They often face language problems, racial and religious discrimination and difficulties in obtaining employment. Even when they feel welcome, the disruption of families, friendships and traditions can be stressful.

Although the impacts of accommodation and protection would be less, they may still be important. The loss of traditional environments, which normally sustain economies and cultures and provide for recreational needs could disrupt family life and create social instability. Regardless of the response eventually chosen, community participation in the decision making process is the best way to ensure that these implications are recognised.

**Reasons for Concern**

Global climate change may raise sea level as much as one metre over the next century and, in some areas, increase the frequency and severity of storms. Hundreds of thousands of square kilometres of coastal wetlands and other lowlands could be inundated. Beaches could retreat as much as a few hundred metres and protective structures may be breached. Flooding would threaten lives, agriculture, livestock, buildings and infrastructures. Salt water would advance landward into aquifers and up estuaries, threatening water supplies, ecosystems and agriculture in some areas.

As a result of present population growth and development, coastal areas worldwide are under increasing stress. In addition, increased exploitation of non-renewable resources is degrading the functions and values of coastal zones in many parts of the world. Consequently, populated coastal areas are becoming more and more vulnerable to sea level rise and other impacts of climate change. Even a small rise in sea level could have serious adverse effects.

The Coastal Zone Management Subgroup has examined the physical and institutional strategies for adapting to the potential consequences of global climate change. Particular attention was focused on sea level rise, where most research on impacts has been conducted. The Subgroup has also reviewed the various responses and has recommended actions to reduce vulnerability to sea level rise and other impacts of climate change.

**Responses**

The responses required to protect human life and property fall broadly into three categories: retreat, accommodation and protection. Retreat involves no effort to protect the land from the sea. The coastal zone is abandoned and ecosystems shift landward. This choice can be motivated by excessive economic or environmental impacts of protection. In the extreme case, an entire area may be abandoned.

Accommodation implies that people continue to use the land at risk but do not attempt to prevent the land from being flooded. This option includes erecting emergency flood shelters, elevating buildings on piles, converting agriculture to fish farming, or growing flood or salt tolerant crops. Protection involves hard structures such as sea walls and dikes, as well as soft solutions such as dunes and vegetation, to protect the land from the sea so that existing land uses can continue. The appropriate mechanism for implementation depends on the particular response. Assuming that land for settlement is available; retreat can be implemented through anticipatory land use regulations, building codes, or economic incentives. Accommodation might be evolved without governmental action, but could be assisted by strengthening flood preparation and flood insurance programmes. Protection can be implemented by the authorities currently responsible for water resources and coastal protection.

In addition to the above responses, there are two possible responses to meet the growing concerns of sea level rise. These are adaptation and mitigation. Adaptation seeks to reduce the adverse effects of sea level rise on living organisms, including human and the environment. The ability to adapt and cope is a function of wealth/income, technology, scientific and technical knowledge and skills, information, infrastructure, policy and management institutions and equity (Chatterjee, R. Huq, 2002). Sea level rise adaptation can be addressed by changes in policies that lessen pressure on resources, improve management of environmental risks, and enhance adaptive capacity.

Coming to the mitigation response it has been well said that, 'Prevention is better than cure'. Emission control is the prevention of climate change and sea level rise. With its massive population, lavish life style and increased and inefficient transport system with rapidly growing industrialisation, India is rapidly contributing to the carbon emission and consequently to the global warming in a big way. The country should, therefore, take necessary steps to cut down its growing emissions level. Control of deforestation and fossil fuel use is essential for the purpose. IPCC (2001a) indicates the main greenhouse gas emission reduction measures which are: (i) demand reduction and/or efficiency improvement, (ii) substitution among fossil fuels, (iii) switch to nuclear energy, (iv) switch to biomass, (v) switch to other renewables, (vi) $CO_2$ scrubbing and removal, and (vii) afforestation. India should follow its level best, the above mentioned seven measures to get rid of the adverse effects of the sea level rise and could able to protect and prevent its coastal population and their livelihood from grave consequences.

**Conclusion**

Sea level rise although not that high for India, but its population explosion and heavy coastal population density will be exposed to the hazards of sea level rise in the coming years and decades given its contribution to carbon emission and global warming. Then also countries surrounding Indian ocean and Bangladesh are especially victims of sea level rise, even the coastal Indian states of Tamil Nadu, Andhra Pradesh, Odisha and West Bengal always been subject to severe loses of life, property and habitations' due to heavy flood, untimely cyclones and many other sea related natural calamities.

Along with the above said protective measures, improving scientific and public understanding of the problem is also a critical component of any response strategy to get rid of the menace. Besides, highest priorities must need to be accorded on basic and applied research for projecting sea level rise, rate of change and the frequency and intensity of storms. Equally important is to collect available information on coastal land elevation.

## References

Church, J. A. and Gregory, J. M. 2001. Changes in sea level. In Climate change 2001. The scientific basis (ed.) J. T. Houghton, Y. Ding, D. J. Griggs, M. Noguer, P. J. van der Linden and D. Xiaosu), Cambridge, UK: Cambridge University Press.

Nicholis, J Robert and Tol, S. J, Richard. 2006. Impacts and responses to sea-level rise: a global analysis of the SRES scenarios over the twenty-first century, Philosophical Translation of Royal Society Association, Vol. 364, No. 1073–95,

Douglas, C. Bruce, Kearney, S. Michael and Leatherman, P. Stephen. 2001. Sea Level Rise, History and Consequences, International Geophysics Series, Volume No. 75, Academic Press, San Diego: California). Sea Level Change, Geophysics Study Committee, National Research Council, 1990, National Academy Press: Washington DC.

Hekstra, G. P. 1988. Prospects of sea level rise and its policy consequences, discussion paper for Symposium on Controlling and Adapting to Greenhouse Warming, June 14-15, 1988, Resources for the Future, Washington, DC.

Unnikrishnan, S.A, Rupa Kumar, K, Fernandes, E. Sheron, Michael, G.S. and Patwardhan, S.K. 2006. Sea level changes along the Indian coast: Observations and projections, Current Science, Vol. 90, No. 3, 10 February.

Shankar, D and Shetye, R. S. 2001. Why is mean sea level along the Indian coast higher in the Bay of Bengal than in the Arabian Sea? Geophysical Research Letters, Vol. 28, No. 4, 15 February.

DEFRA/GoI. 2005a. 'Indian Climate Change Scenarios for Impacts Assessments, Key Sheet 2'. Indian Institute of Tropical Meteorology (IITM), Pune, India. In: *Investigating the Impacts of Climate Change in India*. Report by Department of Environment, Food and Rural Affairs (DEFRA), UK and Ministry of Environment and Forests (MoEF), Government of India (GoI), 2005.

Solomon, S., D. Qin, M. Manning, R. B. Alley, T. Berntsen, N.L. Bindoff, Z. Chen, A. Chidthaisong et al., IPCC. 2007. 'Technical Summary'. In: *Climate Change 2007: The Physical Science Basis. Contribution of Working Group I to the Fourth Assessment Report of the Intergovernmental Panel on Climate Change,* in Solomon, S.,D. Qin., M. Manning, Z. Chen, M. Marquis, K.B. Averyt, M. Tignor and H.L Miller (eds), Cambridge University Press, Cambridge, UK and New York, USA.

National Snow and Ice Data Centre, http://nsidc.org

IPCC. 2001. http://envfor.nic.in/cc/what.html.ippcc01

Observed Effects of Climate Change in India, www.greenpeace.org

M.L. Parry, O.F. Canziani, J.P. Palutikof, P.J. van der Linden, and C.E. Hanson, IPCC. 2007:.Summary for policymakers. In *Climate Change 2007: Impacts, Adaptation and Vulnerability. Contribution of Working Group II to the Fourth Assessment Report of the Intergovernmental Panel on Climate Change.* Eds. Cambridge University Press, pp. 7-22.

Shukla, P.R. Sharma Subodh K., Ravindranath, N.H, Garg, A. and Bhattacharya, S. 2003. 'Vulnerability and Adaptation: Challenges Ahead'. In: Shukla, P.R. Sharma Subodh K., Ravindranath, N.H, Garg, A. and Bhattacharya, S., (eds), *Climate Change and India-Vulnerability Assessment and Adaptation.* Universities Press (India), Hyderabad, India, 389-432.

GoI. 2004b. *India's Initial National Communication (NATCOM) to the United Nations Framework Convention on Climate Change (UNFCCC).* NATCOM 2004. Ministry of Environment and Forests (MoEF), Government of India, 2004.

GoI. 2006. *Crisis Management: From Despair to Hope.* Second Administrative Reforms Commission Third Report, Govt of India, Sep. 2006.

Emanuel, K. 2005. 'Increasing Destructiveness of Tropical Cyclones over the Past 30 Years'. *Nature.* 436: 686-88.

DEFRA/GoI. 2005b. 'Climate Change Impacts on Sea Level in India. Key Sheet 4'. National Institute of Oceanography, Goa, India. In: *Investigating the Impacts of Climate Change in India.* Report by Department of Environment, Food and Rural Affairs (DEFRA), UK and Ministry of Environment and Forests (MoEF), Government of India (GoI), 2005. Available online at http://www.defra.gov.uk/ environment/ climatechange/internat /devcountry/ india2.htm

Ocean and Sea Level Rise, www.climate.org

www.ipcc.ch

IPCC AR4, www.ipcc.ch

Hughes, T., J. L. Fastook, and G. H. Denton. 1979. Climatic Warming and *the Collapse of the West Antarctic Ice Sheet.* Orono, Maine: University of Maine.

Bentley, C. R. 1983. 'West Antarctic Ice Sheet: Diagnosis and Prognosis.' in P*roceedings: Carbon Dioxide Research Conference: Carbon Dioxide, Science* *and Consensus*. Conference 820970. Washington, DC.: Department of Energy, pp. IV.3-IV.50.

www.ipcc-wg1.ucar.edu/wg1/Report

www.npr.org

www.epa.gov/climatechange/effects/eco_animals.html

Bruun, P. 1962. 'Sea Level Rise as a Cause of Shore Erosion'. *Journal of the Waterways and Harbors Division* 88 (WW1): 117-30.

Schwartz, M. L. and J. J. Fisher. eds. 1979. *Proceedings of the Per Bruun* Symposium. Newport, RI.: IGU.

Commission on the Coastal Environments.

Gibbs, J. Michael, Economic Analysis of Sea Level Rise: Methods and Results, http://earth.epa.gov/climatechange/effects.

IPCC. 2007: Climate Change 2007: Impacts, Adaptation and Vulnerability. Contribution of Working Group II to the Fourth Assessment Report of the Intergovernmental Panel on Climate Change [Parry, Martin L., Canziani, Osvaldo F., Palutikof, Jean P., van der Linden, Paul J., and Hanson, Clair E. (eds.)]. Cambridge University Press, Cambridge, United Kingdom, p. 1000.

Dhar, Sujoy, Environment - India: Rising Sea Threatens Bengal's Deltaic People, June 4, 2007/IPS/IFE.

Borroto, R.J. 1998. *Global warming, rising sea level, and growing risk of cholera incidence: a review of the literature and evidence*, Geo Journal 44 (2), pp.111-120.

IPCC. 2001a. *Climate Change 2001: Mitigation*, Contribution of Working Group III to the Third Assessment Report of the Intergovernmental Panel on Climate Change (IPCC), Cambridge University Press, UK.

Chatterjee, R., Huq, S. 2002. *A Report on the Inter-regional Conference on Adaptation to Climate.*

*Change*, Mitigation and Adaptation Strategies for Global Change 7, pp. 403-06.

# 14

# Climate Change Challenges to Human Security in Third World Countries

*Alok Kumar Ray*

## Meaning of Climate Change and its Genesis

Climate Change is a long-term change in the statistical distribution of weather patterns over periods that range from decades to millions of years. It may be a change in the average weather conditions or a change in the distribution of weather events with respect to an average, for example, greater or fewer extreme weather events. Climate change may be limited to a specific region, or may occur across the whole Earth.

In recent usage, especially in the context of environmental policy, climate change usually refers to changes in modern climate. It may be qualified as anthropogenic climate change, more generally known as global warming or anthropogenic global warming... (http://timesofindia indiatimes.com/city/delhi/Climate-change-debate-hots-up/articles).

In the context of climate variation, anthropogenic factors are human activities that change the environment. In some cases, the chain of causality of human influence on the climate is direct and unambiguous (for example, the effects of irrigation on local humidity), while in other instances, it is less clear. Various hypotheses for human-induced climate change have been argued for many years. Presently the scientific consensus on climate change is that human activity is very likely the cause for the rapid increase in global average temperatures over the past several decades. Consequently, the debate has largely shifted onto ways to reduce further human impact and to find ways to adapt to change that has already occurred.

Of most concern in these anthropogenic factors is the increase in $CO_2$ levels due to emissions from fossil fuel combustion, followed by aerosols (particulate matter in the atmosphere) and cement manufacture.

Other factors, including land use, ozone depletion, animal agriculture and deforestation, are also of concern in the roles they play-both" separately and in conjunction with other factors-in affecting climate, microclimate, and measures of climate variables.

In the atmosphere, gases such as water vapour, carbon dioxide, ozone, and methane act like the glass roof of a greenhouse by trapping heat and warming the planet. These gases are called greenhouse gases. The natural levels of these gases are being supplemented by emissions resulting from human activities, such as the burning of fossil fuels, farming activities and land-use changes. As a result, the Earth's surface and lower atmosphere are warming, and this rise in temperature is accompanied by many other changes.

Rising levels of greenhouse gases are already changing the climate. According to the Intergovernmental Panel on Climate Change (IPCC) Working Group I (WGI) Fourth Assessment Report, (UNO) from 1850 to 2005, the average global temperature increased by about 0.76°C and global mean sea level rose by 12 to 22 cm during the last century. These changes are affecting the entire world, from low-lying islands in the tropics to the vast Polar Regions.

Climate change predictions are not encouraging; according to the IPCC WGI Fourth Assessment Report, a further increase in temperatures of 1.4°C to 5.8°C by 2100 is projected. Predicted impacts associated with such temperature increase include a further rise in global mean sea level, changes in precipitation patterns, and more people at risk from dangerous 'vector-borne diseases' such as malaria.

**Climate Change Threat**

Here are some of the most mentioned climate change threats:

1. More frequent extreme weather events such as droughts, floods, hurricanes that will not only create huge environmental but also huge economic damage.
2. Arctic completely free of summer ice, and rapid melting of Antarctica's ice. This could cause serious sea level rise, and threaten many coastal areas around the globe.
3. More hunger in the world, because of new pests and new diseases that will have the negative impact on crop health. More hunger in the world could lead to new wars.
4. Severe water shortages in some parts of the world. This, just like hunger issue, could lead to new wars.
5. More refugees. Climate change has the potential to make living impossible in certain areas of our planet, which could turn lot of people into refugees.

6. Extinction of many animal and plant species. We are not talking only about polar bears but many other animals, which will find it hard to adapt to rapid changes in climate. Biodiversity loss is already a huge problem, and in years to come, it could reach catastrophic proportions

**Climate Change and Anthropogenic Factors**

During the early hunter-gather period that lasted for a long period, human beings did the following:

1. Survived by collecting plants, hunting, fishing and scavenging meat from animals.
2. Moved from place to place.
3. Changed the environment very little.

During the later hunter-gather period, after human beings discovered fire and made tools, they did the following:

1. Converted some forests into grasslands.
2. Caused the extinction of some large animals like the mastodon, mammoth, and the giant sloth.
3. Carried seeds and plants to new areas.

With the beginning of agriculture, people settled as communities. They adapted slash-and-burn cultivation: cutting forests, burning the underbrush, and cultivating wild food plants. Initially they had little impact on the environment. Then the agricultural revolution began and the picture had been changed. The agricultural revolution began about 10,000 years ago. Gradually agriculture became our major activity. In this process, human beings modified the environment in many ways. According to the IPCC (Inter-Governmental Panel of Climate Change), the three main causes of the increase in green house gases observed over the past 250 years have been fossil fuels, land use, and agriculture. Agriculture is itself responsible for an estimated one third of global warming and climate change. It is generally agreed that about 25 per cent of the main green house gas, carbon dioxide, is produced by agricultural sources, mainly deforestation and the burning of biomass. The Food and Agriculture Organisation has estimated that meat production accounts for nearly a fifth of global greenhouse gas emissions. These are generated during the production of animal feeds. Ruminants, particularly cows, emit methane which is 23 times more effective as a global warming agent than carbon dioxide (H. Raj Goutam in the Kurukshetra, July 2009: 5).

The industrial revolution began in England in the 18th century and spread to the USA in the 19th century. The steam engine was invented and factories were set up. Production and trading of goods rapidly increased. Most importantly our relationship with nature changed forever. Humans started believing that:

1. They were the most important species on earth and
2. They could conquer and use nature as they please.

These two ideas have spread to all parts of the world now. In the 19th and 20th centuries, human beings have made a major impact on the environment. Again, scientific discoveries and new technologies led to the rapid exploitation of natural resources. The result is change in the climatic conditions of the world leading to disaster.

Human beings have adopted two methods to increase the Earth's capacity to support it and to satisfy its ever growing needs and desires.

(a) Extension of the inhabited Earth
(b) Intensification of the use of environment to increase productive capacity

Broadly speaking, intensification of man's use of environment thrives upon interaction of four main factors.

(a) Population growth
(b) Rise in aspiration and urge for increasing comfort level
(c) Attitude
(d) Advances in technology

Technology is a factor and a vector outcome of interactions of the three domains mentioned above. Technology affects the environment in two ways.

First, it has increased the ability of man to cause environmental change.

Secondly, the number of ways in which man affects the environment has increased and their effects have become more pronounced and widespread due to technology.

**The Idea of Security**

The traditional idea of security is described in the narrow sense of security within the framework of the Westphalia system of international relations, in which a sovereign state is to defend its political independence and territorial integrity. Its people from any sort of untoward incident and to protect through military and other necessary means, the people and assets within its territory from invasion by outside enemies (Peace of Westphalia in 1648). This sort of narrow sense meaning of security centring round the state, relying on military and coercive power is not suitable to dealing with transnational terrorism, massive population movements, infectious diseases such as HIV/AIDS, and threats to the earth's environment, global issues that transcend the territorial/geographical boundaries of states. The collapse of the Cold War system that was based on the balance of fear between two superpowers along with ideological clash has lessened the sense of imminent danger of a nuclear confrontation. Nuclear deterrent has acted in a positive direction to bring about a condition of no-war in the present world set-up. As a result, other security concerns have become more conspicuous to discuss about.

Human security is an emerging field for understanding global vulnerabilities whose exponents challenge the traditional notion of national security by arguing that the proper focus for security should be the individual rather than the state (Richard Jolly and Deepayan Basu Roy of Institute of Development Studies, Sussex. UNDP, NHDR Occasional paper, National Human Development Report Unit, May 2006). Human security holds that a people-centred view of security is necessary for national, regional and global stability. The concept emerged from a post-Cold War, multi-disciplinary understanding of security involving a number of research fields. The United Nations Development Programme's 1994 Human Development Report is considered a milestone publication in the field of human security, with its argument that insuring "freedom from want" and "freedom from fear" for all persons is the best path to tackle the problem of global insecurity (World Summit Outcome Document, 14-16 September, 2005, United Nations, Paragraph - 143). Thus, a 'more explicit definition' of human security is provided by two main aspects: 'safety from such chronic threats as hunger, disease and repression', and 'protection from sudden and hurtful disruption in the patterns of daily life'.

**Ecology and Human Security**

There are four distinctive views of the relationship between ecology and security: traditional, semi-traditional, alternative, and eclectic. Traditional views are represented by the dominant social paradigm of the realist perspective and the idea of a politics of scarcity. This idea is very similar to the state-centric view of security.

According to the theory of the politics of scarcity, scarcity may develop into international conflict when there is restricted access to vital non-renewable and renewable natural resources such as oil, coal, water as well as food. Thomas Homer-Dixon is representative of this school of thought. The decreasing availability of physical environment and land resources such as clean water, good agricultural land for arable and animal husbandry could create a condition of 'simple scarcity', 'group identity' and 'deprivation' in the area that could provoke violent conflicts of high magnitude due to population movements and scramble for the available resources. Environmental degradation or deficiencies create the conditions that render conflict even more likely, or determine the source of conflict, act as multipliers that aggravate core causes of conflict or acts as a catalytic factor in creating conflict (Homer-Dixon, 1994 : 5—40). Environmental security encompasses both renewable and non-renewable resources and takes into consideration the direct contribution of exploitation of non-renewable resources to the environmental degradation of the renewable resources.

Other prominent environmentally concerned theorists and policy advocates who base their arguments in this conception of ecological security are Lester Brown and Jessica Mathews (Lester R Brown, 1986 and Jessica T Mathews, Spring 1989, Foreign Affairs :162-77). They stress that global environmental threats or social and/or political disorder can be caused by 'environmental refugees' moving across national borders. Richard Ullman(1980, International Security: 162-77) extends this concept of security to include natural disasters including earthquakes.

The alternative view of environmental security is framed from the ecological perspective and is based on Harold and Margaret Sprout's (Sprout and Sprout, 1965, 1971:18) pioneering works in international relations. The ecological perspective envisages 'international politics as a *system of relationships* among *interdependent, earth-related* communities that share with one another an *increasingly crowded planet* that offers *finite* and *exhaustible* quantities of *basic essentials* of human well-being and existence' Regarding the root causes of ecological crisis, this perspective points to 'a transformation both in human capacity to cope with nature *and* in human attitudes toward nature and other men'. In other words, rapid technological advancement and the 'narrower engineering perspective' drive human beings to further exploit natural resources and reshape nature. The 'engineering perspective' is that which results in the invention of new technologies without concern for side effects damaging human health or the natural environment. Similarly, according to the most current advocates of the ecological perspective, the prevailing social perspective reflects an industrial culture that shapes the structure of social institutions. In other words, 'there is a general set of values, attitudes, beliefs, and perceptions that are shared by most members of industrial societies'. The proponents of this view of the ecology-security relationship argue that environmental security will demand changes in attitudes, beliefs, and perceptions if an environmentally secure world is to be realised (The Interlinkage of Climate Security and Human Security, 9 March 1999).

The eclectic view, finally, revolves around the concept of 'sustainable development', as articulated in the 1987 report of the World Commission on Environment and Development (Brundtland Commission). The commission called for radically different approaches to economic development, equity, resource management, and related issues in order to meet the challenge of a rapidly growing world population. Although it has acknowledged the necessity of development in order to break the vicious cycles linking poverty to population growth and environmental destruction, it has rejected unlimited development. The commission defined sustainable development as meeting 'the needs of the present without compromising the ability of future generations to meet their own needs'. However, sustainability is a term that is commonly used and misused when considering development and environmental concerns. One reason for its misuse is a lack of consensus or understanding of what exactly is meant by sustainability. There are also great differences in views on how to achieve sustainable development and how to measure this achievement. When we start to compare ways in which these are viewed across cultural boundaries the diversity and lack of consensus inflate.

## Typology of Human Security

More specifically, human security is considered in seven main categories: economic, food, health, environmental, personal, community, and political security. Economic security requires jobs to secure an assured basic income. Food security means that all people at all times have both physical and economic access to basic foods. Health security is to provide healthy environment and health services to meet the challenges of poor nutrition, infectious diseases, and so on. Environmental security is concerned with lack of access to clean water, deforestation, salinisation, air pollution, natural disasters, and so on. Personal security is to protect human lives from threats of various kinds of violence by states and other groups. This includes categories like crimes, industrial and traffic accidents, threats to women, abuse of children. Community security is about threats like oppressive practices and ethnic clashes in traditional communities. Political security means the protection of human rights and democratisation.

## Freedom from Fear and Freedom from Want

***Freedom from Fear*** — This school seeks to limit the practice of human security to protecting individuals from violent conflicts while recognising that these violent threats are strongly associated with poverty, lack of state capacity and other forms of inequities This approach argues that limiting the focus to violence is a realistic and manageable approach towards human security. Emergency assistance, conflict prevention and resolution, peace-building are the main concerns of this approach. Canada, for example, was a critical player in the efforts to ban landmines and had incorporated the 'Freedom from Fear' agenda as a primary component in its own foreign policy.

***Freedom from Want*** — This school advocates a holistic approach to achieving human security and argues that the threat agenda should be broadened to include hunger, disease and natural disasters because they are inseparable concepts in addressing the root of human insecurity and they kill far more people than war, genocide and terrorism combined· Different from 'Freedom from Fear', it expands the focus beyond violence with emphasis on development and security goals (Human Security Centre: 'What is Human Security'. Retrived on 19th April, 2008 from www.humansecurityreport.info/index.php?).

Despite their differences, these two approaches to human security can be considered complementary rather than contradictory. Franklin D. Roosevelt's famous Four Freedoms speech of 1941 contains 'Freedom from Want' as the third and 'Freedom from Fear' as the fourth such fundamental and universal freedom.

**The Case of Third World Countries**

Scarcities of critical environmental resources— in particular cropland, freshwater, and forests—are contributing to mass violence in several areas of the world. While these 'environmental scarcities' do not cause wars between countries at the first instance, they do sometimes sharply fuel stresses within countries, helping stimulate ethnic clashes, urban unrest, and insurgencies also. In Chiapas, Mexico, Zapatista insurgents recently rose against land scarcity and insecure land tenure produced by longstanding inequalities in land distribution, by rapid population growth among groups with the least land, and by changes in laws governing land access. The insurgency threatened Mexico to the core, helped trigger the peso crisis

In Pakistan, shortages and the maldistribution of good land, water, and forests have encouraged the migration of huge numbers of rural poor into major cities such as Karachi and Hyderabad. The result of this immigration with high fertility rates is causing urban populations to grow at a staggering four to five per cent a year, producing fierce competition and often violence among ethnic groups over land, basic services, and political and economic power. This turmoil incurs a huge cost on the national economy. It may also encourage the Pakistani regime to establish its internal legitimacy by adopting a more belligerent foreign policy on issues such as Kashmir and nuclear proliferation.

In South Africa, severe land, water, and fuel-wood scarcity in the former black homelands 'has helped drive millions of poor blacks into teeming squatter settlements in the major cities. The settlements are often constructed on the worst urban land, in depressions prone to flooding, on hillsides vulnerable to slides, or near heavily polluting industries. Scarcities of land, water, and fuelwood in these settlements provoke interethnic rivalry and violent feuds between settlement warlords and their followers (Nations in Conflict.San Fransisco.Freeman, 1975).

In India, the issue of Narayanpatna in KBK (Koraput-Balangir-Kalahandi) area of Odisha is one such instance where in the process of land grab by the non-tribals and the state, the innocent tribals have been affected more. In order to meet their dire necessities, the tribals become easy prey of mortgaging their land to non-tribals. With the opening of tribal areas to non-tribal people, land of the tribals has become a marketable commodity. The state in the name of development continues to acquire land, over which tribals have a stake, in a large measure. The Naxalite outfits are also taking advantage of the lingering tension between the *Panas* (Scheduled Castes) and tribals in the Gajapati district, as well as of land disputes between tribals and Bangladeshi refugees in the Malkangiri district. Peaceful eastern Indian state Odisha is now hit by naxal violence. Odisha's Koraput district is very much affected by the terror activities. Communist Party of India (Maoist) with the support of the local tribal intensified its movement for land rights in Narayan Patna area. In the banner of Chasi Mulia Adivasi Sangha, (association of tribal farmers and laborers) the tribal raises their voice against Indian Government. They are continuing their attack on non-tribal and forced them to get out from the tribal land. Recently arbitrary arrests of tribals in Narayanpatna of Koraput continue with about 109, including at least 12 children, thrown in jail so far on charges that include criminal conspiracy, rioting, sedition and waging war against the state and police and paramilitary forces have stepped up operations to hunt down activists of the Chasi Mulia Adivasi Sangha (CMAS), a tribal rights organisation active in the area. This is the case not just in Narayanpatna but also across villages and towns in India's tribal areas in Odisha, Chhattisgarh, Jharkhand, Bengal and Andhra Pradesh. Police atrocities against tribals are fuelling support for the Maoists. Tribals are the most marginalised section of Indian society, worse off than even the Dalits (formerly referred to as Untouchables). Around 49.5 per cent of tribals live under the official poverty line, 76.2 per cent are illiterate and almost 30 per cent have no access whatsoever to doctors in clinics. Displaced from their land and discriminated against in the industrial job market they are now fighting to keep their land, their only remaining resource. There are striking similarities between what is happening in Odisha with the controversial *salwa judum* (peace march) in neighbouring Chhattisgarh. Like Odisha, Chhattisgarh's southern districts of Bastar and Dantewada are rich in minerals. And as in Odisha, tribals who have been resisting the takeover of their land by the state and private mining companies are subjected to humiliation and torture.

**Forms of Scarcity**

In the world, nearly 3 billion people rely on agriculture for their main income; perhaps 1 billion of these are subsistence farmers who survive by eating what they grow. More than 40 per cent of the world's people—some 2.2 billion—use fuel-wood, charcoal, straw, or cow dung as their main source of energy: 50 to 60 per cent rely on these biomass fuels for at least some of their energy needs. Over 1.2 billion people lack access to clean drinking water. Basing on this backdrop the growing notion of resource scarcity needs to be discussed.

The cropland, forests, and water supplies that standardise the livelihoods and well-being of these billions are renewable. Unlike non-renewable resources such as oil and iron ore, renewable are replenished over time by natural processes. If used prudently and reasonably they should sustain an adequate standard of living indefinitely. Unfortunately, in the majority of regions where people are highly dependent on renewable resources, these resources are being depleted or degraded faster than they are being renewed due to lopsided development paradigm based on fossil fuel technology and western model of consumerism culture. From Gaza and the Philippines to Honduras, the evidence is established: aquifers are being overdrawn and salinised, coastal fisheries are disappearing, and steep uplands have been stripped of their forests, leaving their thin soils to erode into the sea.

These environmental scarcities usually have complex causes. Resource depletion and degradation are a function of the physical vulnerability of the resource, the size of the resource-consuming population, and the technologies and practices this population uses. The size of the population and its technologies and practices are in turn a result of a wide array of other variables, from the status of women to the availability of human and financial capital.

Moreover, resource depletion and degradation, taken together, are only one of three sources of environmental scarcity. Depletion and degradation produce a decrease in total resource *supply*—that is, a decrease in the size of the total resource 'pie'. But population growth and changes in consumption behaviour can also cause greater scarcity by boosting the *demand* for a resource. Thus, if a rapidly growing population depends on a fixed amount of cropland, the amount of cropland per person—the size of each person's slice of the resource pie—falls inexorably. In many countries, resource availability is being squeezed by both supply and demand pressures.

The third cause of scarcity is a severe imbalance in the distribution of wealth and power, which results in some groups in a society receiving disproportionately large slices of the resource pie while others get slices that are too small to sustain their livelihoods. Often the imbalance is deeply rooted in the institutions and class and ethnic relations inherited from the colonial period. Often it is sustained and reinforced by international economic relations that trap developing countries into dependence on a few raw material exports. It can also be reinforced by heavy external debts that encourage countries to use their most productive environmental resources—such as their best croplands and forests—to generate hard currency rather than to support the most impoverished segments of their populations.

**Interaction between Scarcities**

In the past, scholars and policymakers have usually addressed these three sources of scarcity independently. But supply, demand, and structural scarcities interact and reinforce each other in extraordinarily pernicious ways.

One type of interaction is resource capture. This occurs when powerful groups within a society recognise that a key resource is becoming scarcer (due to both supply and demand pressures) and use their power to shift resource access in their favour. This shift imposes severe structural scarcities on weaker groups. In Chiapas, worsening land scarcity (caused in part by rapid population growth) encouraged powerful landowners and ranchers to exploit weaknesses in the state's land laws in order to seize land from indigenous farmers. Gradually peasants were forced deeper into the state's lowland rain forest, further away from the state's economic heartland and further into poverty.

In the Jordan River basin, Israel's critical dependence on groundwater flowing out of the West Bank—a dependence made acute by a rising Israeli population and salinising aquifers along the Mediterranean coast—encouraged Israel to restrict groundwater withdrawals on the West Bank during the occupation. These restrictions were far more severe for Palestinians than for Israeli settlers. They contributed to the rapid decline in Palestinian agriculture in the region, to the increasing dependence of young Palestinians on day labour within Israel and, ultimately, to rising frustrations in the Palestinian community.

Another kind of interaction, ecological marginalisation, occurs when a structural imbalance in resource distribution joins with rapid population growth to drive resource-poor people into ecologically marginal areas, such as upland hillsides, areas at risk of desertification, and tropical rain forests. Higher population densities in these vulnerable areas—along with a lack of the capital and knowledge needed to protect local resources—causes resource depletion, poverty, and eventually further migration, often to cities.

Ecological marginalisation affects hundreds of millions of people around the world, across an extraordinary range of geographies and economic and political systems. We see the same process in the Himalayas, the Sahel, Central America, Brazil, India's Rajasthan, and Indonesia. For example, in the Philippines an extreme imbalance in cropland distribution between landowners and peasants has interacted with high population growth rates to force large numbers of landless poor into interior upland regions of the archipelago. There, the migrants use slash and bum agriculture to clear land for crops. As millions, more arrive from the lowlands, new land becomes hard to find, and as population densities on the steep slopes increase, erosion, landslides, and flash floods become critical. During the 1970s and 1980s, the resulting poverty helped drive many peasants into the arms of the communist New People's Army insurgency that had a stranglehold on upland regions. Poverty drove countless others into wretched squatter settlements in cities like Manila.

Of course, many factors unique to the Filipino situation have combined with environmental and demographic stress to produce these outcomes. Environmental scarcity is never a determining or sole cause of large migrations, poverty, or violence; it always joins with other economic, political, and cultural factors to produce its effects. In the Filipino case, the lack of clear property rights in upland areas encouraged migration into these regions and discouraged migrants from conserving the land once they arrived.

Analysts often overlook the importance of such contextual factors and, as a result, jump from evidence of simple correlation to unwarranted conclusions about causation. Thus some commentators have asserted that rapid population growth, severe land scarcity, and the resulting food shortfalls caused the Rwandan genocide. In an editorial in August 1994, *The Washington Post* argued that while the Rwandan civil war was 'military, political, and personal in its execution', a key underlying cause was 'a merciless struggle for land in a peasant society whose birthrates have put an unsustainable pressure on it'. Yet, while environmental scarcities in Rwanda were serious, close analysis shows that the genocide arose mainly from a conventional struggle among elites for control of the Rwandan state. Land scarcity played at most a peripheral role by reducing regime legitimacy in the countryside and restricting alternatives for elite enrichment outside the state.

Despite these caveats, in many cases environmental scarcity powerfully contributes to mass violence. Moreover, it is not possible entirely to subordinate its role to a society's particular institutions and policies. Some sceptics claim that a society can fix its environmental problems by fixing its institutional and policy mistakes; thus, they assert, environmental scarcity's contribution to conflict does not merit independent attention. But research shows that such arguments are incomplete at best.

First, environmental scarcity is not only a consequence of institutions and policy: it also can reciprocally influence them in harmful ways. For example, during the 1970s and 1980s the prospect of chronic food shortages and a serious drought encouraged governments along the Senegal River to build a series of irrigation and flood-control dams. Because of critical land scarcities elsewhere in the region, land values in the basin shot up. The Mauritanian government, controlled by Moors of Arab origin, took control of this resource by changing the laws governing land ownership and abrogating the traditional rights of black Mauritanians to farm, herd, and fish along the river.

Second, environmental scarcity should not be subordinated to institutions and policies because it is partly a function of the physical context in which a society is embedded. The original depth of soils in the Filipino uplands and the physical characteristics that make Israel's aquifers vulnerable to salt intrusion are not functions of human social institutions or behaviour. And third, once environmental scarcity becomes irreversible (as when a region's vital topsoil washes into the sea), then the scarcity is, by definition, an external influence on society. Even if enlightened reform of institutions and policies removes the original political and economic causes of the scarcity, it will be a continuing burden on society.

**Resource Scarcity May Cause Interstate War**

Scarcity-induced resource capture by Moors in Mauritania helped ignite violence over water and cropland in the Senegal River basin, producing tens of thousands of refugees. Expanding populations, land degradation, and drought spurred the rise of the Shining Path guerrillas in the southern highlands of Peru. In Haiti, forest and soil loss worsens a chronic economic crisis that generates strife and periodic waves of boat people. And land shortages in Bangladesh, exacerbated by fast population growth, have prompted millions of people to migrate to India—an influx that has, in turn, caused ethnic strife in the Indian states of Assam and Tripura (B.L. Turner II et. al., eds. 1990).

Close examination of such cases shows that severe environmental scarcity can reduce local food production, aggravate the poverty of marginal groups, spur large migrations, enrich elites who speculate on resources, and undermine a state's moral authority and capacity to govern. These long-term stresses can slowly tear apart a poor society's social fabric, causing chronic popular unrest and violence by boosting grievances and changing the balance of power between contending social groups and the state.

The violence that results is usually chronic and diffuse, and almost always sub-national, not international. There is virtually no evidence that environmental scarcity causes major interstate war. Yet among international relations scholars, it has been conventional wisdom for some time that critical scarcities of natural resources can produce international conflict. During the 1970s, for example, Nazli Chourci and Robert North argued that countries facing high resource demands and limited resource availability within their territories would seek the needed resources through trade or conquest beyond their boundaries. Although this 'lateral pressure' theory helped explain some past wars, such as World War I, our more recent research highlights a number of the theory's errors. Most importantly the theory makes no distinction between renewable and non-renewable resources.

There is no doubt that some major wars in this century have been motivated in part by one country's desire to seize another's non-renewable resources, such as fossil fuels or iron ore. For example, before and during World War II, Japan sought to secure coal, oil, and minerals in China and Southeast Asia. But the story is different for renewable like cropland, forests, fish, and freshwater. It is hard to find clear examples from this century of major war motivated mainly by scarcities of renewable.

There are two possible explanations. First, modern states cannot easily convert cropland and forests seized from a neighbour into increased state power, whereas they can quickly use non-renewable like iron and oil to build and fuel the military machines of national aggression. Second, countries with economies highly dependent on renewable tend to be poor, and poor countries cannot easily buy large and sophisticated conventional armies to attack their neighbours. For both these reasons, the incentives and the means to launch resource wars are likely to be lower for renewable than for non-renewable.

The exception, some might argue, is water, especially river water: adequate water supplies are needed for all aspects of national activity, including the production and use of military power, and rich countries are as dependent on water poor countries (often they are more dependent). Moreover, about 40 per cent of the world's population lives in the 214 river basins shared by more than one country. Thus at a meeting in Stockholm in August 1995, Ismail Serageldin, the World Bank's vice president for environmentally sustainable development, declared that the 'wars of the next century will be over water', not oil.

The World Bank is right to focus on the water crisis. Water scarcity and pollution are already hindering economic growth in many poor countries. With global water use doubling every 20 years, these scarcities—and the sub-national social stresses they cause—are going to get much worse. But Serageldin is wrong to declare that we are about to witness a surge of 'water wars'.

Wars between upstream and downstream neighbours over river water are likely only in a narrow set of circumstances: the downstream country must be highly dependent on the water for its national well-being; the upstream country must be able to restrict the river's flow; there must be a history of antagonism between the two countries; and, most importantly the downstream country must be much stronger militarily than the upstream country.

There are very few river basins around the world where all these conditions hold. The most obvious example is the Nile. Egypt wholly dependent on the river's water, has historically turbulent relations with its upstream neighbours, Sudan and Ethiopia, and is vastly more powerful than either. And Egypt has several times threatened to go to war to guarantee an adequate supply of Nile waters. But more common is the situation along the Ganges, where India has constructed a huge dam—the Farakka Barrage—with harsh consequences on downstream cropland, fisheries, and villages in Bangladesh. Bangladesh is so weak that the most it can do is to plead with India to release more water. There is little chance of a water war here between upstream and downstream countries (although the "barrage's" effects have contributed to the migrations out of Bangladesh into India). The same holds true for other river basins where alarmists speak of impending wars, including the Mekong, Indus, Parana, and Euphrates.

**Pivotal States**

The chronic and diffuse sub-national strife that environmental scarcity helps generate is exactly the kind of conflict that bedevils conventional military institutions. Around the world, we see conventional armies pinned down and often utterly impotent in the face of interethnic violence or attacks by ragtag bands of lightly armed guerrillas and insurgents. As yet, environmental scarcity is not a major factor behind most of these conflicts. But we can expect it to become a far more powerful influence in coming decades because of larger populations and higher resource consumption rates.

The world's population is growing by 1.6 per cent a year; on average, real economic product per capita is also rising by 1.5 per cent a year. These increases combine to boost the earth's total economic product by about 3 per cent annually. With a doubling time of approximately 23 years, the current global product of $25 trillion should exceed $50 trillion in today's dollars by 2020.

A large component of this increase will be achieved through greater consumption of the planet's natural resources. Already, as a group of geographers has noted, 'transformed, managed, and utilised ecosystems constitute about half of the ice-free earth; human-mobilised material and energy flows rival those of nature.' Such changes are certain to grow because of the rapidly increasing scale and intensity of our economic activity.

At the level of individual countries, these changes often produce a truly daunting combination of pressures. Some of the worst-affected countries are 'pivotal states'—to use the term recently coined in Foreign *Affairs* by historian Paul Kennedy. These countries—including South Africa, Mexico, India, Pakistan, and China—are keys to international stability in their regions.

India deserves particularly close attention. Since independence, it has often seemed on the brink of disintegration. But it has endured, despite enormous difficulties, and by many measures India has made real progress in bettering its citizens' lives. Yet, although recent economic liberalisation has produced a surge of growth and a booming middle class, India's prospects are uncertain at best.

Population growth stubbornly remains around 2 per cent a year; the country's population of 955 million (of which about 700 million live in the countryside) grows by 17 million people annually, which means it doubles every 38 years. Demographers estimate that India's population will reach 1.4 billion by 2025. Yet severe water scarcities and cropland fragmentation, erosion, and salinisation are already widespread. "Fuelwood" shortages, deforestation, and desertification also affect sweeping areas of countryside.

Rural resource scarcities and population growth have combined with an inadequate supply of rural jobs and economic liberalisation in cities to widen wealth differentials between the countryside and urban areas. These differentials propel huge waves of rural-urban migration. The growth rates of many of India's cities are nearly twice that of the country's population, which means that cities like New Delhi, Mumbai, and Bangalore double in size every 20 years. Their infrastructures are overtaxed: New Delhi has among the worst urban air pollution levels in the world, power and water are regularly unavailable, garbage is left in the streets, and the sewage system can handle only a fraction of the city's waste water.

India's rapidly growing population impedes further loosening of the state's grip on the economy: as the country's workforce expands by 6.5 million a year, and as resentment among the poor rises against those castes and classes that have benefited most from liberalisation, left-wing politicians are able to exert strong pressure to maintain subsidies for fertilisers, irrigation, and inefficient industries and to keep statutory restrictions against corporate layoffs. Rapid population growth also leads to fierce competition for limited status and job opportunities in government and education. Attempts to hold a certain percentage of such positions for lower castes cause bitter inter-caste conflict. The right-wing Bharatiya Janata Party capitalizes on upper and middle-caste resentment of encroachment on their privileges, mobilising this resentment against minorities like Muslims.

These pressures are largely beyond the control of India's increasingly corrupt and debilitated political institutions. At the district and state levels, politicians routinely hire local gang leaders or thugs to act as political enforcers. At the national level, kickbacks and bribes have become common in an economic system still constrained by bureaucracy and quotas. The central government and many state governments are widely seen as unable to manage India's rapidly changing needs, and as a result have lost much of their legitimacy.

The fast expansion of urban areas in poor countries like India may have the dual effect of increasing both the grievances and the opportunities of groups challenging the state: people concentrated in slums can communicate more easily than those in scattered rural villages, which might reinforce incipient economic frustrations and, by reducing problems of coordination, also increase their power in relation to the police and slums. While there is surprisingly little historical correlation between rapid urbanisation and civil strife, India shows that the record may be changing: the widespread urban violence in early 1993 was concentrated in the poorest as strictly communal, between Hindus and Muslims, Hindus directed many of their attacks against recent Hindu migrants from rural areas. B.K. Chandrashekar, a sociology professor at the Indian Institute of Management, has noted that 'the communal violence was quite clearly a class phenomenon. Indian cities became the main battlegrounds because of massive migrations of the rural poor in the past decades.'

Indian social institutions and democracy are now under extraordinary strain. The strain arises from a rapid yet incomplete economic transition, from widening gaps between the wealthy and the poor, from chronically weak political institutions, and—not least—from continued high levels of population growth and resource depletion. Should India suffer major internal violence as a result—or, in the worst case, should it fragment into contending regions— the economic, migratory, and security consequences for the rest of the world would be staggering.

**Banking on Ingenuity?**

As long as market reforms and adequate economic growth continue, India should be able to solve its problems of poverty, population growth, and environmental stress.

The most rigorous representatives of this optimistic position are neoclassical economists. They generally claim that properly functioning economic institutions, especially markets, can provide incentives to encourage conservation, resource substitution, the development of new sources of scarce resources, and technological innovation. Increased global trade allows resource-rich areas to specialise in the production of goods (like grain) that are derived from renewable. These optimists are commonly opposed by neo-Malthusians— often biologists and ecologists—who claim that finite natural resources place strict limits on the growth of human population and consumption both globally and regionally; if these limits are exceeded, poverty and social breakdown result.

The debate between these two camps is now thoroughly sterile. Each grasps a portion of the truth, but neither tells the whole story. Neoclassical economists are right to stress the extraordinary ability of human beings to surmount scarcity and improve their lot. The dominant trend over the past two centuries, they point out, has not been rising resource scarcity but increasing aggregate wealth. In other words, most important resources have become less scarce, at least in economic terms.

The optimists provide a key insight: that we should focus on the supply of human ingenuity in response to increasing resource scarcity rather than on strict resource limits. Many societies adapt well to scarcity, without undue hardship to their populations, and often end up better off than they were before. These societies supply enough ingenuity in the form of new technologies and new and reformed social institutions—like efficient markets, clear property rights, and rural development banks—to alleviate the effects of scarcity.

What determines a society's ability to supply this ingenuity? The answer is complex: different countries, depending on their social, economic, political, and cultural characteristics, will respond to scarcity in different ways and with varying amounts and kinds of ingenuity.

Optimists often make the mistake of assuming that an adequate supply of the right kinds of ingenuity is always assured. However, in the next decade's population growth, rising average resource consumption, and persistent inequalities in resource access guarantee that scarcities of renewable will affect many regions in the developing world with a severity, speed, and scale unprecedented in history. Resource substitution and conservation tasks will be more urgent, complex, and unpredictable, increasing the need for many kinds of ingenuity. In other words, these societies will have to be smarter—socially and technically—to maintain or increase their well-being in the face of rising scarcities.

Simultaneously, the supply of ingenuity will be constrained by a number of factors, including the brain drain from many poor societies, their limited access to capital, and their chronically incompetent bureaucracies, corrupt judicial systems, and weak states. Moreover, markets in developing countries often do not work well: property rights are unclear; prices for water, forests, and other common resources do not adjust accurately to reflect rising scarcity; and thus incentives for entrepreneurs to respond to scarcity are inadequate.

Most important, the supply of ingenuity can be restricted by stresses generated by the very resource crises the ingenuity is needed to solve. In Haiti, for example, severe resource shortages—especially of forests and soil—have inflamed struggles among social groups, struggles that, in turn, obstruct technical and institutional reform. Scarcities exacerbate poverty in rural Haitian communities and produce significant profit opportunities for powerful elites. Both deepen divisions, distrust between rich and poor and impede beneficial change. Thus, for example, the Haitian army has blocked reforestation projects by destroying tree seedlings; the army and the notorious Tonton Macoutes fear such projects will bring disgruntled rural people together and threaten their highly profitable control of forest resource extraction.

Similar processes are at work in many places. In the Indian state of Bihar, which has some of the highest population growth rates and rural densities in the country, land scarcity has deepened divisions between land-holding and peasant castes, promoting intransigence on both sides that has brought land reform to a halt. In South Africa, scarcity-driven migrations into urban areas and the resulting conflicts over urban environmental resources (such as land and water) encourage communities to segment along lines of ethnicity or residential status. This segmentation shreds networks of trust and debilitates local institutions. Powerful warlords, linked to the Inkatha Freedom Party or the African National Congress, have taken advantage of these dislocations to manipulate group divisions within communities, often producing violence and further institutional breakdown.

Societies like these may face a widening 'ingenuity gap' as their requirement for ingenuity to deal with scarcity rises while their supply of ingenuity stagnates or drops. A persistent and serious ingenuity gap boosts dissatisfaction and undermines regime legitimacy and coercive power, increasing the likelihood of widespread and chronic civil violence. Violence further erodes the "society's" capacity to supply ingenuity, especially by causing human and financial capital to flee. Countries with a critical ingenuity gap therefore risk entering a downward and self-reinforcing spiral of crisis and decay.

A focus on ingenuity supply helps us rethink the neo-Malthusian concept of strict physical limits to growth. The limits a society faces are a product of both its physical context and the ingenuity it can bring to bear on that context. If a hypothetical society were able to supply infinite amounts of ingenuity, then that "society's" maximum sustainable population size and rate of resource consumption would be determined by biological and physical laws, such as the second law of thermodynamics. Since infinite ingenuity is never available, the resource limits societies face in the real world is more restrictive than this theoretical maximum. And since the supply of ingenuity depends on many social and economic factors and can therefore vary widely, we cannot determine a "society's" limits solely by examining its physical context, as "neo-Malthusians" do. Rather than speaking of limits, it is better to say that some societies are locked into a 'race' between a rising requirements for ingenuity and their capacity to supply it.

In coming decades, some societies will win this race and some will lose. We can expect an increasing bifurcation of the world into those societies that can adjust to population growth and scarcity—thus avoiding turmoil—and those that cannot. If several pivotal states fall on the wrong side of this divide, "humanity's" overall prospects will change dramatically for the worse (Homer Dixon, 1996 : 359-365).

**Working out a Panacea for Human Security**

Threats to biodiversity, food shortages, urban sprawl...lessons for environmental problems that confront us today may well be found in the past. The archaeological record contains hundreds of situations in which societies developed long-term sustainable relationships with their environments—and thousands in which the relationships were destructive. By discussing archaeological case studies from around the world—from the deforestation of the Mayan lowlands to soil erosion in ancient Greece to the almost total depletion of resources on Easter Island—one unearths the never-ending impact of human culture on the environment. The difference is that human intervention has fastened the process to a great extent during the last half to one century.

The burden of enormous populations is making itself felt: as governments struggle with the need to educate children, create jobs, and deal with the environmental effects of population growth, any new threat-such as AIDS or aquifer depletion-can rapidly escalate to disastrous proportions. The industrialised countries have held HIV infection rates among their adult populations to one per cent or less, but infection rates are as high as one-quarter of the adult population in some African countries. With their rising mortality rates, more reminiscent of the Dark Ages than the bright millennium so many had hoped for, these countries are falling back to an earlier demographic stage with high death rates and high birth rates, and ultimately little growth in population. Events in many countries could spiral out of control, leading to spreading political instability and economic decline. In examining the stakes involved in potentially adding another 3.3 billion people to the world population over the next 50 years, we need a clarion call for immediate expansion of international family planning assistance to the millions of couples who still lack access, and new investment in education young people-especially women-in the Third World, helping to promote a shift to smaller families

It is argued that global sustainability will never be engineered from the top-down, a local vision of sustainable community (Thomas Prugh, Robert Costanza and Herman Daly, 2010). As a prescription for how but can only emerge out of countless numbers of local communities creating and realising communities should go about this, we may look back to 1984 and Benjamin Barber's idea of strong democracy. Strong democracy is a vision of participatory governance rich in discourse and volunteerism. To flesh out what strong democracy might actually look like examples may be cited back in time to Northern Italy circa 1300 and Eastern Switzerland's Republic of Rhaetia circa 1600. Nearer to our present time examples of strong democracy in action were Denmark's consensus conferencing, New England town meetings, and an example of watershed restoration work in Oregon. Aware that mainstream economics still invokes the idea of rational economic "man" introduced by Alfred Marshall, we may echo sociologists such as Max Weber and critical theorists such as Jurgen Habermas who call for a notion of rationality that redeems human dignity and freedom. Marshall's idea of rationality sketched homo economicus, a version of human who does not create alternatives for action, who never contemplates or learns about values or preferences, who never thinks in terms of 'we', but only 'me'. Rational economic 'man' only seeks the most efficient means to produce the highest egoistic utility possible. Such a conception may provide a convenient and powerful hypothesis-testing tool, but it is historically and physiologically inaccurate, and, as critical theorists point out, it leaves out essential aspects of what it means to be human (Barber, B. 1984).

Scientists of all disciplines as well as citizens who gain their knowledge experientially or through tradition have important roles to play in helping provide information about likely consequences of choices. This information one needs when designing and implementing a public discourse about local sustainability. Another glaring point is to pay attention to problems of discursive competence, especially hidden motives of free-riders and ego centred parties who mislead the group and seek to exploit public discussion for private ends. The idea of strong democracy is an ideal that has to take into account the mischievous aspects of people. More than anything else, having a successful process probably means getting all involved to adopt the right attitude, that is, to think 'what is best for all', not 'what is best for me'. How are we to accomplish this when people are socialised in an economised society where the pursuit of hedonistic utilities is strongly condoned from the time one leaves the womb to the time one enters the grave? More research and pulling of brains are necessary to find out a solution for this problem. National and international bodies and forces, corporations and government agencies have their role to play to have a coordinated network to establish political power otherwise the local communities will be vulnerable to the wanton whims of much more powerful global players.

Human beings recognise the importance of nature, its bio-diversity and the need to maintain it. Nature fulfills the survival instinct of all the living beings. Among all the living beings, man only stretches the carrying capacity of the Earth beyond its natural limits. The urge for material prosperity as well as unsustainable exploitation creates an imbalance. The indicators of prosperity envisaged in the development model hinge upon material prosperity which is based on unsustainable exploitation of the ecosystem. Environmental degradation occurs in both the cases of overexploitation for material comfort and over dependence for livelihood. Over dependence on environment for livelihood is a result of lop-sided development that has created enclaves of prosperity by jeopardising the livelihoods of many. Ecological responsibility is to be based on the twin pedestal of curtailment of material need by the rich and promotion of livelihood security for the poor.

## References

Barber, B. 1984. *Strong Democracy: Participatory Politics for a New Age.* Berkeley, CA: University of California Press.

B.L Turner II et al., eds., *The Earth as Transformed by Human Action; Global and Regional Changes in the Biosphere over the Past 300 Years* Cambridge: Cambridge University Press.

Harender Raj Goutam, "Effects of Climate Change on Rural India" in The Kurukshetra, July 2009, New Delhi, page-5 http://times of india indiatimes. com/ city/delhi/Climate-change-debate-hots-up/articles)

Homer-Dixon,T. F(1994), 'Environment Scarcity and Violent Conflict, Evidence from cases' Peace and Conflict Studies Programme. University of Toronto, *International Security*, Vol. 19, No 1, pp. 5-40 (Summer 1994).

Homer Dixon, T. F. (November 1996). Environmental scarcity, mass violence, and the limits to ingenuity. *Current History 95.* 359–365.

Lester R Brown, Redefining National Security in Lester Brown et al. State of World Environment, Newyork, Norton,1986 and Jessica T Mathews, Redefining Security, Foreign Affairs, published by the Council of Foreign Relations, Spring 1989, foreign affairs.com, pp. 162-177).

Richard Jolly and Deepayan Basu Roy of Institute of Development Studies, Sussex, The Human Security Framework and National Human Development Reports: A Review of Experiences and Current Debates. UNDP, NHDR Occasional paper, National Human Development Report Unit, May 2006.

Richard Ullman, Redefining Security, International Security,Vol. 8, No.1, pp. 162-77(1980) Sprout and Sprout,The Ecological Perspective of Human Affairs;With Special Reference to International Politics, Princeton, Princeton University Press,1965-Towards a Politics of Planet Earth, New York,Van Nostard Reinhold,1971, p. 18.

The Interlinkage of Climate Security and Human Security; The Convergence of Policy Requirements.Heroshi Ohta.Waseba University. 9 March 1999, prepared for –Climate/Security, University of Copenhagen, Copenhagen, Denmark.

Human Security Centre. "What is Human Security". Retrived on 19th April, 2008 from www.humansecurityreport.info/index.php?

Thomas Prugh, Robert Costanza and Herman Daly,The Local Politics of Global Sustainability, by Washington D.C; Island Press, 2010).

World Summit Outcome Document, 14-16 September, 2005, United Nations, Paragraph-143.

# 15

# Politics of Development in North-East India: Its Impact upon the Bio-diversity, Environment and Human Security

*Sasanka Ghosh*

The phrase 'India's North East' has a magnetic attraction among the academic community that is involved with social science research. Due to some typical characteristics of the region, it has got such interest among the academicians. This region is often described as 'underdeveloped area', 'ethnic conflict haunted zone', 'terrorist trodden area', etc. The area is known for its underdevelopment, ethnic chaos, terrorism, border-dispute, trans-border illegal migration of foreign people and illegal trans-border trade. But very recently, another aspect of the region has cast the attention of both the social scientists, social activists and decision makers. It is about the importance of bio-diversity protection/preservation of the region (Fernandes 2010). The area is one of the 25 mega-biodiversity zones of the world and one of its 18 biodiversity hotspots (Fernandes, 2010).

## Geo-Politics of North East India

The North Eastern Region of India lies between 22°N and 29°5 North Latitude and 88°E and 97°30 East Longitude. The region covers nearly 2,62,379 sq. km area and has been divided into two bio-geographic zones, i.e., Eastern Himalaya and North East India, based on floristic composition, the naturalness of the flora and the local climate (Rodgers and Panwar 1988). The Eastern Himalaya comprises both Arunachal Pradesh and Sikkim, whereas the North East India bio-geographic zone consists of six out of the 'seven sisters' i.e. Assam, Nagaland, Manipur, Meghalaya, Mizoram and Tripura (WWF, 2006).

The region is isolated from the rest of India by our neighbouring country Bangladesh and the Bay of Bengal except the slim 'Siliguri bottle neck'. The North Eastern Region of India is virtually an island with only a 20 km wide land connection with the rest of India, i.e., the Siliguri Bottle neck (Chaube 2010). The region occupies 7.7 per cent of India's total geographical area. It shares an international boundary of more than 4,500 kilometers with the neighbouring countries like Bangladesh, China, Bhutan and Myanmar. The Region has about 1,956 km boundary with Bangladesh, 1,643 km with Myanmar, about 1,000 km with China and about 650 km with Bhutan (Bandyopadhyay 2005).

Various climatic conditions, difference in altitude and numerous water bodies like streams, lakes, rivers and swamps have created a rich biodiversity over the region. Out of nine important vegetation types in India, six are found in the North Eastern Region. Nearly half of Indian variety of plants is found there. Various endemic plants spontaneously grow over the region. Nearly 31.58 per cent of the flora found over there is endemic (Yumnam 2008). A large variety of medicinal plants is also available over the region.

The Region is also a suitable abode of numerous birds and animal species. High rate of endemism is there among the birds, mammals, reptiles and vertebrates.

| *Birds* | | *Mammals* | | *Reptiles* | | *Amphibians* | | *Vertebrates* | |
|---|---|---|---|---|---|---|---|---|---|
| Total | Endemic | Total | Endemic | Total | Endemic | Total | Endemic | Total | Endemic |
| 1170 | 140 | 329 | 73 | 484 | 201 | 202 | 114 | 2185 | 528 |

*Source:* (WWF 2006).

Like the above mentioned higher proportion of endemism, variety of species is found from different categories of fauna. As for example, three families of primates occur in India with 15 known species, nine of which occur in North East India. Assam and Arunachal Pradesh are the safe homes for the big cats like the tiger, Leopard, Snow Leopard and the Clouded Leopard. Out of 28,000 wild elephants in India, about 33 pert cent are found in North East India. Fifty-eight species of snakes have been recorded in Assam, thirty four from Manipur and ninety two from Arunachal Pradesh (WWF 2006).

## Demography of North East India

Demographic situation of North East India is also interesting. Firstly, because the Region hosts a large number of tribal groups (nearly 225 tribes live in the region out of 450 in the country) and secondly, it is a safe area for a large number of illegal migrants from Bangladesh. According to the 2001 Census the North Eastern States have tribal population in the following figures:

| *Name of the state* | *Total Population* | *Tribal Population* | *Percentage of Tribal Population* |
|---|---|---|---|
| Arunachal Pradesh | 1,097,968 | 705,158 | 64.2 |
| Assam | 26,655,528 | 3,308,570 | 12.4 |
| Manipur | 2,166,788 | 741,141 | 34.2 |
| Meghalaya | 2,318,822 | 1,992,862 | 85.9 |
| Mizoram | 888,573 | 839,310 | 94.5 |
| Nagaland | 1,990,036 | 1,774,026 | 89.1 |
| Tripura | 3,199,203 | 993,426 | 31.1 |

The ratio of the tribal population in some states of the region like Tripura and Assam is rapidly decreasing due to migration and other socio-economic causes. The following table shows the tragic picture of the demographic change in Tripura:

| *Census Year* | *Total Population* | *Tribal Population* | *Percentage of Tribal Population* |
|---|---|---|---|
| 1931 | 3,82,450 | 1,92,249 | 50.26 |
| 1941 | 5,13,010 | 2,56,991 | 50.09 |
| 1951 | 6,45,707 | 2,37,953 | 36.85 |
| 1961 | 11,42,005 | 3,60,070 | 31.53 |
| 1971 | 15,56,342 | 4,50,544 | 28.95 |
| 1981 | 20,53,058 | 5,83,920 | 28.44 |
| 1991 | 27,57,205 | 8,53,345 | 30.95 |
| 2001 | 3,199,203 | 993,426 | 31.1 |

*Source*: (Mohanta 2004).

Similarly, in Assam the demographic balance has been seriously disturbed due to the problems mentioned above. However, the tribal population of the region is heterogeneous in character. A large number of tribes and sub-tribes live in the region with their own distinct cultural identity, traditions and value-system. Although the wave of modernisation has brought a considerable change among the elite sections of them, yet both the common people and the intelligentsia belonging to those groups are conscious to protect their culture and heritage.

However, the changes that are brought about by migration and demographic change in the states like Assam, Tripura and Meghalaya have seriously affected the economic interests of the tribal people living over the region. The indigenous people of the region have been deprived from the possession of plain land and forest area. The problem of indebtedness is concomitant with that of landlessness and land alienation, resulting from the growing influx of people and increasing pressure (Bose 1996).

**Economy of the Region and Rationale behind Development**

The economy of the region is basically dependent upon agriculture (Roy 2010). Both the *jhum* cultivation (a primitive mood of shifting cultivation based on slush and burn) and settled cultivation are practiced here. Gogoi (2007) has examined the agriculture, industry, infrastructure and social sector of the region and said, 'All the North Eastern States could place themselves under 'low' category of development in one sector or the other.'

Sound agricultural and industrial base along with strong infrastructural growth and solid social sector are essential prerequisites for an over all development of the Region. But most of them are not sufficiently present over there. Low productive investment and high unemployment are the two most visible characteristics of the economy of the Region. Naturally the Region is haunted by the problems like unemployment and underemployment. An analysis of the Annual Economic surveys of the States shows that the North East has an unemployed backlog of 25 per cent of its active workforce. (Fernandes and Pereira 2005). To remove these problems, it is generally advised that 'a heavy dose of investment and government expenditure in social and infrastructural sectors in particular' is among the most urgent.

Jairam Ramesh (2008), one of the important Ministers in the present UPA Government has examined thoroughly the policy of the Indian Government towards the North Eastern Region and pointed out four paradigms that have been adopted so far. The first among them is the 'Culture Paradigm'. The region is a phenomenally diverse mosaic of cultures which have to be preserved and enriched. According to Mr Ramesh in 1950s and 60s New Delhi believed that 'exotic and endangered' cultures of North East had to be protected from encroachment. But after the Chinese invasion of 1962, the 'security paradigm' was tried over the region. Again in 1970s, with the formation of new states in North East, the New Delhi Government applied the 'politics paradigm'. Finally, in 1980s, the new *mantra* of 'development paradigm' was introduced. According to Mr Ramesh, the latest paradigm wants to spread the feel good factor of development. If new schools, bridges, internet centres, IITs, refineries, hydro-electric projects etc. are established rapidly over the region, the people will 'forget about their problems of identity, problems of assertion, problems associated with creating a nation out of essentially tribal community'.

In this way development initiatives are superimposed upon the North Eastern Region. The region is already tormented by some issues that have been generated by the nation-building process. Contradictions such as insiders vs. outsiders, customary rights vs. legal rights, ethnic autonomy vs. national integration and unification created unrest and violence in the North Eastern States like Nagaland, Assam, Mizoram and Tripura. Such contradictions have started to interfere with the so-called development missions. In addition, these development initiatives are designed in such a manner that they would never pose a threat to market interests of investors, capitalists and their dependents. The entire development strategy creates an extraeconomic coercion on local producers and is legitimised by an ideology of rapid growth of economy in terms of industries and massive capacity building (Biswas 2007). Such developmental efforts are put under question by many people on the ground that they go clearly against the welfare of micro level communities like indigenous people and other marginal sections of society.

**Development of North East and its Bio-diversity Significance**

As stated in the very beginning of this chapter that the North Eastern Region is one of the 25 mega biodiversity zones of the world and one of its 18 biodiversity hotspots. Therefore, it is very clear the existing variety and richness of the biodiversity of North East India is not only important for India but also for the world as a whole. The world is continuously being threatened by the danger of green house effect and climate change. Due to its floristic diversity and density of green forest cover the North East India has also been identified as 'one of the carbon sinks' of the world to protect Europe and other parts of the world from the threat of climate change (Fernandes 2010).

However, the so called developmental activities that are going on in this region hardly care for biodiversity and the issues related with human security. A brief review of some of the developmental projects that are going on in this region can establish the statement.

**Mining and Development**

The region has an abundance of mineral resources. Coal, limestone, dolomite, natural gas, petroleum etc. have been discovered in different parts of the Region. It is also estimated that the Region has a rich reserve of those minerals. A vast reserve of uranium also has been discovered at Kyelleng-Pyndengsohiong, Mawthabah in Meghalaya.

But, extracting of some minerals like coal, limestone etc. is not at all favourable for the rich biodiversity of the region. Environment-friendly and sustainable mining is yet to be introduced in this area.

**KPM Uranium Mining Controversy**

Kyelleng-Pyndengsohiong, Mawthabah uranium mining project is located in the west of Khasi Hills district of the state of Meghalaya. It is estimated that the site has 16 per cent of the total uranium reserve in India. The Uranium Corporation of India Limited (UCIL) received clearance from the Government of India in 2007. It also managed the approval of the Government of Meghalaya. However, a strong wave of protest from different NGOs and civil society organisations including the Khasi Students' Union opposed the project. The antagonists of the project hold that the uranium extraction would invite massive health hazard in the area due to the release of large quantity of radioactive waste. Secondly, it would also invite environmental hazards due to the toxic nature of uranium ore and the waste associated with it. Finally, the UCIL proposed to acquire 10 sq. km of area and because of it nearly 30,000 residents of the area would be displaced (Kumar 2010).

**Development and Deforestation**

The region is very rich in forest resources, i.e., floral and faunal diversity. North East India has 64 per cent of its total area under forest cover and is often referred as a forest surplus region (WWF 2006). Side by side, there has been a traditional symbiotic relationship between the indigenous people and the eco-system of the region. Economic activities, particularly traditional agricultural practices are eco-friendly and sustainable (Jeeva, Laloo and Mishra 2006).

However, the so called developmental activities around the region hardly care for the importance of forest and the need for the protection/ preservation of traditional eco-friendly indigenous knowledge. Large-scale deforestation for promoting developmental activities has become a familiar tendency in the region. Such type of deforestation also leads to displacement of the tribal people. The displacement of the Hmars and the Kukies from the North Cachar and Karbi Anglong districts of Assam due to deforestation is an indirect cause of violence over that area of Assam (Biswas 2007).

Moreover, in some cases the areas that are selected for the camps/ headquarters of the security forces are the forests that are occupied by the tribes. Previously, the tribal people used those areas for *jhum* cultivation, collection of roots, shoots, fire-wood, etc. However, the Government deliberately deprives the people from the possession over forest and forest resources by establishing such camps of the security forces (Debbarma 2007).

**Urbanisation and Displacement**

Although the rate of urbanisation is not remarkable in this area, yet some government initiatives like the construction of East-West corridor in the outskirts of Guwahati have resulted in the displacement and subsequent exodus of the Lalungs from their usual habitats (Biswas 2007). The towns and cities of the Region are mainly surrounded by forest area. Therefore, whenever there is an initiative of expanding them, it certainly destroys the forest, biodiversity and livelihood of marginal section of tribal people.

**Construction of Big Dams and Its Consequences**

Mighty Rivers of the North Eastern Region like Brahmaputra, Barak and their large number of tributaries along with other rivers have created a huge potential for generating hydroelectric power. It is estimated that the region has an exploitable potential of 63,257 MW of electricity whereas the region currently generates only 1011 MW (Bandyopadhyay 2005).

Therefore, to exploit such a huge potential, the Government has adopted a plan to construct as many dams as possible. There is a project to construct 48 hydroelectric dams in the region in the present decade only and in future, there will be another 168 dams to be constructed. 69 of them will be constructed only in Arunachal Pradesh, which has the highest component of biodiversity over the region (Fernandes 2010). These projects will certainly destroy most of the components of biodiversity along with the sustenance of local people.

Same examples also can be drawn from other parts of the region. The case of the large project like Tipaimukh dam for the generation of an estimated 1500 MW of power has already generated fear of massive displacement of people from an area of 275.50 sq. km, i.e., the area of permanent submergence by the dam. It is estimated that the project is going to displace over 15,000 people (Hussain and Phanjoubam 2007). The human and environmental cost of the project is taken into account for the supposed quantity of power generation (Biswas 2007). Similarly, the Dumbur hydropower project of Tripura has submerged 4,562 hectares of fertile land, which used to feed 24,472 families of tribal people (Debbarma 2007). In addition, in both the cases the victims are yet to be compensated. Perhaps these symptoms of the so-called developmental projects promote such doubts. 'By allowing the dams to supply power to the rest of India and to South East Asia will the region be deprived of its biodiversity and sustenance and remain supplier of one more raw material in the form of power?' (Menon 2008). Moreover, the region is still geologically active. It is located in a high seismic zone (Chakma 2010). Therefore, the mega dams that are being constructed over the area may be dangerous for the security of common people.

**Need for Alternative Method of Development**

Nobody denies the need for an over all economic development. Not all the debates and disputes are intended to replace the strategy of economic growth by the so-called myth of 'zerogrowth' in the name of 'greenpolitics'. Rather, they are against the tendency of imposing an artificial and shortsighted mission of development. North East India has a unique character in the country, so the region requires special care and attention to implement developmental programmes. While introducing any development initiative, the policy-makers must consider the following issues that are inherently associated with the region:

Firstly, as stated earlier, the region is one of the 25 mega-biodiversity zones of the world and one of its 18-biodiversity hotspots. So while implementing different developmental policies, it must be verified that none of them would do any harm to the richness of biodiversity of the region. Sacrificing biodiversity means imposing a partial initiative of welfare upon the present society at the cost of future of not only the region or the Indian Nation but also of the entire world!

Secondly, the people of the area basically belong to different tribal groups. Civil and political rights as assured by the Indian Constitution are not of great importance for the tribes like the other sections of the Indian population. The problems of the tribes stemmed not so much from within as from outside. The tribes had hardly aspired to become members of the larger Indian society. That is why the tribes' struggle has always been for the greater control over their land, forest and other resources through articulation of political autonomy than share for greater participation and control of the state structure and institution (Xaxa 2006). Each of the tribes has its own structure of customary law that defines almost all aspects of its life. Therefore, it must be deliberated carefully that these laws should not be contradicted or violated without any serious or sufficient ground.

Thirdly, the indigenous people of the region have their own pattern of economic activities which are eco-friendly and sustainable. Use of plant and animal parts as pesticide and also as the medicine for various crops diseases, growing of location-specific nitrogen fixing trees for increasing soil fertility, providing artificial shelter amid the crop field for birds which prey on harmful pests are some of the examples of such activities (Chhetry and Belbahri 2009). Those techniques should be popularied and given sufficient importance.

Fourthly, as stated earlier, the demographic structure of the Region has seriously affected by some historical causes like the partition and the consequent wave of migration from the then East Pakistan (present Bangladesh). Those immigrants who settled over the region were habituated with the practice of settled cultivation. However, most of the indigenous people of the region were comfortable with their traditional *jhum* cultivation. So the 'outsiders' have captured maximum of plain land of the region. Side by side, due to increasing pressure of population and the government's decision of reserving a large forest area as the 'reserved' forest, the indigenous tribal groups of the region are alienated from the ownership of land. Perhaps this is the most important cause for which there is such a sharp division between the so-called 'insiders' and 'outsiders'.

Fifthly, the region is haunted by inter-ethnic chaos, insurgency and terrorism. The existing and emergent civil society organisations of the region circulate essentially within the confines of their own allied ethnic communities. So in most of the cases, they are either exclusively committed to the cause of their own communities, or strong enough to deny the authority of the state (Das 2007). Therefore, the common people very often lose the favour of them. Finally, it is predicted that an unchecked trend of climate change could potentially lead to a greater influx of environmental refugees to the North Eastern part of India. It is estimated that 130 million people would be displaced from different coastal area of South East Asia because of climate change, out of which 70 million are expected from Bangladesh only. This large number of refugees may further complicate the already existed complexities in the region (Bhoumik 2010).

## References

Bandyopadhyay. P.K. 2005. *The North East Saga:* Ministry of Information and Broadcasting, Govt. of India, New Delhi.

Basu, Pradip Kumar. 1996. *The Communist Movement in Tripura*. Kolkata: Progressive Publishers.

Bhaumik Subir. 2010. Seminar Proceeding as published in *Look East,* October 2010 Vol. 1, Issue: 7.

Biswas P. 2007. 'Insurgency and Regional Underdevelopment: Philosophical Issues in Youth Movements of North East India'. in Bora and Goswami (ed). *Youth at the Crossroads: A Study of North East India*. Guwahati, DVS Publishers.

Chakma, M.K. 2010. 'Environmental thwart will wreck North East' in *Look East* October 2010; Vol. 1, Issue: 7.

Chaube, S.K. 2010. 'North East India: A Continental View' in '*Man and Society: A Journal of North East Studies'* Vol. VII, Summer 2010.

Chhetry, Gopal Kumar Niroula and Belbahri, Lassaad. 2009. 'Indigenous Pest and Disease Management Practices in Traditional Farming System in North East India: A Review' in *Journal of Plant Breeding and Crop Science,* Vol. 1(3), May 2009.

Das, S.K. 2007. 'Youth and the Making of Civil Society in North East India'in Bora and Goswami (ed.). *Youth at the Crossroads: A Study of North East India* Guwahati, DVS Publishers.

Debbarma, S. 2007. 'Genesis of Youth Unrest in Tripura: A Case Study' in Bora and Goswami (ed.). *Youth at the Crossroads: A Study of North East India* Guwahati, DVS Publishers.

Fernandes, Walter. 2010. 'Regionalism, Globalisation and Constraints to Research in North East' in *Man and Society: A Journal of North East Studies,* Vol. VII, Summer 2010.

Fernandes and M. Pereira. 2005. *Land Relations and Ethnic Conflicts: The Case of North East India.* Guwahati, North Eastern Social Research Centre.

Gogoi, J.K. 2010. 'Regional Disparity in the Levels of Development in India and Youth Unemployment Problem with Special Reference to the North Eastern Region' in Bora and Goswami (ed.). *Youth at the Crossroads: A Study of North East India* Guwahati, DVS Publishers.

Hussain, Monirul and Phanjoubam, Pradip. 2007. *A Status Report on Displacement in Assam and Manipur.* Mahanirvan Calcutta Research Group, Kolkata.

Jeeva, Solomon Retna Dhas Nadar, Laloo,Christopher and Mishra, Bhanu Prakash. 2006. 'Traditional Agricultural Practices in Meghalaya, North East India' in *Indian Journal of Traditional Knowledge,* Vol. 5(1) January, 2006.

Kumar, Abhay. 2010. *KPM Uranium Mining in Meghalaya: A Controversial Project* in http://www.pragoti.in/node/4002.

Menon, Manju. 2008. 'Land Alienation Due to Large Hydro-Power Projects in Arunachal Pradesh' in Fernandes and Barbora (ed.) *Land, People and Politics: Contest for Tribal Land in North East India.* Guwahati: North Eastern Social Research Centre and IWGIA.

Mohanta, Bijan. 2004. *Tripura in the Light of Socio-Political Movements since 1945* Kolkata, Progressive Publishers.

Ramesh, Joyram. 2008. 'North East India in a New Asia' in *North East Quarterly* Vol. 1, No. 2 (January 2008).

Rodgers W.A. and Panwar, H.S. 1988. *Planning a Wildlife Protected Area Network in India.* Dehradun, Wildlife Trust of India.

Roy, Niranjan and Debnath, Abijit. 2010. 'Performance of the States of North East India: A Comparative Analysis of Post and Pre-Reform Periods' in *Man and Society: A Journal of North East Studies,* Vol. VII, Summer 2010.

WWF. 2006. *Background Paper on Biodiversity Significance of North East India* 30 June, 2006 from mdoner.gov.in/writereaddata/.../13Biodiversity 9583338441pdf.

Xaxa, Virginius. 2006. 'Reading Tribes in the Light of Citizenship Rights' in M.N. Karna (ed.) *Democrac, Pluralism and Conflicts.* Jaipur: Rawat Publications.

Yumnam, J.Y. 2008. 'Rich Biodiversity of North East India Needs Conservation' in *Current Science,* Vol. 95, No 3 (August, 2008).

# 16

# Reconceptualising Security in the Context of Global Climate Change

*Sonali Mohapatra*

In the traditional analysis security and state are one and in unity. State is legitimised to define what security threat is and respond appropriately to the extraordinary situation created by the threat. This is only true when people and state are held synonymous and the source of security threats remains only with the state. With the emergence of multifarious security threats not necessarily from state but from other non-state sources and segmentation of people into various sectors such as vulnerable, rich and poor, male and female the credibility of traditional and classical notion of security built around state and its military apparatus to give protection to people against all types of threats was challenged.

With the end of cold war the relative dominance of the military-political security plummeted reflecting the widespread recognition of the sources of security being diversified away from the centre stage, i.e., the state. In recent years what have catapulted into the centre stage of both policy and strategic thinking are the environmental apocalypse, poverty, economic decline and other social crisis of identity, drug and human trafficking and politics of racism, minority, human rights and feminist indignation and the issues relating to the very survival and dignity of human beings brushing aside the political-military construct of security hovered around state and its apparatus to the fringe. The growing disenchantment with the inadequacy of state to meet all these security challenges from diverse sources making a mince meat of the realist and neorealist understanding of security gave rise to the 'wide' versus 'narrow' debate about security. The widening of the base of security threats as outlined above and confinement to the use of force blurs the distinction between the traditional and wide concept of security, endangers the intellectual coherence of security, and dilutes the meaning. If sources of security threats have been diversified, to cling to Westphalia thematic of security built around state and its military apparatus seems not in congruence with the old logic and the changing and plurality of security threats. As a response and solution to these threats it is not always the state and its military apparatus but the some collective security thinking appears logical as found required in the case of climate change.

The traditionalists demonstrated strong proclivity toward putting primacy on military security despite a vast array of non-military security threats affecting humanity at large. They are not averse to looking at these non-military security threats and conflict in the international system yet are prompt to loosen their state centrism without making an attempt to defend and stick to their conventionally understood centrality of state in security terrain when many non state actors are in the fray along with state to claim equal status in the military game.

As a result of the end of the cold war a global reconceptualisation of security has emerged, that geographically widened due to globalisation. With the approaching shift in earth history from the 'Holocene' to the 'Anthropocene' new objective security dangers have been socially constructed since the 1980s. Now global problems pose fundamental threats, challenges, vulnerabilities and risks for individual human beings and humankind.

Security is a basic term and key concept in the social sciences that is used in intellectual traditions and schools, conceptual frameworks, and approaches. The term 'security' is associated with many different meanings that refer to frameworks and dimensions, apply to individuals, issue areas, societal conventions, and changing historical conditions and circumstances. Thus, security as an individual or societal political value has no independent meaning and is always related to a context and a specific individual or societal value system and its realisation.

Security is a societal value or symbol that is used in relation to protection, lack of risks, certainty, reliability, trust and confidence, predictability in contrast with danger, risk, disorder and fear. As a social science concept, 'security is ambiguous and elastic in its meaning. Arnold Wolfer pointed to two sides of the security concept: "Security, in an objective sense, measures the absence of threats to acquired values, in a subjective sense, the absence of fear that such values will be attacked."

For the constructivists, security is inter subjective referring to 'what actors make of it'. Thus, security depends on a normative core that cannot simply be taken for granted. Political constructions of security have real world effects, because they guide action of policymakers, thereby exerting constitutive effects on political order. Security is the result of a speech act ,according to which an issue is treated as: 'an existential threat to a valued referent object to allow' 'urgent and exceptional measures to deal with the threat'. Thus, the 'securitising actor' points 'to an existential threat' and 'thereby legitimises' 'extraordinary measures'.

'Security in an objective sense' refers to specific security dangers, i.e. to 'threats, challenges, vulnerabilities and risks' to specific security dimensions and referent objectives as well as sectors while 'security in a subjective sense' refers to security concerns that are expressed by government officials, media representatives, scientists or 'the people' in a speech act or in written statements by those who securitise 'dangers' as security 'concerns' being existential for the survival of the referent object and that require and legitimise extraordinary measures and means to face and scope with these concerns. Thus, security concepts have always been the product of orally articulated or written statements by those who use them as tools to analyse, interpret, and assess past actions or to request or legitimie present or future activities in meeting the specified security threats, challenges, vulnerabilities, and risks. The Copenhagen school did not review the sectorialisation of security from the perspective of national (international, regional) and human security.

Influenced by different worldviews, rival theories and mindsets, security is a key concept of competing schools of (a) war, strategic or security studies from a realist perspective, and (b) peace and conflict research from an idealist or pragmatatic view. Since 1990, interparadigm debates emerged between traditional, critical, and constructive approaches. Within the UN and NATO, different concepts coexist, a state-centred political and military concept, and an extended security concept with economic, societal, and environmental dimensions. A widening and depending of the security concept prevailed in OECD countries, while other countries adhered to a narrow military concept.

Not only the scope of 'securitisation' has changed, but also the referent object from a 'national' to a human-concerned' security concept, both within the UN system, and in the academic security community. In European security discourses, an 'extended' security concept is used by governments and in scientific debates distinguishing a 'national' and three expanded security concepts of 'social', human and environmental security.

While since the 19th century the key 'actor' has been the state, it has not necessarily been a major 'referent object' of security which is often referred to as 'the people' or 'our people' whose survival is at stake.

Whether a threat, challenge, vulnerability and risk become an 'objective security danger' or a 'subjective security concerns' also dependents on the political context. While in Europe climate change has become a major security issue, in the US, during the administration of

George W. Bush this problem was downgraded. Labelling climate change a security issue implies different degrees of urgency and means for coping with it.

The traditional understanding of security 'as the absence of existential threats to the state emerging from another state' has been challenged both with regard to the key subject (the state) and carrier of security needs, and its exclusive focus on the 'physical or political dimension of security of territorial entities' that are behind the suggestions for a horizontal and vertical widening of the security concept. The meaning of security was also interpreted as a reaction to globalisation and to global environmental change.

**Securitisation Theory**

This reconceptualisation of security has impacts on international agendas and thus on political action on many different levels. UNDP (1994) introduced a 'people-centred human security concept that was subsequently promoted by the human security network' (as 'freedom from fear') and by the human security commission (as 'freedom from want'), to which Kofi Annan added as a third pillar: 'freedom to live in dignity' and the united Nations University (UNU) as the fourth pillar: Freedom from hazard impacts.

The process of rethinking or 'reconceptualisation security concepts' and 'redefinition of security interests' that was triggered by the global turn of 1989-1991 and slightly modified by the events of 11 September 2001. The sub-sequent US-led 'war on terror' has become a truly global process.

Trends in the reconceptualisation of security:

- Widening, deepening, and sectoralisation of security concepts;
- Shift of referent object from the state to human beings or humankind (human security);
- Perception of new security dangers (threats, challenges, vulnerabilities, and risks) and securitisation of new security concerns due to articulations by national and international organisations, scientific epistemic communities, and an attentive public with a progressing decentralisation and diversity of information control through the internet;
- Search for new non-military strategies to face and cope with these newly perceived security dangers and concerns and new environmental dangers, hazards, and disasters that pose no classical security dilemma for states but a 'survival dilemma' for people.

**Securitisation of Global Environmental Change**

He defined peace 'as security and the secure access to resources that are essential for living' where climate change affects some populations to access (a) clean water (water security) (b) sufficient food (food security) (c) stable health conditions (health security) (d) ecosystem resources (environmental and ecological security), and (e) security of settlements (urban security).

AI Gore in 2007 emphasised the close conceptual relationship of peace and security with development and environment or among the four components of a 'conceptual quartet'. During 2007, climate change has been perceived and addressed-in many parts of the world although not by all governments, as a major security threat, challenge, vulnerability and risk and in this process of 'securitiation' of problems of global environmental change, the IPCC has indirectly—in the public perception worldwide—become a major 'securitising actor' by upgrading climate change to an 'existential threat' to different referent objects from the international community (global, international and regional security), the state (state or national security) and humankind (human and gender security).

Addressing global environmental change (GEC) and climate change (CC) as a new objective security danger and subjective security concern takes note of the fundamental reconceptualisation of security that is reflected in the widening (from a narrow political and military to economic, social and environmental dimensions), deepening (from a state-centred to a human-centred perspective or from the 'state' to the 'people' as referent objects of securitisation; thus from 'national' to 'human' security) and the sectoralisation of security with applications of this concept to energy, food, health, livelihood and water as well as those related to climate, soil, population and urbanisation.

## Global Environmental and Climate Change: The IPCC as a Securitising Actor

At least three standpoints exist on environmental issues:

- A pessimist or Neo-Malthusian view stimulated by Malthus' essay on population (1798) that stressed the limited carrying-capacity of the Earth to feed the growing population:
- An optimist or cornucopian view that believed.

Thus, since 1988, climate change has increasingly become an urgent policy issue and was thus 'politicised' and since the turn of the century, climate change has gradually been perceived and discussed as an international and human security issue. During the past two decades both the scientific and the policy debates on problems of global environmental change and on the reconceptualisation of security were pursued by different scientific and policy communities, by environmental epistemic communities and by the debates on peace and security.

The United Nations University Institute on environment and human security in Bonn has developed the environmental dimension of human security further. From its perspective, the improvement of human security, particularly the improvement of the environmental dimension of human security, requires a better understanding of the various forms of vulnerability in different societies, their economies and of the environmental conditions for hazards of natural origin as well as with regard to creeping environmental degradation that impact on the vulnerability and the hazard components. Bogardi and Brauch suggested that human security should rest on three pillars reflecting the corresponding pillars of sustainable development:

- 'Freedom from want';
- 'Freedom from fear';
- 'Freedom from hazard impacts'.

**The Hot Issue: Securitisation of Climate Change**

Especially since 2007, several policy-oriented studies have securitised climate change from different advantage points and concepts of security by analysing climate change as:

- An international security threat, challenge, vulnerability and risk;
- A national security threat for the United States and as
- A human security challenge that will affect the highly socially vulnerable poor population.

In analysing climate change as a security issue the policy efforts for securing a stable climate or maintaining a rate of change below the levels for human and ecological systems should be made.

**Climate Change as an International Security Danger and Concern**

Global climate change will potentially alter agricultural productivity, freshwater availability and quality, access to vital minerals, costal and island flooding, and more. Among the consequences of these impacts will be challenges to political relationship.

**Realignment of Energy Markets and Regional Economies and Threats to Security**

There is debate about the extent to which resource constraints or environmental problems alone can lead to conflict. However, it is widely acknowledged that resource constraints can lead to economic pressures and tensions or as triggers to conflicts when other tensions exist between states or political actions. These challenges, together with the long history of political frictions and disputes worsened by environmental stresses, suggest that global climatic changes have the potential to exacerbate international and sub national tensions and conflicts. The Pacific Institute has published a series of analyses of the relationship between climate and security beginning in the late 1980's. In an ongoing effort to understand the connections between climate change and international security, the Pacific Institute will continue to publish and analyse the links between climate and conflict.

From a state-centred international security perspective, the German Advisory Council on Global Change (WBGU 2007/2008) reviewed the Scientific Research on Climate Change as a Security Risk. The report argued that:

Without resolute counteraction, climate change will overstretch many 'societies' adaptive capacities within the coming decades. This could result in destabilisation and violence, jeopardising national and international security to a new degree. However, climate change as a threat to humankind and soon sets the course for the avoidance of dangerous anthropogenic climate change and will draw ever deeper lines of division and conflicts between and within countries over the distribution of resources ,especially water and land, over the management of migration, or over compensation payments between the countries mainly responsible for climate change and those countries most affected by its destructive effects [this report] is based on the findings of research into environmental conflicts, the causes of war, and of climate impact research. It appraises past experience but also ventures to cast a glance far into the future in order to assess the likely impacts of climate change on societies, nation-states, regions and the international (see WBGU website) society.

Climate change poses significant security risks due to an increased occurrence of severe weather patterns, degradation of vital natural resources and threats to the livelihoods and threats to the livelihoods and safety of populations on every continent. Pressure on resources, natural disasters and humanitarian crises-including flooding, drought, desertification and loss of arable land, massive and rapid migration and refugee flows-has the potential to threaten economic, political and social stability while increasing the risk of internal civil unrest. Poor and underdeveloped communities and countries are particularly vulnerable and yet are marginalised in international negotiations on climate policy. An absence of political will on the part of developed nations to accommodate more comprehensive policy changes may meet with increasing resentment from those countries most vulnerable to the effects of climate change. Thus, it is imperative to take action now to decrease the momentum of climate change. Women's leadership must help increase political will at the national and global level, guide the private sector away from voluntary initiatives and toward legally required changes in practice, give voice to affected communities most in need of assistance to mitigate and cope with the effects of climate change. 'An integrated human, gender and environmental security approach is needed for dealing with the growing threat of climate change, in order to develop appropriate adaptation and mitigation strategies.

The links between climate change, peace and war were analysed in a report by international Alert, an independent peace building organisation on: A climate of conflict: The Links between Climate Change, peace and war 'consequences of consequences', i.e., The effects climate change has in interacting with socio-economic and political problems in poorer countries. It highlighted four key elements of risk—political instability, economic weakness, food insecurity and large-scale migration and it makes twelve recommendations for addressing climate change in fragile states.

With regard to the impacts of climate change on the Middle East, Eco-peace/friends of the Earth Middle East tabled a report on 'Climate Change: A New Threat to Middle East Security' to the COP in Bali in December 2007 in which the authors from Israel, Palestine and Jordan argued that 'the climate crisis and its potential physical and socioeconomic impacts are likely to exacerbate this cross-border instability' and that 'climate change is likely to act as a 'threat multiplier'—exacerbating water scarcity and tensions over water within and between nations linked by hydrological resources; geography; and shared political boundaries.' However, this crisis offers opportunities 'for local, cross-border and international cooperation to ameliorate the problems that are already occurring and that are projected to intensify.'

Sir Crispin Tickell, the former UK Permanent Representative to the UN, highlighted the environmental factors behind societal collapse. Professor John Mitchell, the chief scientist, forecasted that the coming decades would see a 30 per cent increase in severe drought and that Africa will experience increased desertification, water stress and diseases.

In a speech in Berlin in late 2006, the British foreign secretary Margaret Beckett considered climate change as a serious threat to international security' that 'must not be dealt with using guns and tanks, but through dialogue and the sharing of new technologies between developed and developing countries'.

When people were faced with the stresses of overpopulation, including a lack of key resources such as water, energy and food, 'so we see the slide down the spectrum from stability to instability'. 'A global temperature increase of just two or three percent would cut the crop yields in Africa, the Middle East and South Asia by up to 40 per cent causing millions of people to be denied food even more than they are now. Climate change also affected issues such as immigration. A 50 cm rise in sea levels would displace 25 million in Bangladesh. In short, a failing climate means more failed states. And that has implications for everything we want to achieve from conflict prevention and resolution to counter-terrorism.' Ms

Becktt said, by tackling climate change, we can help address the underlying securities that feed and exacerbate conflicts and instability. By ignoring it we resign ourselves to the same crises flaring up again and again. And new ones, are emerging. She said the world was on the verge of 'global irresponsibility of massive and irreversible scale' by refusing to tackle climate change, but warned: 'The greatest security threat we face as a global community would not be met by guns and tanks'. It will be solved by investment in the emerging techniques of soft power-building avenues of trust and opportunity that will lead to a low-carbon economy. There is no backstop-politics and diplomacy has to work (see British Embassy Berlin Speech, 2006).

He concluded: 'Climate change is a security issue because if we do not deal with it, people will die and states will fail'. In his view, defense and security planners must face a paradox when assessing their responses to the problem. Most security threats in today's world are amenable to some extent to a "hard power" or 'conventional reaction,' he said, 'and demand will rise for such responses to climate change-related security problems'. However, there is no hard power solution to climate change you cannot force your neighbour to change its carbon emissions at the barrel of a gun (Chris Littlecott, 2007).

There are few greater potential threats. She argued that 'an unstable climate will exacerbate some of the core drivers of conflict, such as migratory pressures and competition for resources'. Japan's Ambassador Kenzo Oshima said that 'it is clear that climate change can pose threats to national security... [and] in the foreseeable future climate change may well create conditions or induce circumstances that could precipitate or aggravate international conflicts' (International Herald Tribune, 18 April 2007: 2).

A stark account of the threat is posed by global warming. It argues that the world is drifting towards a 'tipping point' that could lock the world's poorest countries and their poorest citizens in a downward spiral, leaving hundreds of millions facing malnutrition, water scarcity, ecological threats, and a loss of livelihoods. 'Ultimately, climate change is a threat to humanity as a whole. But it is the poor, a constituency with no responsibility for the ecological debt we are running up, who face the immediate and most severe human costs,' commented UNDP Administrator Kendal Derive.

Climate change also poses major obstacles to progress in meeting the Mugs and maintaining progress raising the HDI: 'There is a clear and present danger that climate change will roll back human development for a large section of humanity, undermining international cooperation aimed at achieving the Millennium Development Goals in the process' (UNDP, 2007).

**Climate Change as a National Security Danger and Concern**

The CNA's Military Advisory Board drew five policy recommendations from its analysis:

1. The national security consequences of climate change should by fully integrated into national security and national defense strategies.
2. The US should commit to a stronger national and international role to help stabilise climate change at levels that will avoid significant disruption to global security and stability.
3. The US should commit to global partnerships that help less developed nations build the capacity and resiliency to better manage climate impacts.
4. The Department of Defence should enhance its operational capability by accelerating the adoption of improved business processes and innovative technologies that result in improved US combat power through energy efficiency.
5. The Department of Defense should assess the impact on US military installations worldwide of rising sea levels, extreme weather events and other projected climate change impacts over the next 30 to 40 years.

On 29—31 March 2007 the strategic studies Institute and the Triangle Institute for security studies conducted a colloquium on 'Global climate change: National Security impactions', that reached the following key insights:

- Climate change is underway. The effects will vary according to a broad variety of circumstances and interactions, some of which are not well understood. Likewise, mitigation is not well understood, and will not take place quickly.
- The national security implications of climate change are proportional both to the speed and extent of change. Public awareness should follow a coordinated strategic communication plan that focuses on maintaining credibility.
- Threats to national survival stemming from catastrophic change must be anticipated, evaluated, and neutralised to the greatest degree possible.
- The entire range of plausible threats needs to be delineated, then analysed and early warning criteria established. The alternative approaches and cost-benefit analyses must be run to establish what can be done, when and at what cost.
- While military forces have roles in disasters relief, the broader impact of serious climate change will require multinational, multi-agency cooperation on a scale hitherto unimaginable and could provide no-fault ground for global cooperation.

- Effective interagency action may require new legislation and better definition of Department of Homeland Security authority.
- Should global cooperative measures fail, the first impact will likely come from large numbers of displaced people who, by the very nature of their displacement, will become subject to malnutrition and disease; agricultural dislocation could be aggressive or spark displacement and broader security issues could arise as well.

In heating up the planet, Alan Dupont and Graeme Pearman (2006) analysed the linkages between climate change and security arguing that climate change will complicate Australia's security environment in several ways:

- First, weather extremes and greater fluctuations in rainfall and temperatures have the capacity to refashion the region's productive landscape and exacerbate food, water and energy scarcities in a relatively short time span. Sea level rise is of particular concern because of the density of coastal populations and the potential for the large-scale displacement of people in Asia.
- Secondly, climate change will contribute to destabilising, unregulated population movements in Asia and the Pacific. Most of these flows are likely to be internal, but the ripple effects will be felt beyond the borders of the states most affected, requiring cooperative regional solutions.
- Thirdly, more extreme weather patterns will result in greater death and destruction from natural disasters, adding to the burden on poorer countries, stretching the resources, and coping ability of even the most developed nations.
- Fourthly, extreme weather events and climate-related disasters will not only trigger short-term diseases but also have more enduring health security consequences, since some infectious diseases will become more widespread as the planet heats.
- Fifthly, even if not catastrophic in themselves, the cumulative impact of rising temperatures, sea levels and more mega droughts on agriculture, fresh water and energy could threaten the security of states in Australia's neighbourhood by reducing their carrying capacity below a minimum threshold, thereby undermining the legitimacy and response capabilities of their governments and jeopardising the security of their citizens. Where climate change coincides with other transnational challenges to security, such as terrorism or pandemic diseases, or adds to pre-existing ethnic and social tensions, then the impact will be magnified. However, state collapse and destabilising internal conflicts are more likely outcomes than interstate war. For a handful of small, low-lying pacific nations, climate change is the ultimate security threat, since rising sea levels will eventually make their countries uninhabitable.

### Linking Anthropocene, HUGE and HESP: A New Dimension to Environmental Security

While the first three phases of environmental security research primarily focused on the 'nation state' as the key referent of Environmental Security concepts and policy, this chapter suggests that during the fourth phase the referent object of securitisation should be both widened and deepened. This implies that the environmental dimension of security should include both societal, human and gender issues but also sectoral approaches such as water, food, health, and livelihood security. Furthermore, this includes widening of the referents of securitisation–besides the narrow focus on 'national security'.

### Human, Gender and Environmental Security (HUGE)

The deplorable conditions of physical, environmental, human societal and gender security of a large part of humanity have inspired researchers (Oswald 2001, 2006, 2008; Moller 2001, 2003: 278-79) to ask security for whom (the nation state, society, human beings or humankind or the environment); for what (eternal or earthly peace, social, justice, humane livelihood, sustainable development, gender equity): from whom (state and sub-state actors, terrorists, migrants, alien cultures, multilateral organisation, globalisation, elites, nature, humankind, patriarchy, totalitarian integrity, national unity and identity, survival and quality of life, sustainability, equity, identity, social representations and solidarity).

In synthesis, HUGE integrates social, environmental, human, cultural, and political and identity concerns, aiming at solidarity, resilience, participatory democracy, peace building and equity in a world that is facing many new security dangers and concerns due to the gradual transition from the Holocene to the Anthropocene.

### Human and Environmental Security and Peace (HESP)

#### Policy Goals

- Policy Goals on the Societal and individual levels: Environmental Security studies should contribute to strategies for reducing the impact of environmental stress, decreasing the vulnerability and strengthening the coping capacities and resilience.
- Policy Goals on the communal, sub-national, national and international levels: strategies for coping with outcomes of environmental stress should be developed by improving disaster preparedness and response and by integrating disaster reduction into development planning and resilience building. The resolution, prevention and avoidance of resulting violence should become a major policy goal.

HESP combines a normative orientation at the dual policy goals of sustainable development and sustainable peace with a 'people-centred' human security perspective from the individual to the global level to develop strategies for adaptation and mitigation and to reduce the likelihood and impact of vulnerability to these outcomes by strengthening resilience. Both concepts and policy goals of sustainable development and sustainable peace are as the highly contested as the security concept.

### Facing and Coping with Global Environmental Change by Responding with a New Peace Policy for the 21st Century

From 1974 to 2003 more than 5 billion people were affected by human induced natural hazards most of them by climate related events, such as floods, drought, famine, forest fire and heat waves. Thus climate change is already killing people and the likelihood will increase that climate change will lead to forced migration, domestic crises and even violent conflicts. Neither the military nor any national security strategy can cope with this challenge unilaterally that often poses a 'survival dilemma' for the most vulnerable poor populations in the global south. What is needed is a new global multilateral strategy for coping with climate change by adoption and mitigation measures.

## References

See: "press Conference by Security council president,4 April 2007";at: >http://www.un.org/News/briefings/docs//2007/070404 Parry.doc.htm>; 6 Bloomberg news: "UN attacks climate change as threat to peace", in: International Herald Tribune, 18 April 2007:2;UN Security council,SC/9000,5663rd meeting,17 April 2007: "Security Council holds First-ever debates on impact of Climate change on peace, security, hearing 50 speakers", at: http://un.org/news/press/docs/2007/sc9000.doc.htm; Reuters: "UN Council Hits Impasse over debate on Warming", in: New York Times,18 April 2007; Edith M.Lederer: "Security Council Tackles Climate Change", in: Washington Post,18 April 2007.

See : Chirrs Spence, edited by Pamela Chasek: Summary of the Informal Thematic Debates of the UN General Assembly on climate change as a Global Challenge"; at: <http://www.iisd.ca/climate/unga/UNGA%20Climate%20Change%20briefing%20note.pdf>.

For the documents of the G 8 Meeting in Heiligendamm, Germany on 8 June 2007; at: <http:www.g-8.de/webs/G8/EN/G8Summit/Summitdocuments/Summit-documents.html> and the chair's conclusions; at: http://www.g-8. de/nsc_true/Content/EN/Artikel/_g8-summit/anlagen/chairs-summary.templateId=raw, property=publicationFile.pdf/chairs-summary.

The negotiations are documented at the website of the UNFCCC secretariat; at: http://unfccc.int/2860.php.

See "water security in the 21st century", Ministerial Declaration, second World water Forum, the Hague, 22 March 2000, part. 3, available at: http://

www.waterpage.com/hangue_declaration.htm; Ministerial Declaration, Third world water forum, Kyoto 23 March 2003, para.11, available at: http://www.mofa.go.jp/policy/environmental/wwf/declaration.html.

See the African development Bank; at.http://www.afdb.org/pls/portal/docs/PAGE/ADB_ADMIN_PG/DO-CUMENTS/NEWS/CONCEPT%20NOTE-ENG.PDF.

See the testimony of peter H. Gleick to the United States congress, Committee on Governmental Reform subcommittee on National security, Emerging Threats, and International Relations, Hearing on Energy as a weapon: implications for U.S. Security: "The implications of global climatic changes for international security", May 2006; Peter H.Gleick: "Global Climate Change and international security" (1989);at:http://www.pacinst.org/reports/climate_repots/Global_Climate_change_and_international_security.pdf; peter H.Gleick: "the implications of global climate changes for international security", in: climate change,15 (1989): 309-325.

See for details the WBGU website at: http://www.wbgu.de/wbgu_jg2007_engl.html, where several expert studies are also available for download in English and German. A summary for policy–makers is at: http://www.wbgu.de/wbgu_jg2007_Kurz_engl.html and the full report can be downloaded at: http://www.wbgu.de/wbgu_jg2007_engl.pdf.

Quoted in: Ben Vogel (2007) "climate change creates security challenge" 'more complex than cold war', in: Janes.com; at <http://www.janes.com/security/international_security/news/misc/janes070130_1_n.shtml>.

Quoted by Chris Littlecott (2007) "climate change: The global security impact" 5 February at: http://www.e3g.org/index.php/programmes/climate-articles/climate-change-the-global-security-impact/>.

"Press Conference by security Council President, 4 April 2007"; at: http://www.org/News/brifings/docs//2007/070404 Parry.doc.htm;

Blooming news: "UN attacks climate change as threat to peace", in International Herald Tribune, 18 April 2007:2;

UN Secretary Council, SC/9000, 5663rd meeting, 17 April 2007: "Security Council holds first-ever debate on impact of climate change on peace, security, hearing 50 speakers", at : <http//un.org/news/press/docs/2007/sc9000.doc.htm>; Reuters: "UN Council Hits Impasse over Debate on Warming" in Washington Post, 18 April 2007.

At the G8 meeting in Gleneaagles plans of action on "climate change, Clean Energy and Sustainable Development"; at: <http://www.fco.gov.uk/files/kfile/postG8_Gleneagles_ccchangePlanofAction.pdf>:

For the documents of the G 8 Meeting in Heiligendamm, Germany on 8 June 2007; at: http://www.g-8.de/webs/G8/EN/G8Summit/SummitDocuments/summit-documents.html and the chair's councils; at: <http://www.g-b.de/nsc_true_true/content/EN/Artikel/_g8-summit/anlagen/chairs-summery, templated=raw, property=publicationFile.pdf/chairs-summery >.

Chris Spence; Pamela Chasek: "Summary of the Informal Thematic Debates of the UN General Assembly on Climate Change as a Global Challenge, New York,

31 July – 2 August 2007,in:Earth Negotiations Bulletin (New York: IISD, 7 August 2007).

See: “Background note by the Secretary-General,” at: <http://www.un.org/climatechange/2007highlevel/background.shtml>.

UNDP, 2007: Human Development Reports; at: http://hdr.undp.org/; UNDP/UNEP/World Bank/ADB/AfDB/GTZ/DFID/OCED/EC (2003).

UNEP News Centre: “Climate change and conflict – New Report weighs the risks and Pinpoints Likely Hotpots”; at: <http://www.unep.org/Documents.Multilingual/default.asp?DocumentID=523&ArticleID=57207&l-en>.

Andrew Bounds: “Climate change poses ‘security risk’”, in: FT.Com. 3 March 2008; Ian Traynor; “EU told to prepare for flood of climate change Migrants”, in: The Guardian, 10 March 2008:This report was discussed at a meeting on “national Security and the threat of Climate Change”, by the Environmental change and security Program (ECSP) of the Wilson Center on 14 May 2007.

U.S. Army War College and Triangle Institute for Security Studies, Strategic Studies Institute, Colloquium Brief, and Compiled by Douglas V. Johnson II; at: <http://www.strategic studies Institute, army.mil/pdffiles/PUB779.pdf>.

On 21-23 June 2005, The Global Environmental Change and Human Security (GECHS) project of IHDP organised a workshop in Oslo on ‘climate change and human security’; at: http://www.cicero.uio.no/humsec/; papers at: http://www.cicero.uio.no/humsec/list_participants.html; Six papers have been published in a special issues on “climate change and Human Security”, of: Erde,137,3:155-270.

See the Greek concept paper “Human Security and the Climate Change Impact on Vulnerable Groups” of 8 May 2007; at: <http://www.humansecuritynetwork.org/docs/2007-ministerial-meeting-04-greek% 20paper.doc>.

See for the conference documents at: <http://www.sidym2006.org/eng/eng_ponencias_conclusiones.asp>.

See for a brief report at: http://www.afes-press.de/pdf/UNCCD_journal_050511.pdfand a documentation of the CRIC-meeting at: <http://www.afes-press.de/pdf/Earth_negot_bulletin_04_175.pdf> ;see also the opening speech by Brauch; at: http://www.afes-press.de/pdf/Brauch_Almeria_Valencia.pdf and the major speech by Oswald/Brauch; at: <http://www.afes-press.de/pdf/Oswald_Brauch_lang.pdf>.

# Index


**A**

A Climate of Conflict: The Links between Climate Change, 104
Accommodation, 276
Ad Hoc Working Group-Kyoto Protocol, 171
Adapt well to scarcity, 297
Afghan War of 1979-1989, 69
Afghan War of 2001, 71, 72
Africa, 30
African National Congress, 298
African Nations, 194
Agricultural Productivity Loss, 99
Air pollution regulations, 152
Al Gore, 192
Amartya Sen defined human security, 82
An integrated human, gender and environmental secu, 318
Anthony Zinni, 192
anthropogenic climate change, 29
AOSIS, 168, 169, 173, 176, 180
Approches to Environmental Peace-building, 60
Arctic, 281
armed conflicts, 31
army camps, 235
Arnold Wolfer's famous discussion of security, 25
Attacks on the land, 69
Attacks on the water infrastructure, 73

**B**

Baldwin, 25
Bali Action Plan, 170
Bangladesh, 80, 86, 88, 89, 95, 99, 101, 106, 107
Bangladesh Climate Change Strategy and Action Plan, 91
Bangladesh coastal zone, 91
Bangladesh has export interest, 218
Bangladesh Jute Mills Corporation, 219
Bangladesh on high alert, 104
Bangladesh's Environmental Goods and Services Industry, 213
Bangladesh's Imports of OECD Classified Environment, 213
BASIC, 169
group, 165
group of countries, 180
partners, 176
Basudheiba Kutumbakam, 46
beach erosion, 266
Ben Wisner, 30
Benjamin Barber's idea of strong democracy, 299
Bihar, 298
Bio-fuels, 197
Biomass, 152
Blending of ethanol, 178
Bogardi and Brauch, 317
Brazil, 176, 213
British initiative, 31
Brundtland Commission, 285

**C**

Canada, 179
Canadian Government, 25
Accords, 176
Agreement, 177
Cancun Summit, 175, 176, 181
capita equity, 169

carbon emission, 160
carbon sink projects, 156
Carrying capacity, 56
Causes of Sea Level Rise, 261
CCAPS programme, 200
CDM, 160, 174
project, 160
Challenges of climate change, 46
Changing Energy Strategy, 149
changing temperature and rainfall patterns, 191
Chasi Mulia Adivasi Sangha, 288
Chhattisgarh, 288
China, 173, 176, 177
China's entry, 183
China's international environment policy, 173
Chris Littlecott, 320
civil environmental disruption, 76
classic security policy, 41
classical formulation, 80
Clean Development Mechanism, 156, 170
clean energy, 178, 230
Climate and Sea Level, 259
Climate change, 32, 79, 117, 191, 207, 318
Climate change adaptation, 229
Climate change and agriculture, 191
climate change and conflict, 30
Climate Change and Human Security, 29, 84
Climate Change and Security, 31, 32, 195
Climate Change as a National Security Danger, 321
Climate Change Challenges, 137
Climate Change Council, 237
Climate change impacts, 86
Climate change impacts and violent conflict, 132
Climate change issue, 189
Climate change mitigation, 221
Climate change negotiations, 168, 181
Climate Change Policy, 2011, 230, 236
Climate Change Secretariat of the Ministry of Envi, 127
Climate justice, 166
Climate negotiations, 173
Climate records, 137
Climate related risks, 271
Climate security, 224
Climate Technology Network, 177
Climate-food connection, 35
Climate-induced stress, 32
Climate-sensitive Diseases, 107
Climatic anomalies, 88
Club Rome group, 23
$CO_2$ emission reduction, 149
Coast Conservation, 128
Coast Conservation Department, 128
Coastal erosion, 267, 271
Coastal flooding, 195
Coastal wetland ecosystems, 273
coastal zone, 39
Coastal Zone Management Plan (CZMP) and the Master, 127
Coastal Zone Management Subgroup, 275
Coastal zones, 258
Coasts, 193
Cold war era, 56
Commercial energy, 151
Commission on Human Security, 26
Commissioned report on climate change and conflict, 30
Community Forestry User Group, 227
Community Water Supply and Sanitation Project, 127
Concept of G-2, 173
Concept of human security, 21, 25, 81
Concept of security, 187
Concern for India, 147
Conference of Parties, 169
Conference of Parties (COP), 168, 169
Conference on 'Climate Change: The Global Security, 30
Conflict Constellations, 33
Conflict over water, 234
Consequences of climate change, 45
Conventional military institutions, 294
COP-1 in Berlin, 170
COP-15, 171
COP-16 at Cancun, 175

Copenhagen, 171, 207
Copenhagen Accord, 172, 175
Copenhagen and Cancun agreements, 167
Copenhagen Summit, 174, 175, 207
Coral reefs, 251
Cornucopian view, 316
Costs, 149
Costs associated with a low GHG energy strategy, 149
Costs of mitigation, 155
Countries accountable for all emissions, 160
Countries with populations in the low-elevation co, 39
Crispin Tickell, 30, 319
Cyclones, 272
Cyclonic storms, 272
Cyril Obi, 245

**D**

D. Deudney, 57
Damage to housing and infrastructure, 124
Dams and climate change, 196
Darfur, 203
David A. Baldwin, 53
David King, 246
Declaration of the Stockholm Conference on the Hum, 166
Decline in Food Production and Conflict, 34
Definition of security, 27
Deglaciation, 261
Degradation of Freshwater Resources and Conflict, 33
Demographic stress, 202
Dengue cases, 121
Dependent on the river's water, 293
Desertification and drought, 203
destructive changes in temperature, 46
Developed countries, 166, 182
Developed countrycomplacency syndrome, 182
Developing countries, 166, 168
Development, 46
Development based on fossil fuel technology, 43
Differences between the MoEF and the MEA, 184
Differentiated responsibility, 159
Disaster Management Center (DMC), 126
Disaster-Development Nexus, 98
Disasters Threatening Livelihood—Governance Secur, 38
Dispute over 'climate justice', 166
Distress Migration, 140
divided, 174
Doha Ministerial Declaration, 208, 209
Domestic rainwater harvesting, 127
Downstream inundation, 38
drinking water, 33
dro-climatic hazards, 132
drought and desertification, 198
Droughts, 92, 124
Droughts and flood events, 117
Dubash's terminologies, 170
Durban Platform for Enhanced Action, 179
Durban Summit, 179, 180

**E**

Earth, 45
Earth Summit, 170
Eastern Himalaya, 302
Ecological Institute, 82
Ecological marginalisation, 290
Economic Impacts of sea level rise, 266
Economic wisdom, 143
Effects of global warming, 191
Effects of sea level rise, 265
EGS liberalisation, 208
Egypt, 293
Emigration from coastal zones, 148
Emission mitigation, 154
Emission of developing countries, 159
Emission quotas, 160
Emission Reduction, 154
Emission standards, 126
Emission trading system, 160
Emissions reductions strategies, 167
Empowerment, 27
ENCOP, 59

ENCOP's research, 59
end of cold war, 24, 312
Energy conservation and increased efficiency, 151
Energy efficiency, 230
Energy efficiency in vehicles, 152
Energy savings, 151
Environmental, 203
Environmental and social disruptions, 68
Environmental Award Programme, 254
Environmental battlefield disruption, 68
Environmental cooperation, 60
Environmental debate, 54
Environmental Degradation, 55, 83, 84, 107, 140, 228, 284, 300
Environmental degradation and threats to military, 56
Environmental degradation in South Asia, 233
Environmental factors behind societal collapse, 319
Environmental Goods, 208
Environmental Goods and Services (EGSs), 208, 218
Environmental impact, 72
Environmental impacts of sea level, 266
Environmental insecurity, 247
Environmental issues on the security agenda, 56
Environmental Peacemaking, 61
Environmental problems, 45
Environmental projects, 212
Environmental Protection Agency estimates, 265
Environmental refugees, 75, 254
Environmental scarcities, 289
Environmental scarcity, 291
Environmental security, 53, 188, 224, 284, 286
Environmental Security Agenda, 55
Environmental security and human security, 29
Environmental Security Approaches, 57
Environmental security aspect of human security, 84
Environmental security threats, 54
environmental service, 211
Environmental Services Industry, 214
Environmental stress, 202
Environmental warfare, 68
Environmentally concerned theorists, 285
Environmentally preferable products, 211
Environmentally Preferable Products (EPPs), 219
Equitable climate regime, 159
Erosion and Accretion on Bangladesh, 93
Examples of environmental disruption, 69
Excepted Affects of Climate Change, 134
Explosive remnants of war, 76
Extinction of a significant number of species, 125
Extreme Events, 149
Extreme events due to climate change, 136
Extreme weather events, 38

**F**

Factors contributing to violent conflicts, 58
Failed States Index 2011, 104
Failure of the Copenhagen Summit, 172
Fairness' in adaptation to climate, 36
FCCC objective, 147
Filipino situation, 291
Filipino uplands, 292
First Climate Conference at Geneva, 165
Flood-and drought-resistant rice, 128
Floods, 89
Floods in Bangladesh, 102
Food, 194
Food and livelihood, 196
Food and livelihood pressure, 34, 37
Food insecurity, 196
Food Security, 119, 199, 251
Food security in Bangladesh, 100
Food—Fuel Conflict and Human Security, 37

Forced energy savings' by the poor, 151
Forced Migration in Bangladesh, 107
Forcing mechanisms, 134
Forum for China-Africa Cooperation, 183
Forward-looking treaty regimes, 33
Framework Convention on Climate Change (FCCC), 146
Franco-Dutch War, 73
Freedom, 82
Freedom from Fear, 286
Freedom from Want, 286
freshwater systems, 193
Fur-Arab war, 202

**G**

G-77, 168, 173, 174, 179
gender impact analysis, 37
German Advisory council, 318
German Ministry, 30
Ghenghis Khan, 69
GHG concentrations, 147
GHG emissions, 116, 230
Glaciers, 261
Global carbon price floor, 157
Global Climate Change, 45, 189
Global climate regime, 182
Global Environmental and Climate Change: The IPCC, 316
Global Environmental Changes and Human Security, 83, 84
Global governance system, 42
Global Humanitarian Forum, 86
Global issue, 132
Global sustainability, 299
Global warming, 68, 75, 76, 320
Global warming proceeds, 75
Global World Energy Council, 178
globalisation, 62
Globalisation, 241
Green Fund, 173, 174
Green growth strategy, 230
Green technology investments, 178
Greenhouse gas concentration levels, 153
Greenhouse gas concentrations, 166
Greenhouse gases, 207, 281
Groundwater, 290
Growing international distributional conflicts, 40
Guadalupe, 200
Gulf War, 71

**H**

Hardest hit by climate change, 80
Harvesting seasons, 200
Health, 193
Health and hygiene, 121
Health Hazards and Conflict, 36
Health Impacts, 197
hierarchy of needs, 54
high yielding rice varieties, 128
Himalayan Glaciers, 108
Holland Water, 73
Human and Environmental Security and Peace (HESP)P, 323
Human development category, 101
Human Development Index, 31, 225
Human Development Report, 225
Human, Gender and Environmental Security, 323
Human Health, 251
Human living conditions, 194
Human medical tragedies, 70
Human rights, 41
Human security, 27, 28, 82, 117, 132, 199, 205
Human security paradigm, 28
Human-induced climate change, 259
hunger, 281
Hurricane Katrina, 182
hydropower generation, 260

**I**

IBSA (India, Brazil and South Africa), 178
ice, 281
imbalance in the distribution of wealth and power, 289
impacts of cattle industry, 182
impacts of climate change on the Middle East, 319
Impacts of Sea Level Rise, 263
Impacts of sea-level rise, 260
Implications of climate change, 109
Importance of nature, 300

Increase in Storm and Flood Disasters and Conflict, 35
Increase in the number of poor, 149
Independent Commission on Disarmament and Security, 24
India, 287
India's demands, 180
India's emissions, 159
India's emissions profile, 150
India's food production, 159
India's position and perception, 174
India's rapidly growing population, 295
Industrialised countries, 166
Industrialised and the non-industrialised countries, 166
Ingenuity in the form of new technologies and new, 297
Inkatha Freedom Party, 298
Input-Output Model, 150
Intellectual Property Rights, 176
Intellectual Property Rights (IPR) control, 176
Intentional attacks, 68
Inter-ethnic chaos, insurgency and terrorism, 310
Interface between freshwater and saline water, 96
Intergovernmental negotiations, 166
Intergovernmental Panel on Climate Change, 165, 205, 281
Intergovernmental Panel on Climate Change, 165
Internal Forcing Mechanisms, 134
International aid, 203
International Institute for Environment and Development, 84
International Peace and Security, 139
International resolutions, 132
International Security Assistance Force, 71
Interrelated building blocks of human – and national security, 83
Interstate war, 292
Intervention on human security, 25
Inventories of their GHG emissions, 168
Investments in bio-fuels, 37
IPCC draft report, 44
IPCC third assessment report, 153
Irrigation and flood-control dams, 291

**J**

Jairam Ramesh, 175, 179, 305
Japan's Ambassador Kenzo Oshima, 320
John Ashton, 30
John Reid, 139
John Reid MP, 30, 196
Jordan River basin, 290

**K**

Kofi Annan, 31, 243
Koraput-Balangir-Kalahandi, 287
Korean Demilitarised, 74
Korean Demilitarised Zone [DMZ], 74
Korean War, 69
KPM Uranium Mining Controversy, 307
Kyoto Protocol, 136, 156, 167, 170, 171, 173, 176, 179, 181, 207, 254
Kyoto Protocol's Clean Development Mechanism, 36

**L**

Land grab, 287
Land scarcity, 298
Land use planning, 128
Land-use change in developing countries, 152
Landslides, 38
Landslides and cyclones, 118
Lateral pressure, 292
Lateral pressure' theory, 292
LDCs, 168, 180
Link between Carrying Capacity and Warfare, 43
Linkages between climate change and security, 322
Links between climate change, peace and war, 319
Livelihood security, 34, 38
Livelihoods, 124
Locke, 45
Lorraine Elliott, 28
Lower Agricultural Production, 147

**M**

Magnitude of climate change, 98
Mahbub-ul-Haq, 25
Malaria, 199
Malaria epidemics, 121
Maldives, 247
Maldives' socio-economic development, 253
Malte Meinshausen, 44
Mangroves of Sunderbans, 273
Maoist group and the Government, 233
Margaret, Beckett, 319
Marrakesh talks, 171
Mean sea level, 270
Measuring, Reporting and Verification, 174
Megha Drinking Water Project, 225
Melting of glaciers, 261
Methane emissions, 182
Methane Emissions and Subsistence Emissions, 147
Migrants, 68
Migration, 41, 203
Migration and Conflict, 35
Millennium Development Goals (MDGs), 29, 98, 138
Ministry of Disaster Management, 126
Mitigation Costs and Benefits, 155
    mitigation response, 276
Mitigation-adaptation-inequity and conflict, 36
Mohammed Nasheer, 240
Montreal (COP-11), 171
Montreal Protocol, 146

**N**

National Adaptation Plan of Action (NAPA), 253
National Climate Change Strategy and Action Plan, 98
National Communications (NC), 253
National Dengue Eradication Month, 129
National Environment Act, 126
National Environmental Satellite, Data and Information, 132
National inventories of their GHG emissions, 168
National policy of clean development mechanism (CD, 116
National Policy on Agriculture, 127
National policy on rainwater harvesting, 127
National Security, 243
National Water Supply and Drainage Board, 120
Nationally Appropriate Mitigation Actions, 171
Natural resources, 194
    neo-Malthusians, 296
    'neo-realist' paradigm, 242
    'neo-realist' paradigm of security, 242
    neoclassical economists, 296
Nepal, 225
Nepalese government, 235
    new challenges, 224
    new kind of security agenda, 27
Newtonian and Cartesian epistemology, 43
Nile Basin Initiative, 33
Nobel Peace Prize 2007, 244
Non-commercial energy, 151
Non-proliferation Treaty (NPT), 183
North East India, 302
    notion of rationality, 300

**O**

Ocean acidification, 76
OECD, 160
OECD countries, 155
Oil and gas exploration, 141
Organisation for Economic Cooperation and Development, 211
Organisation of Petroleum Exporting Countries, 168
Osama bin Laden was killed, 71

**P**

Pakistan, shortages, 287
Paul Collier and A. Hoeffler, 59
Peloponnesian War, 69
Per capita equity, 169
Personal security, 286
Pest and disease occurrence, 119
Pest control methods, 128

Phases of environmental security research, 323
Pivotal states, 298
Polar ice cover, 141
Policy of the Indian Government towards the North, 305
Political economy of energy, 142
Political Instability, 226
Political security, 286
Political stability in Bangladesh, 104
Pollution control measures, 152
Population growth, 295
    post-Copenhagen scenario, 179
    post-Kyoto period, 171
    potential threats, 320
Poverty and climate change, 97
Poverty and resource management, 195
Poverty with environment, 225
Power Generation and Energy Supply, 120
President Narsheed, 253
Preventing Conflict, 60
Problems troubling men of all nations, 23
Projected sea-level, 39
Protagonists of the idea of human security, 23
Protection, 27
Protocol I on International Armed Conflicts 1977, 76
Protocol II on Non-international Armed Conflicts 1977, 76

**R**

Rand underdeveloped communities, 318
    raising, the, HDI, 320
    rapid melting, 108
Rational economic 'man', 300
Realignment of Energy Markets, 317
Realist School, 80
Rebellion of 1850—64, 70
Receding sea ice, 141
Redefining Security, 26, 53
Redrawing global trade routes, 141
Reforestation and afforestation activities and res, 128
    refugees, 281
Refugees, environmental and other, 74
Relative sea level, 262
Replacement of fossil fuel, 46
Report of the Commission, 28
Report of the Commission on Global Governance (CGG), 28
Request-offer approach, 213
Resource capture, 290
Resource constraints, 191
Resource depletion and degradation, 289
Resource Scarcity, 140
Reviewed the scientific research on Climate Change, 318
Rio Summit, 165
Rising global temperatures, 79
Rising Temperatures, 198
Risks for global economic development, 40
River Bank and Coastal Erosion, 93
River basins, 293
Rural and urban poor, 32
Rural resource scarcities and population growth, 295
Rwandan civil war, 291

**S**

Salinity intrusion, 96
Salinity intrusion in surface water, 96
Salinity of seawater, 274
Salwa judum, 288
School of thought, 59
Science, 43
Scientific or technological 'solutions', 143
Sea level changes, 270
Sea Level Rise, 94, 136, 148, 250, 258, 259
Sea level rise and coastal erosion, 125
Sea Level Rise and Displacement Security Threats, 38
Sea level Rise in Bay of Bengal, 270
    sea-level rise, 141
Second Indo-China War, 69
Second Sino-Japanese War, 74
Second Sino-Japanese War of 1937—45, 74
Securitisation, 234

Securitisation and democratisation, 235
Securitisation debate, 169
Securitisation of Climate Change, 317
securitisation of the environment, 56
Securitisation of Global Environmental Change, 315
Securitisation Theory, 315
Security, 80, 81, 283
Security challenges, 84
Security dimensions of climate change, 139
Security implication of climate change, 30
Security of the Livelihoods of People, 124
Security Paradigm, 83
security risks, 318
security threats, 30
self-inflicted scorched-earth, 73
Senegal River, 291
Settlements, 193
Share of responsibilities, 207
Short-term fluctuations, 134
shrinking water resources, 108
Shyam Saran, 184
Simon Dalby, 55
Sink Projects through CDM, 156
situation, 30
social factors, 247
Social Impact, 72
Social Implications, 275
Soil Resource Development Institute, 96
Solution to climate change, 30
South Africa, 287
South Africa, scarcity-driven migrations, 298
South-South cooperation, 164, 166, 172, 175
South-South cooperation movement, 179
Southern African countries, 38
sovereignty, 28
Sri Lanka, 115
Sri Lanka's forest cover, 115
Sri Lanka's forest cover had declined, 115
State actor adaptations, 32
state security, 28
Stockholm Initiative on Global Security and Governance, 24
Storm Surges, 267
storms and flooding, 38
strategic partnership' with the Afghan, 71
Strategy for Accelerated Poverty Reduction, 98
Strong democracy, 299
Study on environment and security, 27
Sub-Saharan Africa, 198
Sunderbans face a threat, 273
Sustainable Community Development and Community Forestry Programmes, 230
Sustainable development, 61, 285
Sustainable Development in India, 150
Sustainable energy-consumption, 77
Swiss Peace Foundation's Environment and Conflicts, 58

## T

Targets to tackle climate change, 44
Technology cooperation, 183
Technology Transfer (TT) and CDM, 157
Temporary Movement of Natural Persons (TMNP), 218
Temporary Service Providers of Bangladesh, 220
The National Environment Policy, 126
The United Nations Conference on Trade and Development, 211
Theory of the politics of scarcity, 284
Third World Countries, 287
Thirty Years' War of 1618, 48, 70
Thomas Homer-Dixon, 57, 284
threats, 81, 312
Threats to Economic Security, 101
Threats to Health Security, 106
Threats to International Stability and Security, 39
Three pillars, 317
Tipaimukh dam, 308
Top-down approach and the bottom-up approach, 167

Toronto School, 57, 58
Tourism, 125
Tourism and fisheries, 250
Trade Routes, 141
Traditional concept of security, 27
Traditional technologies, 183
Traditional view of security, 80
Trans-boundary hydro-conflicts, 108
Triangle Institute for security studies, 321
Tsunamis, 274

**U**

UK's emissions, 45
  ultimate security, 53
Under-nourishment in 2010, 101
UNDP's Human Development Report, 26
United Nations Conference on Human Environment, 165
United Nations Development Report, human security, 243
United Nations Environment Programme, 71
United Nations Framework Convention, 254
United Nations Framework Convention on Climate Change, 165
United Nations Intergovernmental Panel on Climate, 274
Unsustainable consumption, 146
Upstream and downstream neighbours, 293
Upstream and downstream neighbours over river water, 293
Urban air pollution, 152
Urbanisation and Displacement, 307
US War of Independence of 1775—83, 70
USA, 71

**V**

Variations in atmospheric pressure, 263
Variations in sea level, 263
Vector-borne diseases, 129
Vietnam Conflict of 1961—75, 70,76
Violence and political conflict, 196
Vulnerable, 106
Vulnerable areas in terms of sea level rise, 94

**W**

Walter Lippman (1943) in his classical formulation, 80
  Warfare (armed conflict), 67
  Wars, 67
  Wars in this century, 293
  Wartime environmental disruption, 75
Water, 193
  availability, 260
  management and forestry projects, 32
  policy of Nepal, 236
  poverty index, 34
Water Resources, 194, 251
Water scarcity, 120, 246, 293
Water Security in Bangladesh, 108
  water shortages, 132, 281
Weak and fragile states, 39
Weaponisation, 47
Wetlands, 273
Wheat yields, 199
Willy Brandt, 24
Wind-driven waves, 262
Women, 32, 35
Working Group-Long-Term Cooperative Action, 171
World Bank, 293
World Commission on Environment and Development, 285
World Environment Day, 254
World Trade Organisation, 181
World Trade Organisation negotiations, 181
World War I, 68
World War II, 29, 69, 70
  worst affected, 46